Series on Analysis, Applications and Computation – Vol. 7

ISAAC

The "Golden"
Non-Euclidean Geometry

Hilbert's Fourth Problem, "Golden" Dynamical
Systems, and the Fine-Structure Constant

Series on Analysis, Applications and Computation

Series on Analysis, Applications and Computation – Vol. 7

ISAAC

The "Golden" Non-Euclidean Geometry

Hilbert's Fourth Problem, "Golden" Dynamical Systems, and the Fine-Structure Constant

∘ Alexey Stakhov

International Club of the Golden Section, Canada
& Academy of Trinitarism, Russia

∘ Samuil Aranson

The Russian Academy of Natural Sciences, Russia

∘ Assisted by Scott Olsen

College of Central Florida, USA

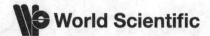

NEW JERSEY · LONDON · SINGAPORE · BEIJING · SHANGHAI · HONG KONG · TAIPEI · CHENNAI · TOKYO

Published by

World Scientific Publishing Co. Pte. Ltd.

5 Toh Tuck Link, Singapore 596224

USA office: 27 Warren Street, Suite 401-402, Hackensack, NJ 07601

UK office: 57 Shelton Street, Covent Garden, London WC2H 9HE

Library of Congress Cataloging-in-Publication Data

Names: Stakhov, A. P. (Alekseĭ Petrovich) | Aranson, S. Kh. | Olsen, Scott Anthony

Title: The "golden" non-Euclidean geometry : Hilbert's fourth problem, "golden" dynamical
　　systems, and the fine-structure constant / Alexey Stakhov (International Club of the Golden
　　Section, Canada & Academy of Trinitarism, Russia), Samuil Aranson (The Russian Academy
　　of Natural Sciences, Russia) ; assisted by Scott Olsen (College of Central Florida, USA).

Other titles: Non-Euclidean geometry

Description: New Jersey : World Scientific, 2016. | Series: Series on analysis, applications,
　　and computation ; volume 7 | Includes bibliographical references.

Identifiers: LCCN 2016011605 | ISBN 9789814678292 (hardcover : alk. paper)

Subjects: LCSH: Geometry, Non-Euclidean.

Classification: LCC QA685 .S794 2016 | DDC 516.9--dc23

LC record available at http://lccn.loc.gov/2016011605

British Library Cataloguing-in-Publication Data

A catalogue record for this book is available from the British Library.

Printed in Singapore

In fond memory of
Evgeniya Alexandrovna Leontovich-Andronova,
Yuri Alexeyevich Mitropolsky,
Dmitriy Viktorovich Anosov,
Alexander Andreevich Volkov

Preface

In mathematics, there are the objects, which, at first glance, seem unrelated to one another. But upon deeper examination of such objects some general features are revealed which can be a source for new mathematical discoveries and theories. This is the central idea of our scientific collaboration which began in 2007. This collaboration led to a number of fundamental results, in particular, to an unusual solution to Hilbert's Fourth Problem and the new recursive (or "golden") non-Euclidean geometry.

The objective of this study is to trace the connection of the golden ratio, Fibonacci numbers [1–4], Lucas sequences [5] and Binet's formulas [1–4] with the "golden" hyperbolic Fibonacci and Lucas functions (Alexey Stakhov, Ivan Tkachenko, Boris Rozin [6, 7]), Fibonacci λ-numbers [8, 9] and metallic proportions. These latter were obtained simultaneously and independently by the Argentinean mathematician Vera W. de Spinadel [10], French mathematician Midhat Gazale [11], American mathematician Jay Kappraff [12], Russian engineer Alexander Tatarenko [13], Armenian philosopher and physicist Hrant Arakelyan [14], Russian researcher Victor Shenyagin [15], Ukrainian physicist Nikolai Kosinov [16], Spanish mathematicians Sergio Falcon and Angel Plaza [17], and others.

In 2006 Stakhov expanded the general theory of recursive hyperbolic functions, based on Fibonacci λ-numbers and metallic proportions, by introducing hyperbolic Fibonacci and Lucas λ-functions [18].

The primary result of our studies during the 7-year period 2007–2014 is that the metallic proportions of Vera W. de Spinadel and the Fibonacci λ-numbers together provide the basis for Stakhov's general theory of recursive hyperbolic functions [18, 19]. This led to an unusual solution of Hilbert's Fourth Problem and, ultimately, to the justification of recursive or "golden" non-Euclidean geometry (Stakhov and Aranson) [20–34].

This discovery challenges those in the theoretical natural sciences to discover these (recursive) non-Euclidean geometries in Nature. One of these geometries, based on *"golden" hyperbolic functions*, was already discovered in Nature by Ukrainian researcher Oleg Bodnar (Bodnar's geometry in the phenomenon of phyllotaxis [35–38]). The search will continue throughout Nature for other recursive hyperbolic geometries.

One object of this study is to trace the development of non-Euclidean geometry. This study began with a critical analysis of Euclid's Fifth Postulate and the creation of what Nikolai Lobachevsky called "imaginary geometry". However, the primary purpose is to present the scientific community with recursive or "golden" non-Euclidean geometry and stimulate the search for these geometries throughout Nature.

The present study introduces the ancient idea of harmony (Pythagoras, Plato, Euclid) into hyperbolic and spherical geometry and to the theory of the fine-structure constant that may lead to a radical shift in the scientific paradigm of modern mathematics and theoretical physics towards the ancient idea of harmony rooted in the golden ratio.

Introduction

The primary purpose of this book is to present in a concise manner the fundamentals of "golden" non-Euclidean geometry (hyperbolic and spherical), following from our original solution to Hilbert's Fourth Problem, the "golden" qualitative theory of differential equations and dynamical systems, and the new theory of fine-structure constant, relating to the Physics Millennium Problem. It is intended to be accessible to a wider audience, including not only mathematicians, but also representatives of other areas of the theoretical natural sciences, such as physics, chemistry, biology, botany, and genetics.

The book consists of 5 chapters.

Chapter 1 is a compact introduction to the theory of the "golden" hyperbolic Fibonacci and Lucas functions, based on the golden ratio and Fibonacci numbers, and their applications in the natural world (Bodnar's geometry).

Chapter 2 is a brief introduction to the theory of the Fibonacci λ-numbers, the metallic proportions, and the harmonic hyperbolic Fibonacci and Lucas λ-functions. These are generalizations of the classic Fibonacci numbers and golden ratio, and the "golden" hyperbolic Fibonacci and Lucas functions. Stakhov's harmonic hyperbolic Fibonacci and Lucas λ-functions are the primary results of Chapter 2.

Chapter 3 is a concise introduction to Stakhov and Aranson's two fundamentally new mathematical results. Following an analysis of David Hilbert's famous article "Mathematical problems" and his approach to mathematical solutions ("Hilbert philosophy"), we discuss a solution to Hilbert's Fourth Problem based on hyperbolic Fibonacci λ-functions (introduced in Chapter 2). Next we discuss our solution to Hilbert's Fourth Problem based upon *spherical Fibonacci λ-functions*. Our results introduce a new class of elementary functions and demonstrate the existence of fundamental relations between hyperbolic and spherical Fibonacci functions which are presented in a comparative table (Table 3.3).

Chapter 4 unites the Mathematics of Harmony by Stakhov [39] with the qualitative theory of differential equations and dynamical systems as developed by Aranson in his doctoral dissertation (1990) and outlined in several mathematical works [40–46].

Chapter 5 presents for the first time in modern science an original solution to the Physical Millennium Problem, formulated in 2000 by David Gross. We describe in Chapter 5 the dependence of the fine-structure constant α and another fundamental constant, the ratio of the proton mass M to the electron mass m, from the age of the Universe, starting from the Big Bang.

Contents

About the Authors

Alexey Stakhov and Samuil Aranson were already well-known scholars at the beginning of their scientific cooperation in 2007. A Doctor of Physics and Mathematics, Samuil Aranson is a recognized mathematician in differential equations, geometry and topology. The theory of qualitative dynamical systems is his scientific specialty. Aranson's development as a scientist and his scientific activities are inextricably linked with the Russian school of nonlinear oscillations created by academician A. A. Andronov in Gorky (Nizhny Novgorod), and refers to the classic area of mathematics, the qualitative theory of ordinary differential equations. The prominent Russian mathematicians Evgeniya Leontovich-Andronova, Dmitry Anosov and Yakov Sinai (who was awarded the Abel Prize in 2014) played very important roles in the development of Aranson as a scientist and mathematician.

A Doctor of Computer Science, Alexey Stakhov is the creator of many original theories in computer science and mathematics, including algorithmic measurement theory, codes of the golden proportions, hyperbolic Fibonacci and Lucas functions, Fibonacci Q_p-matrices, a generalization of the Cassini formula for Fibonacci λ-numbers, and a theory of golden matrices, the Mathematics of Harmony.

The development of Stakhov as a scientist is inextricably linked with the Kharkov Institute for Radio Electronics, where he was a postgraduate student of the Technical Cybernetics Department from 1963 to 1966. There he defended his Ph.D. thesis in the field of Technical Cybernetics under the leadership of Professor Akexander Volrov. In 1972 Stakhov defended (at age 32) his doctoral dissertation "Synthesis of optimal algorithms for analog-to-digital conversion" (Computer Science specialty). Although the dissertation had an engineering character, Stakhov touched upon two fundamental problems of mathematics, measurement theory and numeral systems.

The prominent Ukrainian mathematician and head of the Ukrainian Mathematical School, Yuri Mitropolsky, gave high praise for the Mathematics of Harmony by Stakhov. Academician Mitropolsky organized Stakhov's speech at the meeting of the Ukrainian Mathematical Society in 1998. Based upon his recommendation, Stakhov's articles were published in Ukrainian academic journals, in particular, the *Ukrainian Mathematical Journal*. Under his direct influence, Stakhov started writing the book *The Mathematics of Harmony* [39] which was published by World Scientific in 2009 following the death of academician Mitropolsky in 2008.

Acknowledgments

This book is dedicated to the memory of four eminent scientists: Evgeniya Alexandrovna Leontovich-Andronova, Yuri Alexeyevich Mitropolsky, Dmitriy Viktorovich Anosov, and Alexander Andreevich Volkov. These scientists played important roles in the scientific development of both Stakhov and Aranson.

The authors are grateful to Professor M. W. Wong (York University, Canada) for the invitation to contribute a book to the Series on Analysis, Applications and Computation published by World Scientific.

The authors express their gratitude to the members of their families, in particular, their wives Antonina Stakhova and Inna Aranson for providing excellent conditions for the authors to work on the book.

The authors also express their gratitude to Yuri Gotsman for participation in the preparation of the manuscript for publication, in particular, the covers and drawings of the book.

Finally, the authors thank the American philosopher and expert in the field of the "Golden Section", Professor Scott Olsen for the English editing of the book.

References to the Front Matter

[1] Coxeter, H. S. M. *Introduction to Geometry*. New York: John Wiley and Sons (1961).

[2] Vorobyov, N. N. *Fibonacci Numbers*. Moscow: Nauka (1984), (first edition 1961) (Russian).

[3] Hoggat, Jr. V. E. *Fibonacci and Lucas Numbers*. Boston, MA: Houghton Mifflin (1969).

[4] Vajda, S. *Fibonacci & Lucas Numbers, and the Golden Section. Theory and Applications*. Ellis Horwood Limited (1989).

[5] "Lucas sequence", *Wikipedia, The Free Encyclopedia*, http://en.wikipedia. org/wiki/Lucas_sequence (accessed November 4, 2015).

[6] Stakhov, A. P. and Tkachenko, I. S. "Hyperbolic Fibonacci trigonometry", *Reports of the Ukrainian Academy of Sciences* (1993) Vol. 208, No. 7: 9–14 (Russian).

[7] Stakhov, A. P. and Rozin, B. N. "On a new class of hyperbolic functions", *Chaos, Solitons & Fractals* (2004) Vol. 23: 379–389.

[8] Stakhov, A. P. "A Theory of Fibonacci λ-numbers", Academy of Trinitarism, Moscow: Electronic number 77-6567, publication 17407 (2012) (Russian). http://www.trinitas.ru/rus/doc/0232/009a/02321250.htm (accessed November 4, 2015).

[9] Stakhov, A. P. "A generalization of the Cassini formula", *Visual Mathematics* (2012) Vol. 14, No. 2. http://www.mi.sanu.ac.rs/vismath/stak-hovsept2012/cassini.pdf (accessed November 4, 2015).

[10] de Spinadel, V. W. *From the Golden Mean to Chaos*. Nueva Libreria (first edition 1998); Nobuko (second edition 2004).

[11] Gazale, M. J. *Gnomon. From Pharaohs to Fractals*. Princeton, NJ: Princeton University Press (1999).

[12] Kappraff, J. *Beyond Measure. A Guided Tour Through Nature. Myth and Number*. Singapore: World Scientific (2002).

[13] Tatarenko, A. "The golden T_m-harmonies and D_m-fractals is an essence of soliton-similar m-structure of the world", Academy of Trinitarism, Moscow: Electronic number 77-6567, publication 12691 (2005) (Russian). http:// www.trinitas.ru/rus/doc/0232/009a/02320010.htm (accessed November 4, 2015).

[14] Arakelyan, H. *Numbers and Magnitudes in Modern Physics*. Armenian Academy of Sciences (1989) (Russian).

[15] Shenyagin, V. "Pythagoras, or how every man creates its own myth. The fourteen years after the first publication of the quadratic mantissa s-proportions", Academy of Trinitarism, Moscow: Electronic number 77-6567, publication 17031 (2011) (Russian). http://www.trinitas.ru/rus/doc/0232/013a/02322050.htm (accessed November 4, 2015).

[16] Kosinov, N. V. "The golden ratio, golden constants, and golden theorems", Academy of Trinitarism, Moscow: Electronic number 77-6567, publication 14379 (2007) (Russian). http://www.trinitas.ru/rus/doc/0232/009a/02321049.htm (accessed November 4, 2015).

[17] Falcon, S. and Plaza, A. "On the Fibonacci k-numbers", *Chaos, Solitons & Fractals* (2007) Vol. 32, Issue 5: 1615–1624.

[18] Stakhov, A. P. "Gazale's formulas, a new class of hyperbolic Fibonacci and Lucas Functions and the improved method of the 'golden' cryptography", Academy of Trinitarism, Moscow: Electronic number 77-6567, publication 14098 (2006) (Russian). http://www.trinitas.ru/rus/doc/0232/004a/02321063.htm (accessed November 4, 2015).

[19] Stakhov, A. P. "On the general theory of hyperbolic functions based on the hyperbolic Fibonacci and Lucas functions and on Hilbert's Fourth Problem", *Visual Mathematics* (2013), Vol. 15, No. 1. http://www.mi.sanu.ac.rs/vismath/2013stakhov/hyp.pdf (accessed November 4, 2015).

[20] Stakhov, A. P., Aranson, S. Kh. and Khanton, I. V. "The golden Fibonacci goniometry, the resonance structure of the genetic code of DNA, Fibonacci–Lorentz transformations and other applications. Part I", Academy of Trinitarism, Moscow: Electronic number 77-6567, publication 14778 (2008) (Russian).

[21] Stakhov, A. P., Aranson, S. Kh. and Khanton, I. V. "The golden Fibonacci goniometry, the resonance structure of the genetic code of DNA, Fibonacci–Lorentz transformations and other applications. Part II. Resonance structure of the genetic code of DNA", Academy of Trinitarism, Moscow: Electronic number 77-6567, publication 14779 (2008) (Russian).

[22] Stakhov, A. P., Aranson, S. Kh. and Khanton, I. V. "The golden Fibonacci goniometry, the resonance structure of the genetic code of DNA, Fibonacci–Lorentz transformations and other applications. Part III. Fibonacci–Lorentz transformations and their relationship with the 'golden' universal genetic code", Academy of Trinitarism, Moscow: Electronic number 77-6567, publication 14782 (2008) (Russian).

[23] Stakhov, A. P., Aranson, S. Kh. and Khanton, I. V. "The golden Fibonacci goniometry, the resonance structure of the genetic code of DNA, Fibonacci–Lorentz transformations and other applications. Part IV. Other applications of Fibonacci numbers, the golden ratio and the Fibonacci golden goniometry," Academy of Trinitarism, Moscow: Electronic number 77-6567, publication 14783 (2008) (Russian).

[24] Stakhov, A. P. and Aranson, S. Kh. "The golden Fibonacci goniometry, Fibonacci–Lorentz transformations and Hilbert's Fourth Problem", Academy of Trinitarism, Moscow: Electronic number 77-6567, publication 14816 (2008) (Russian).

[25] Stakhov, A. P. and Aranson, S. Kh. "The 'golden' Fibonacci goniometry, Hilbert's fourth problem, Fibonacci–Lorentz transformations and the 'golden' interpretation of the special theory of relativity", Academy of Trinitarism, Moscow: Electronic number 77-6567, publication 15225 (2009) (Russian).

[26] Stakhov, A. P. and Aranson, S. Kh. "The mathematics of harmony and Hilbert's fourth problem: the way to the harmonic hyperbolic and spherical worlds of nature", Academy of Trinitarism, Moscow: Electronic number 77-6567, publication 18814 (2014) (Russian).

[27] Stakhov, A. P. and Aranson, S. Kh. "The 'golden' Fibonacci goniometry, Fibonacci–Lorentz transformations, and Hilbert's fourth problem", *Congressus Numerantium* (2008) Vol. 193: 119–156 (International Journal, Canada, USA).

[28] Stakhov, A. P. and Aranson, S. Kh. "Hyperbolic Fibonacci and Lucas functions, 'golden' Fibonacci goniometry, Bodnar's geometry, and Hilbert's fourth problem. Part I. Hyperbolic Fibonacci and Lucas functions and 'golden' Fibonacci goniometry", *Applied Mathematics* (2011) Vol. 2, No. 1: 74–84.

[29] Stakhov, A. P. and Aranson, S. Kh. "Hyperbolic Fibonacci and Lucas functions, 'golden' Fibonacci goniometry, Bodnar's geometry, and Hilbert's fourth problem. Part II", *Applied Mathematics* (2011) Vol. 2, No. 2: 181–188.

[30] Stakhov, A. P. and Aranson, S. Kh. "Hyperbolic Fibonacci and Lucas functions, 'golden' Fibonacci goniometry, Bodnar's geometry, and Hilbert's fourth problem. Part III", *Applied Mathematics* (2011) Vol. 2, No. 3: 283–293.

[31] Stakhov, A. P. "A history, the main mathematical results and applications for the mathematics of harmony", *Applied Mathematics* (2014) Vol. 5, No. 3: 283–293.

[32] Stakhov, A. P. "Hilbert's fourth problem: Searching for harmonic hyperbolic worlds of nature", *Journal of Applied Mathematics and Physics* (2013) Vol. 1: 60–66.

[33] Stakhov, A. P. and Aranson, S. Kh. "The mathematics of harmony, Hilbert's fourth problem and Lobachevsky's new geometries for physical world", *Journal of Applied Mathematics and Physics* (2014) Vol. 2, No. 7: 457–494.

[34] Stakhov, A. P. and Aranson, S. Kh. *The Mathematics of Harmony and Hilbert's Fourth Problem. The Way to the Harmonic Hyperbolic and Spherical Worlds of Nature.* Germany: Lambert Academic Publishing (2014).

[35] Bodnar, O. Y. *The Golden Section and Non-Euclidean Geometry in Nature and Art.* Lvov: Publishing House "Svit" (1994) (Russian).

[36] Bodnar, O. Y. "Dynamic symmetry in nature and architecture", *Visual Mathematics* (2010) Vol. 12, No. 4. http://www.mi.sanu.ac.rs/vismath/BOD2010/index.html (accessed November 4, 2015).

[37] Bodnar, O. Y. "Geometric interpretation and generalization of the non-classical hyperbolic functions", *Visual Mathematics* (2011) Vol. 13, No. 2. http://www.mi.sanu.ac.rs/vismath/bodnarsept2011/SilverF.pdf (accessed November 4, 2015).

[38] Bodnar, O. Y. "Minkovski's geometry in the mathematical modeling of natural phenomena", *Visual Mathematics* (2012) Vol. 14, No. 1. http://www.mi. sanu.ac.rs/vismath/bodnardecembar2011/mink.pdf (accessed November 4, 2015).

[39] Stakhov, A. P., assisted by S. Olsen. *The Mathematics of Harmony. From Euclid to Contemporary Mathematics and Computer Science.* Singapore: World Scientific (2009).

[40] Anosov, D. V., Aranson, S. Kh., Arnold, V. I., Bronshtein, I. U., Grines, V. Z. and Il'yashenko, Yu. S. *Dynamical Systems - I. Ordinary Differential Equations. Smooth Dynamical Systems.* Moscow: Publisher VINITI (1985) (Russian).

[41] Anosov, D. V., Aranson, S. Kh., Arnold, V. I., Bronshtein, I. U., Grines, V. Z. and Il'yashenko, Yu. S. "Dynamical systems I. Ordinary differential equations and smooth dynamical systems", in *Encyclopedia of Mathematical Sciences.* Berlin: Springer-Verlag (1988).

[42] Anosov, D. V., Aranson, S. Kh., Grines, V. Z., Plykin, R. V., Sataev, E. A., Safonov, E. A., Solodov, V. V., Starkov, A. N., Stepin, A. M. and Shlyachkov, S. V. *Dynamical Systems with Hyperbolic Behavior.* Moscow: Publisher VINITI (1991) (Russian).

[43] Anosov, D. V., Aranson, S. Kh., Grines, V. Z., Plykin, R. V., Sataev, E. A. Safonov, E. A., Solodov, V. V., Starkov, A. N., Stepin, A. M. and Shlyachkov, S. V. "Dynamical systems with hyperbolic behavior", in *Encyclopaedia of Mathematical Sciences. Dynamical Systems IX.* Berlin: Springer (1995).

[44] Aranson, S. Kh., Belitsky, G. R. and Zhuzhoma, E. V. *Introduction to the Qualitative Theory of Dynamical Systems on Surfaces.* Providence, RI: American Mathematical Society (1996).

[45] Anosov, D. V., Aranson, S. Kh., Arnold, V. I., Bronshtein, I. U., Grines, V. Z. and Il'yashenko, Yu. S. *Ordinary Differential Equations and Smooth Dynamical Systems.* Berlin: Springer (1997).

[46] Aranson S. Kh., Medvedev, V. and Zhuzhoma, E. "Collapse and continuity of geodesic frameworks of surface foliations", in *Methods of Qualitative Theory of Differential Equations and Related Topics.* Providence, RI: American Mathematical Society (2000).

Chapter 1

The Golden Ratio, Fibonacci Numbers, and the "Golden" Hyperbolic Fibonacci and Lucas Functions

1.1. The Idea of Harmony and the Golden Section in the History of Science

Scientific and technological progress has a long history of development passing through ancient Babylon, Egypt, China, India, Greece, the Middle Ages, Renaissance, Enlightenment, and the 19th century. Scientific and technological developments have continued into the 20th and 21st centuries with a new but very ancient paradigm springing forth — one centered on Harmony.

In the ancient world a series of outstanding mathematical discoveries were made influencing both material and spiritual development. Fundamental developments included the Babylonian sexagesimal notation, the positional principle of number representation underlying our decimal and binary number systems, trigonometry, Euclidean geometry, the foundations of number theory, incommensurable line segments, a theory of irrational numbers, and most importantly the golden ratio and Platonic solids.

These developments tend to build upon one another. Continuity in the development of science can be expressed in various ways. Fundamental scientific ideas permeate all stages of progress, technological, philosophic and aesthetic. Disregarding prior discoveries can stifle future progress.

The principle of Harmony based upon the golden ratio is part and parcel of such fundamental scientific ideas. According to B. G. Kuznetsov, Albert Einstein believed that science, physics in particular, had as its fundamental goal "to find objective harmony in the labyrinth of the observed facts" [1]. Furthermore Einstein stated that "the scientists' religious feeling takes the form of a rapturous amazement at the harmony of natural law ..." [2]. This

quote demonstrates Einstein's deep faith in the existence of universal laws
of the Harmony of the Universe.

Alexey Losev and Johannes Kepler on the golden section.   What is the
subtle underlying idea of ancient Greek science? Some researchers give the
following answer: the idea of Harmony associated with the golden ratio.
As is known, in ancient Greek philosophy, Harmony was in opposition to
Chaos and meant organization and order of the Universe. Alexey Losev,
Russian philosopher of the aesthetics of antiquity and the Renaissance, evaluates the main achievements of the ancient Greeks in this area as follows [3]:

> *"From Plato's point of view, and in general in terms of the entire ancient
> cosmology, the Universe is determined as a certain proportional whole
> that obeys the law of harmonic division — the golden ratio .... The ancient
> Greek system of cosmic proportions in the literature is often interpreted
> as a curious result of unrestrained and wild imagination. In such expla-
> nations we can see the unscientific helplessness of those who claim it.
> However, we can understand this historical and aesthetic phenomenon
> only in connection with a holistic understanding of history, that is, by
> using a dialectical view of culture and looking for the answer in the pecu-
> liarities of ancient social life."*

Here Losev formulates the "golden" paradigm of ancient cosmology. It
is based upon the most important ideas of ancient science that are sometimes treated in modern science as a "curious result of an unrestrained
and wild imagination." First of all, we are talking about the Pythagorean
Doctrine of the Numerical Harmony of the Universe and Plato's Cosmology based on the Platonic solids. Referring to the geometrical structure of
the Cosmos and its arithmetical relations expressing Cosmic Harmony, the
Pythagoreans anticipated the modern mathematical basis of the natural
sciences, which began to develop rapidly in the 20th century. Pythagoras'
and Plato's idea about Cosmic Harmony has proven to be immortal.

Thus, the idea of Harmony underlies the ancient Greek mathematical
doctrine of Nature, and therefore ancient Greek mathematics was the Mathematics of Harmony. This is directly related to the golden ratio, perhaps
one of the most important "harmonic" discoveries in ancient mathematics.

Johannes Kepler, brilliant astronomer and author of Kepler's Laws,
expressed his admiration for the golden ratio in the following words [4]:

> *"Geometry has two great treasures: one is the theorem of Pythagoras;
> the other the division of a line into extreme and mean ratio. The first we
> may compare to a measure of gold; the second we may name a precious
> jewel."*

You will recall that the ancient problem of the division of a line segment in extreme and mean ratio was Euclid's language for the golden ratio!

The ancient Greek mathematical doctrine of Nature. According to the American historian of mathematics, Morris Kline [5], the main contribution of the ancient Greeks, "which had a decisive influence on all subsequent culture, was the fact that they began to study the laws of Nature." The main conclusion to be drawn from Kline's quote was that the ancient Greeks proposed an innovative concept of the Cosmos, where everything was subordinated to mathematical laws. The question is: what were the main ideas and mathematical objects chosen by the ancient Greeks as the basis for this mathematical doctrine of Nature? The answer is contained in the aforementioned quotes of Losev and Kepler, and also in Plato's cosmology and doctrine of the structure of matter. It follows that the main idea of the Greek mathematical doctrine of Nature was "the idea of Harmony", which is associated with the golden ratio and Platonic solids. According to Kline [5], this innovative mathematical view on the Cosmos was developed in ancient Greece during the 6th to 3rd centuries BCE. But according to Andrey Kolmogorov [6], in the same period in ancient Greece, "mathematics arose as an independent science with a clear understanding of the originality of its method and the necessity for systematic development of its basic concepts and proposals in a rather general way."

But then the question arises: is there any relationship between their developments of a doctrine of Nature and that of mathematics, or were they two separate developments? It turns out that such a relationship really did exist. It is in fact arguable that Greek mathematics and their doctrine of Nature is one and the same scientific discipline. Euclid's *Elements*, written in the 3rd century BCE, are the most striking confirmation of this. Proclus' hypothesis, outlined below, is a novel view on Euclid's *Elements*, which radically alters our understanding of the history of mathematics.

1.2.   Proclus' Hypothesis: A New View on Euclid's *Elements* and the History of Mathematics

1.2.1.   *Regular polyhedra*

Polyhedra are "solid" three-dimensional figures similar to two-dimensional polygons (triangles, squares, pentagons, etc.) (see Fig. 1.1).

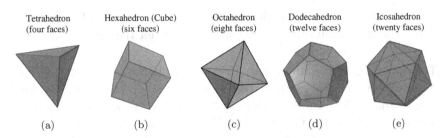

| Tetrahedron
(four faces) | Hexahedron (Cube)
(six faces) | Octahedron
(eight faces) | Dodecahedron
(twelve faces) | Icosahedron
(twenty faces) |

| (a) | (b) | (c) | (d) | (e) |

Figure 1.1.   Platonic solids: (a) tetrahedron, (b) cube, (c) octahedron, (d) dodecahedron, (e) icosahedron.
Source: Wikipedia, The Free Encyclopedia, http://en.wikipedia.org/wiki/Platonic_solid

Polyhedra have vertices, edges, and faces. If a polyhedron has faces, which are regular polygons, and if at each vertex exactly the same number of faces meet, then we have a regular polyhedron. How many regular polyhedra exist? In Euclid's *Elements* we find a strict proof that only five convex regular polyhedra exist, and their faces can be composed of only three types of regular polygons: triangles, squares and pentagons (Fig. 1.1).

Many books are devoted to the theory of polyhedra. The book *Polyhedron Models* by English mathematician M. Wenninger [7] begins with a description of convex regular polyhedra, that is, polyhedra formed from a single type of polygon. These polyhedra are named Platonic solids in honor of Plato, who used them in his cosmology in the *Timaeus*. We begin our consideration with the simplest regular polyhedron, the tetrahedron, whose faces are equilateral triangles (Fig. 1.1(a)). One key observation is that the sum of the interior angles of the polygons meeting at every vertex is always less than 360°. In the tetrahedron (Fig. 1.1(a)), three equilateral triangles meet at one vertex (the sum of their interior angles is equal to 180° = 3 × 60°); thus, their common base forms a new equilateral triangle. The tetrahedron has the least number of faces amongst the Platonic solids and is the three-dimensional analog to the simplest two-dimensional regular polygon, the equilateral triangle. And it has the least number of edges among the regular polygons.

The next geometric solid formed from equilateral triangles is referred to as an octahedron (Fig. 1.1(c)). In the octahedron four equilateral triangles meet at one vertex (the sum of their interior angles is equal to 240° = 4 × 60°); resulting in a pyramid with a square base. If one connects two such pyramids at their bases, then the polyhedron with eight triangular faces, the octahedron, emerges.

Next we can connect 5 equilateral triangles (the sum of the interior angles is equal to $300° = 5 \times 60°$) at one vertex. Combining four of these will result in a polyhedron with 20 triangular faces, namely the icosahedron (Fig. 1.1(e)).

A square is the next regular polygon (whose interior angle is equal to 90°). If we unite 3 squares at one vertex (the sum of their interior angles is equal to $270° = 3 \times 90°$) and then add three new squares, we obtain a perfect polyhedron with 6 faces called a hexahedron or cube (Fig. 1.1(b)).

Finally, there is one more opportunity to construct a regular polyhedron, one based on the pentagon with an interior angle of 108°. If we connect 12 pentagons so that 3 regular pentagons meet at each vertex (the sum of their interior angles is equal to $324° = 3 \times 108°$), then we will get the last Platonic solid, the dodecahedron (Fig. 1.1(d)).

A hexagon is the next regular polygon after the pentagon. A hexagon has an interior angle of 120°. If we connect 3 hexagons at one vertex, it lies flat because the sum of their interior angles equals $360° = 3 \times 120°$. It means that it is impossible to construct a spatial geometric figure from only hexagons. Other regular polygons with more than six edges have interior angles greater than 120°. It means that we cannot form spatial geometric figures from them. It follows from this consideration that there are only 5 convex regular polyhedra, all of whose faces can be formed from either equilateral triangles, squares or pentagons.

There are surprising geometrical connections between all regular polyhedra. For example, the cube and the octahedron are dual to each other. That is to say they can be obtained from one another, if the centroids of the faces of the first solid are taken as the vertices of the other and conversely. Similarly, as the cube (Fig. 1.1(b)) is dual to the octahedron (Fig. 1.1(c)), the icosahedron is dual to the dodecahedron (Fig. 1.1(e)). The tetrahedron is dual to itself. That only 5 convex regular polyhedra exist is surprising since there are an infinite number of regular polygons!

1.2.2. *Three "key" problems of ancient mathematics*

Proclus' hypothesis, formulated in the 5th century of our era by the Greek philosopher and mathematician Proclus Diadochus (412–485), contains a surprising view on Euclid's *Elements*.

According to Proclus, Euclid's goal was not simply to set forth geometry itself, but to build a complete theory of the regular polyhedra (Platonic solids). This theory was outlined by Euclid in the concluding or thirteenth book of the *Elements*. This fact in itself is an indirect confirmation of

Proclus' hypothesis. Typically the most important scientific information is placed in the concluding part of a scientific book.

To attain this goal, Euclid included the necessary mathematical information in his *Elements*. It is significant that he introduced the golden ratio in Book II, later used for the creation of the geometric theory of the dodecahedron (see below). In Plato's Cosmology, the regular polyhedra were associated with the harmony of the Universe. This means that Euclid's *Elements* are based on the harmonic ideas of Pythagoras and Plato, that is, Euclid's *Elements* are historically the first variant of the Mathematics of Harmony. This surprising view on the *Elements* may lead us to conclusions which radically alter our view on the history and structure of mathematics.

In his book [6] the Russian mathematician Andrey Kolmogorov identified the "key" problems, which stimulated the origin and development of Classical Mathematics: namely the counting problem and the measurement problem. However, it follows from Proclus' hypothesis that a third "key" problem, the problem of harmony, which underlies Euclid's *Elements*, stimulated the origin and development of Harmony Mathematics.

Kolmogorov's "key" problems and following from them Classical Mathematics are presented in Fig. 1.2(a).

Harmony Mathematics and its applications, following from Proclus' hypothesis, are presented in Fig. 1.2(b).

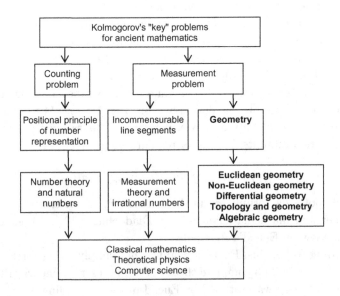

Figure 1.2(a).   Kolmogorov's "key" problems for ancient mathematics.

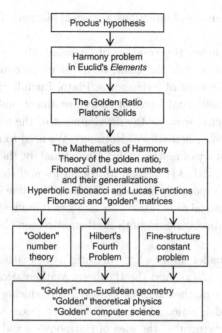

Figure 1.2(b). Harmony Mathematics and its applications, following from Proclus' hypothesis.

The counting problem resulted in the discovery of natural numbers and the measurement problem in the discovery of irrational numbers (Fig. 1.2(a)). The natural and irrational numbers underlie Classical Mathematics. The harmony problem, connected with the division of a line segment in extreme and mean ratio (the golden ratio) resulted in the ancient Mathematics of Harmony (Fig. 1.2(b)), which can be looked upon as a new interdisciplinary direction of contemporary science and mathematics [8].

This result may be surprising to many mathematicians. As it turns out, parallel to the traditional view of Classical Mathematics, another mathematical approach — that of the Mathematics of Harmony — was also developing alongside it.

Similar to Classical Mathematics, Harmony Mathematics has its origin in Euclid's *Elements*. However, Classical Mathematics focuses on the axiomatic approach and other ancient achievements (number theory, theory of irrationals, and so on), while the Mathematics of Harmony is based on the golden ratio (Proposition II.11) and Platonic solids, described in Book XIII of Euclid's *Elements*. Thus, Euclid's *Elements* are a source of the two independent mathematical directions: Classical Mathematics and Harmony Mathematics.

For many centuries, the creation of Classical Mathematics, Queen of the Natural Sciences, was the main focus of mathematicians. However, developing with the work of Pythagoras, Plato, Euclid, Fibonacci, Pacioli, and Kepler, the intellectual forces of many prominent mathematicians and thinkers were directed towards the development of the basic concepts and applications of the Mathematics of Harmony. We need to recall this important fact in the history of mathematics. Unfortunately, these two important mathematical approaches (Classical Mathematics and the Mathematics of Harmony) evolved separately from one another. The time has come to unite these two mathematical directions. This important unification can now lead to novel scientific discoveries in both mathematics and the theoretical natural sciences.

A new and broader approach to the origins of mathematics, Classical Mathematics (Fig. 1.2(a)) and the Harmony Mathematics (Fig. 1.2(b)), is very important for mathematical education, introducing the idea of harmony and the golden ratio. This provides pupils access to ancient science and its main achievement — the idea of Harmony — and tells them about the most important architectural and sculptural works of the ancient world based upon the golden ratio (including Khufu's pyramid (Cheops), Nefertiti, Parthenon, Doryphoros, Venus, etc.).

1.2.3.   A discussion of Proclus' hypothesis in the historical-mathematical literature

The analysis of Proclus' hypothesis is found in many mathematical books, such as [9–11]. In the book *Pythagoras and the Pythagoreans: A Brief History* [9] C. H. Kann states: "*According to Proclus, the main objective of the 'Elements' was to present the geometric construction of the so-called Platonic solids.*"

L. Zhmud in his book [10] argues that this idea got a further development:

"*Proclus, by mentioning all previous mathematicians of Plato's circle, said: 'Euclid lived later than the mathematicians of Plato's circle, but earlier than Eratosthenes and Archimedes .... He belonged to Plato's school and was well acquainted with Plato's philosophy and his cosmology; that's why he put the creation of the geometric theory of the so-called Platonic solids as the main purpose of the Elements.*"

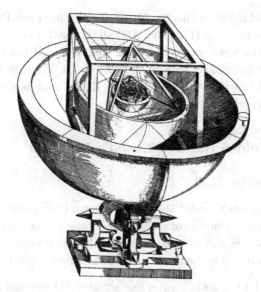

Figure 1.3.　Cosmic Cup — Kepler's model of the Solar system based on Platonic solids.
Source: Wikipedia, The Free Encyclopedia, http://en.wikipedia.org/wiki/Platonic_solid

This important comment draws attention to the deep connection between Euclid and Plato. Euclid, who spent time in the Academy in Athens, fully shared Plato's Pythagorean philosophy and cosmology, based upon the Platonic solids. That is why Euclid suggested that the creation of the geometric theory of Platonic solids was the underlying purpose of his *Elements*.

C. Smorinsky in his book *History of Mathematics* [11] discussed the influence of Plato and Euclid on Kepler's design of the Cosmic Cup (Fig. 1.3):

> "*Kepler's project in 'Mysterium Cosmographicum' was to give 'true and perfect reasons for the numbers, quantities, and periodic motions of celestial orbits.' The perfect reasons must be based on the simple mathematical principles, which had been discovered by Kepler in the Solar system, by using geometric demonstrations. The general scheme of his model was borrowed by Kepler from Plato's 'Timaeus', but the mathematical relations for the Platonic solids (pyramid, cube, octahedron, dodecahedron, icosahedron) were taken by Kepler from the works by Euclid and Ptolemy. Kepler followed Proclus and believed that 'the main goal of Euclid was to build a geometric theory of the so-called Platonic solids.' Kepler was fascinated by Proclus and often quotes him calling him a 'Pythagorean'.*"

We can conclude from this quote that Kepler used the Platonic solids to create the Cosmic Cup (Fig. 1.3). And all the mathematical relations for the Platonic solids were borrowed by him from Book XIII of the *Elements*, unifying Plato's Cosmology with Euclid's *Elements*. Thus, he fully believed Proclus' hypothesis that the main goal of Euclid was to give a complete geometric theory of the Platonic solids (Fig. 1.1).

1.3.   The Golden Ratio in Euclid's *Elements*

1.3.1.   *Proposition II.11*

In Euclid's *Elements* we meet the concept which later plays such a significant role in the development of science. This concept is called the "division of a line in extreme and mean ratio" (DEMR). In *Elements* this concept occurs in two forms. The first is formulated in Proposition 11 of Book II.

Proposition II.11. *Divide a given line segment* AD *into two unequal parts* AF *and* FD *so that the area of the square, which is built on the larger segment* AF *would be equal to the area of the rectangle, which is built on the segment* AD *and the smaller segment* FD.

Depict this problem geometrically (Fig. 1.4).

Thus, according to Proposition II.11, the area of the square AGHF should be equal to the area of the rectangle ABCD. If we denote the length of the larger segment AF as b (which is equal to the side of the square AGHF), and the side of the smaller segment FD as a (which is equal to the vertical side of the rectangle ABCD), then the condition for Proposition II.11 can be written as follows:

$$b^2 = a \times (a + b). \tag{1.1}$$

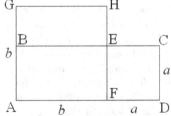

Figure 1.4.   A division of a line segment in extreme and mean ratio (the golden ratio).

A •————————————•————————•————————————• B
                                  C

Figure 1.5.   The golden ratio.

In Euclid's *Elements,* we meet another form of the golden ratio. This form follows from the first one, given by (1.1), if we make the following simple transformations. By dividing both parts of (1.1) at first by a, and then by b, we get the following proportion:

$$\frac{b}{a} = \frac{a+b}{b}. \tag{1.2}$$

The proportion (1.2) has the following geometric interpretation (Fig. 1.5). Divide line segment AB at point C such that the larger part CB is related to the smaller part AC in the same way that segment AB is related to its larger part CB (Fig. 1.5), that is:

$$\frac{AB}{CB} = \frac{CB}{AC}. \tag{1.3}$$

This is the definition of the golden ratio, used in modern science.

We depict the proportion (1.3) by x. Then taking into consideration the fact that AB = AC + CB, the proportion (1.2) can be written as follows:

$$x = \frac{AC + CB}{CB} = 1 + \frac{AC}{CB} = 1 + \frac{1}{\frac{CB}{AC}} = 1 + \frac{1}{x}. \tag{1.4}$$

The following algebraic equation is derived from (1.4):

$$x^2 - x - 1 = 0. \tag{1.5}$$

Real world applications of the proportion (1.2) imply that we should use the positive root of the equation (1.5). We name this root the golden ratio or golden section and denote it by Φ:

$$\Phi = \frac{1 + \sqrt{5}}{2}. \tag{1.6}$$

1.3.2.   *How did Euclid use the golden ratio?*

The question arises: why did Euclid introduce different forms of the golden ratio in *Elements* which we can find in Books II, VI and XIII? To answer this question, we again return to the Platonic solids (Fig. 1.1). Only three types of regular polygons can be the faces of the Platonic solids: the equilateral triangle (tetrahedron, octahedron, icosahedron), the square (cube) and the

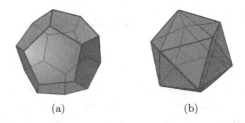

Figure 1.6.   (a) Dodecahedron and (b) icosahedron.
Source: Wikipedia, The Free Encyclopedia, http://en.wikipedia.org/wiki/Platonic_solid

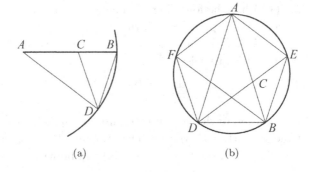

Figure 1.7.   (a) The "golden" isosceles triangle, and (b) the pentagon.

regular pentagon (dodecahedron). In order to construct the Platonic solids, we must first of all be able to build the two-dimensional faces of Platonic solids geometrically (i.e. with a ruler and compass).

It is for this purpose, that Euclid introduced the golden ratio in Book II (Proposition II.11), which is presented in two forms in *Elements*. By using the golden ratio, Euclid constructs the "golden" isosceles triangle, whose base angles are each equal to double the vertex angle (Fig. 1.7(a)).

First we construct the "golden" isosceles triangle by using a ruler and compass (see Fig. 1.7(a)). The triangle ABD has equal sides AB and AD and equal angles B and D at the base BD. These angles are equal to double the angle at the vertex A.

By using the "golden" isosceles triangle ABD (Fig. 1.7) we can construct the regular pentagon in Fig. 1.7(b).

Then only one step remains to construct the dodecahedron (Fig. 1.6(a)), which for Plato is one of the most important regular polyhedra symbolizing the universal harmony of his cosmology.

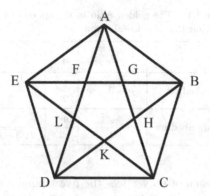

Figure 1.8. The regular pentagon and pentagram.

Note that the regular pentagon (Fig. 1.8), which is the basis of the dodecahedron, was a very important geometric figure in Greek science because the golden ratio is the basis of the construction of the regular pentagon. The word "pentagon" is derived from the Greek word "pentagonon".

If we draw all the diagonals in Fig. 1.8 then we get the pentagonal star, also known as a pentagram or pentacle. The name "pentagram" is derived from the Greek word "pentagrammon" (pente — five and grammon — line).

The points F, G, H, K, L of the intersection of the pentagon's diagonals are always golden section cuts. At the same time, they form a new regular pentagon FGHKL. In the new pentagon we may draw further diagonals. The points of their intersection produce another smaller regular pentagon and can be continued indefinitely. Thus, the pentagon ABCDE contains an infinite number of regular pentagons, which are formed each time by the intersection points of the diagonals. This endless repetition of the same geometric figure (the regular pentagon) creates an aesthetic sense of rhythm and harmony.

The Pythagoreans were delighted with the pentagram, which was considered their distinctive symbol of recognition.

1.3.3. *The golden ratio in the dodecahedron and icosahedron*

The dodecahedron (Fig. 1.6(a)) and its dual the icosahedron (Fig. 1.6(b)) have a special role among the Platonic solids. First of all, we emphasize that the geometry of the dodecahedron and the icosahedron relate directly to the golden ratio. Indeed, all faces of the dodecahedron are regular pentagons, based on the golden ratio. If we look closely at the icosahedron in Fig. 1.6(b),

Table 1.1.   The golden ratio in the spheres of the dodecahedron and icosahedron.

	R_c	R_m	R_i
Icosahedron	$\dfrac{1}{2}\Phi\sqrt{3-\Phi}$	$\dfrac{1}{2}\Phi$	$\dfrac{\frac{1}{2}\Phi^2}{\sqrt{3}}$
Dodecahedron	$\dfrac{\Phi\sqrt{3}}{2}$	$\dfrac{\Phi^2}{2}$	$\dfrac{\Phi^2}{2\sqrt{3-\Phi}}$

we can see that in each of its vertices the five triangles come together and their outer edges form a regular pentagon. These facts alone reveal the crucial role the golden ratio plays in the geometric construction of these two Platonic solids.

There are other deep mathematical confirmations of the fundamental role of the golden ratio in the icosahedron and dodecahedron. For example the Platonic solids have three specific spheres. The first (inner) sphere is inscribed inside the Platonic solid and touches the centers of its faces. We denote the radius of the inner sphere as R_i. The second or middle sphere of the Platonic solid touches its ribs. We denote the radius of the middle sphere as R_m. Finally, the third (outer) sphere is circumscribed around the Platonic solid and passes through its vertices. We denote its radius as R_c. In geometry, it is demonstrated that the values of the radii of these spheres for the dodecahedron and the icosahedron with an edge of unit length is expressed through the golden ratio (Table 1.1).

Note that the ratio of the radii $\frac{R_c}{R_i} = \frac{\sqrt{3(3-\Phi)}}{\Phi}$ is the same for both the icosahedron and dodecahedron. Thus, if the dodecahedron and icosahedron have the same inner spheres, their outer spheres are equal as well. This is a reflection of the hidden harmony in both the dodecahedron and icosahedron.

1.3.4.   Icosahedron as the main geometrical object of mathematics

In the late 19th century, prominent German mathematician Felix Klein drew a great deal of attention to the Platonic solids. He predicted an outstanding role for them, particularly that of the icosahedron, in the future development of science and mathematics. In 1884 Felix Klein published the book *Lectures on the Icosahedron* [12], dedicated to the geometric theory of the icosahedron.

Figure 1.9.   Felix Klein (1849–1925).
Source: *Wikipedia, The Free Encyclopedia, http://en.wikipedia.org/wiki/Felix_Klein*

According to Klein, the theories of mathematics, extending widely and freely like sheets of fabric, are united by geometric objects allowing for their broader and more general understanding. In Klein's opinion the icosahedron is precisely just such a mathematical object [12]: "*Each unique geometrical object is somehow or other connected to the properties of the regular icosahedron.*" In fact, Klein considers it to be the one mathematical object out of which five of the branches of mathematical theory follow, namely, geometry, Galois' theory, group theory, invariant theory and differential equations. Thus Klein, in following Pythagoras, Plato, Euclid, and Kepler, directed attention to the fundamental role of the Platonic solids for the development of science and mathematics.

Unfortunately, Klein's contemporaries could not understand and appreciate the revolutionary importance of his idea. However, its significance was appreciated one century later in 1982 when the Israeli scientist Dan Shechtman discovered a special alloy called quasi-crystals. And in 1985 researchers Robert F. Curl, Harold W. Kroto and Richard E. Smalley discovered a special class of carbon they named fullerenes. It is important to emphasize that quasi-crystals are based on Plato's icosahedron and the fullerenes on the Archimedean truncated icosahedron. This means that in the 19th century Klein made a brilliant prediction which was verified a century later.

1.4.   The Algebraic Identities for the Golden Ratio

1.4.1.   *The simplest identities*

We start with the simplest algebraic properties of the golden ratio. Beginning with equation (1.5) as follows:

$$x^2 = x + 1. \tag{1.7}$$

If we substitute the root Φ (the golden ratio from (1.6)) in place of x in the equation (1.7), we then get the first remarkable property of the golden ratio:

$$\Phi^2 = \Phi + 1. \tag{1.8}$$

If we divide all terms of (1.8) by Φ, we get the following expression for Φ:

$$\Phi = 1 + \frac{1}{\Phi}. \tag{1.9}$$

If we multiply each side of (1.8) by Φ, and then divide each side of (1.8) by Φ, we get two new identities:

$$\Phi^3 = \Phi^2 + \Phi \tag{1.10}$$

and

$$\Phi = 1 + \Phi^{-1}. \tag{1.11}$$

If we now continue to multiply the terms of the identity (1.10) by Φ, and divide the terms of (1.11) by Φ and continue the process to infinity, we get the following elegant identity, which connects the powers of the golden ratio:

$$\Phi^n = \Phi^{n-1} + \Phi^{n-2}, \quad n = 0, \pm 1, \pm 2, \pm 3, \ldots. \tag{1.12}$$

1.4.2.   *The "golden" geometric progression*

Consider the sequence of the powers of the golden ratio:

$$\left\{ \ldots, \Phi^{-n}, \Phi^{-(n-1)}, \ldots, \Phi^{-2}, \Phi^{-1}, \Phi^0 = 1, \Phi^1, \Phi^2, \ldots, \Phi^{n-1}, \Phi^n, \ldots \right\}. \tag{1.13}$$

The sequence (1.13) has some very interesting mathematical properties. On the one hand, the sequence (1.13) is a "geometric progression", in which each term is equal to the previous one multiplied by the golden ratio Φ, i.e.,

$$\Phi^n = \Phi \times \Phi^{n-1}. \tag{1.14}$$

On the other hand, in accordance with (1.12), the terms of the geometric progression (1.13) have the so-called "summing" property (1.12), because each term of the progression (1.13) is the sum of the two previous terms. Note that the property (1.12) is characteristic only for the geometric progression with the denominator Φ, and therefore such a geometric progression is called the golden progression.

As every geometric progression corresponds to some logarithmic spiral, in the opinion of many researchers, the property (1.12) singles out the golden progression (1.13) amongst the other geometric progressions. This is the cause of the widespread belief that the golden progression (1.13) is the underlying basis of spiral forms in Nature.

1.4.3. *A representation of the golden ratio in "radicals"*

Let us once again consider the identity (1.8). If we take the square root of the right and left parts of (1.8), we get the following expression for Φ:

$$\Phi = \sqrt{1 + \Phi}. \tag{1.15}$$

If we next substitute the right-hand side of (1.15) in place of Φ in the same expression we obtain the following:

$$\Phi = \sqrt{1 + \sqrt{1 + \Phi}}. \tag{1.16}$$

If we continue this substitution into the right-hand side of (1.16) *ad infinitum*, we then get the following astonishing representation of the golden ratio in a "nested radical":

$$\Phi = \sqrt{1 + \sqrt{1 + \sqrt{1 + \sqrt{1 + \cdots}}}}. \tag{1.17}$$

1.4.4. *A representation of the golden ratio in the form of a continued fraction*

In order to represent the golden ratio as a continued fraction, we use the identity (1.9). If we substitute into the right-hand part of (1.9) in place of Φ the same expression (1.9), then we come to the following representation of Φ:

$$\Phi = 1 + \frac{1}{1 + \frac{1}{\Phi}}. \tag{1.18}$$

If we continue this substitution in the right-hand part of (1.18) *ad infinitum*, then we get the following wonderful representation of the golden ratio in the form of a continued fraction:

$$\Phi = 1 + \cfrac{1}{1 + \cfrac{1}{1 + \cfrac{1}{1 + \cfrac{1}{1 + \cdots}}}}.$$ (1.19)

The expression (1.19) has a profound mathematical sense. The Russian mathematicians A. Ya. Khinchin [13] and N. N. Vorobyov [14] drew attention to the fact that this continued fraction expression (1.19) singles out the golden ratio as the last irrational number to be approximated by rational numbers. This fact reveals the uniqueness of the golden ratio amongst all irrational numbers from the point of view of continued fractions.

From this we now can derive fractional approximations for the golden ratio. To do this we employ (1.19) in the following manner:

$$1 = \frac{1}{1} \qquad \text{(first approximation)};$$

$$1 + \frac{1}{1} = \frac{2}{1} \qquad \text{(second approximation)};$$

$$1 + \cfrac{1}{1 + \frac{1}{1}} = \frac{3}{2} \qquad \text{(third approximation)};$$

$$1 + \cfrac{1}{1 + \cfrac{1}{1 + \frac{1}{1}}} = \frac{5}{3} \qquad \text{(fourth approximation)}.$$

By continuing this process, we find the sequence of continued fractions approximating the golden ratio, which is the sequence of the ratios of adjacent Fibonacci numbers [14, 15]. That is,

$$\frac{1}{1}, \frac{2}{1}, \frac{3}{2}, \frac{5}{3}, \frac{8}{5}, \frac{13}{8}, \frac{21}{13}, \cdots \to \Phi = \lim_{n \to \infty} \frac{F_n}{F_{n-1}} = \frac{1 + \sqrt{5}}{2}.$$ (1.20)

This sequence (1.20) expresses the famous law of phyllotaxis [16], according to which Nature constructs pine cones, pineapples, cacti, heads of sunflowers and other botanical objects. In other words, Nature uses these unique mathematical properties of the golden ratio (1.19) and (1.20) in its wonderful creations!

In conclusion, a few words about the aesthetic aspects of the identities (1.17) and (1.19). Mathematicians intuitively seek to express their mathematical results in the simplest, most compact form. And if one finds such an "aesthetic form", it leads to "aesthetic pleasure". In this respect mathematical creativity is similar to the creativity of composers or poets, whose main task is to obtain perfect musical or poetic forms, which also lead to "aesthetic pleasure".

Note that the endless repetition of the same simple mathematical elements in the formulas for Φ, given by (1.17) and (1.19), produce within us an "aesthetic pleasure" replete with rhythm and harmony.

1.5. Fibonacci Numbers

1.5.1. *Fibonacci's role in the development of western mathematics*

The Middle Ages are associated in our consciousness with the horrors of the inquisition, torture racks and flames on which heretics, midwives, herbal healers, and so-called witches were burned alive, and the atrocities of the Crusades were perpetrated in their purported quest for the "Ark of the Covenant". The science of those times was obviously not the focus of public attention. Under such circumstances, the appearance of the mathematical book *Liber Abaci*, written in 1202 by the Italian mathematician Leonardo Pisano (nicknamed Fibonacci), ironically found its way onto fertile ground and became an important event in the scientific life of society.

Although Fibonacci was one of the brightest mathematical minds in the history of western mathematics, his contributions have received less recognition than deserved. In his 1919 book, *Integer Number*, Russian mathematician Professor A. V. Vasil'ev estimated the significance of Fibonacci's mathematical influence on the west as follows:

> "*The work of the learned merchant from Pisa was so far above the level of mathematical knowledge of even the scientists of his time, that its influence on mathematical literature first became noticeable two centuries after his death, at the end of the 15th century. That is when Luca Pacioli, professor at several Italian universities and Leonardo da Vinci's friend, employed many of Fibonacci's theorems and problems in his works. And then again at the beginning of the 16th century, when a group of talented Italian mathematicians, Ferro, Cardano, Tartalia, and Ferrari, formulated the beginnings of higher algebra thanks to the solution of cubic and biquadrate equations.*"

It follows from this that for almost two centuries Fibonacci surpassed the western mathematicians of the time. Fibonacci's historical role for western

Figure 1.10.   Fibonacci (c.1170–c.1250).
Source: *Wikipedia, The Free Encyclopedia, http://en.wikipedia.org/wiki/Fibonacci*

science is similar to that of Pythagoras, who received scientific education from the ancient Egyptians and Babylonians and then transferred this scientific knowledge to the Greeks. Also educated by Arab mathematicians, Fibonacci acquired, in particular, the Arabic–Hindu decimal notation, which he then introduced into western mathematics. In a manner similar to Pythagoras, Fibonacci transferred Arabic mathematical knowledge into western science through his mathematical works. By this he laid the foundation for further development of western mathematics.

1.5.2.   *Fibonacci's recurrence relation*

By solving the "rabbit reproduction" problem, Fibonacci derived the following important recurrence relation:

$$F_n = F_{n-1} + F_{n-2}. \tag{1.21}$$

Note that the concrete values of the numerical sequences, generated by the recurrence relation (1.21), depend on the initial values (the seeds) of the sequences F_1 and F_2. Here it is assumed that

$$F_1 = F_2 = 1. \tag{1.22}$$

For this case the recurrence formula (1.21) with initial conditions (1.22) "generates" the following numerical sequence:

$$1, 1, 2, 3, 5, 8, 13, 21, 34, 55, 89, 144, \ldots \qquad (1.23)$$

In mathematics, this numerical sequence (1.23) is called, as a rule, Fibonacci numbers. Fibonacci numbers (1.23) have a variety of remarkable mathematical properties (see below).

1.5.3. *Variations on the Fibonacci theme*

Variations on a given theme are a well-known genre in music. A distinctive feature of such musical compositions is that they begin, in most cases, with one simple essential musical theme, which thereafter undergoes considerable changes in tempo, mood and nature. If we follow this example of a musical piece and select a simple mathematical subject, such as the Fibonacci numbers, we can consider it together with its numerous variations.

The sum of the first *n* consecutive Fibonacci numbers. Fibonacci numbers have a range of delightful mathematical properties which for many centuries have stimulated the mathematical imagination. For example, calculate the sum of the first n Fibonacci numbers beginning with the simplest sums as follows:

$$
\left.
\begin{aligned}
1 + 1 &= 2 = \mathbf{3} - 1 \\
1 + 1 + 2 &= 4 = \mathbf{5} - 1 \\
1 + 1 + 2 + 3 &= 7 = \mathbf{8} - 1 \\
1 + 1 + 2 + 3 + 5 &= 12 = \mathbf{13} - 1
\end{aligned}
\right\}. \qquad (1.24)
$$

If in these sums (1.24) we consider the numbers marked by bold type: **3, 5, 8, 13,** ..., then it is easy to see that they are Fibonacci numbers! We can then write the sums (1.24) using the following general form:

$$F_1 + F_2 + \cdots + F_n = F_{n+2} - 1. \qquad (1.25)$$

The sum of consecutive Fibonacci numbers with odd indices. We start with the simplest sums:

$$
\left.
\begin{aligned}
1 + 2 &= \mathbf{3} \\
1 + 2 + 5 &= \mathbf{8} \\
1 + 2 + 5 + 13 &= \mathbf{21} \\
1 + 2 + 5 + 13 + 34 &= \mathbf{55}
\end{aligned}
\right\}. \qquad (1.26)
$$

The following general formula is derived from (1.26):

$$F_1 + F_3 + F_5 + \cdots + F_{2n-1} = F_{2n}. \qquad (1.27)$$

The sum of consecutive Fibonacci numbers with even indices. This sum is

$$F_2 + F_4 + F_6 + \cdots + F_{2n} = F_{2n+1} - 1. \qquad (1.28)$$

The sum of the squares of n consecutive Fibonacci numbers. Next we calculate the sum of the squares of n sequential Fibonacci numbers:

$$F_1^2 + F_2^2 + \cdots + F_n^2. \qquad (1.29)$$

We start from an analysis of the simplest sums of the kind (1.29):

$$
\left.
\begin{array}{l}
1^2 + 1^2 = 2 = \mathbf{1 \times 2} \\
1^2 + 1^2 + 2^2 = 6 = \mathbf{2 \times 3} \\
1^2 + 1^2 + 2^2 + 3^2 = 15 = \mathbf{3 \times 5} \\
1^2 + 1^2 + 2^2 + 3^2 + 5^2 = 40 = \mathbf{5 \times 8}
\end{array}
\right. \qquad (1.30)
$$

This analysis of (1.30) leads us to the following general formula:

$$F_1^2 + F_2^2 + \cdots + F_n^2 = F_n F_{n+1}. \qquad (1.31)$$

The sum of the squares of two adjacent Fibonacci numbers. Next let us consider the sum of the squares of two adjacent Fibonacci numbers:

$$F_{n-1}^2 + F_n^2. \qquad (1.32)$$

We start from an analysis of the simplest sums of the kind (1.32):

$$
\left.
\begin{array}{l}
1^2 + 1^2 = 1 + 1 = \mathbf{2} \\
1^2 + 2^2 = 1 + 4 = \mathbf{5} \\
2^2 + 3^2 = 4 + 9 = \mathbf{13} \\
3^2 + 5^2 = 9 + 25 = \mathbf{34}
\end{array}
\right. \qquad (1.33)
$$

The analysis of (1.33) leads us to the following general formula:

$$F_{n-1}^2 + F_n^2 = F_{2n-1}. \qquad (1.34)$$

The "extended" Fibonacci numbers. Fibonacci numbers (2.5) can be extended for negative values of the indices of n (see Table 1.2).

Table 1.2. The "extended" Fibonacci numbers.

n	0	1	2	3	4	5	6	7	8	9	10
F_n	0	1	1	2	3	5	8	13	21	34	55
F_{-n}	0	1	−1	2	−3	5	−8	13	−21	34	−55

The "extended" Fibonacci numbers are connected by the following simple relation:

$$F_{-n} = (-1)^{n+1} F_n. \tag{1.35}$$

Cassini's formula for Fibonacci numbers. The history of science does not reveal why French astronomer Giovanni Domenico Cassini (1625–1712) took such a great interest in Fibonacci numbers. Most likely it was simply an object of rapture for the great astronomer. At that time many serious scientists were fascinated by Fibonacci numbers and the golden ratio. We recall that these numbers were also an aesthetic object for Johannes Kepler, who was a contemporary of Cassini.

Consider now the Fibonacci series: 0, 1, 1, 2, 3, 5, 8, 13, 21, 34, ... Take the Fibonacci number 5 and square it, that is, $5^2 = 25$. Next consider the product of the two Fibonacci numbers 3 and 8 that are adjacent to each side of the number 5, that is, $3 \times 8 = 24$. Then we can write:

$$5^2 - 3 \times 8 = 1.$$

Note that the difference obtained is equal to (+1).

Now we follow the same procedure with the next Fibonacci number 8, that is, we first square it ($8^2 = 64$), then we calculate the product of the two Fibonacci numbers 5 and 13, which lie adjacent on either side of the Fibonacci number 8, that is, $5 \times 13 = 65$. After a comparison of the product $5 \times 13 = 65$ to the square $8^2 = 64$ we get:

$$8^2 - 5 \times 13 = -1.$$

Note that the difference obtained is equal to (−1).

Furthermore we have:

$$13^2 - 8 \times 21 = 1;$$
$$21^2 - 13 \times 34 = -1$$

and so on.

We notice that the square of any Fibonacci number F_n always differs from the product of its two adjacent Fibonacci numbers F_{n-1} and F_{n+1}

by 1. This difference is alternately $+1$ or -1 to infinity and depends on the index n of the initial Fibonacci number F_n. If the index n is even, then the result is -1, and if odd, $+1$. This property of Fibonacci numbers can be expressed by the following mathematical formula:

$$F_n^2 - F_{n-1}F_{n+1} = (-1)^{n+1}. \tag{1.36}$$

This wonderful formula evokes an aesthetic if not spiritual thrill when one realizes that it is valid for any value of n (where n is any integer from $-\infty$ up to $+\infty$). Thus this endless alternation of $+1$ and -1 in the expression (1.36) evokes a genuine aesthetic feeling of rhythm and harmony.

Fibonacci Q-matrix.   In recent decades the theory of Fibonacci numbers has been supplemented by the theory of the so-called Fibonacci Q-matrix [15]. This is a (2×2)-matrix of this form:

$$Q = \begin{pmatrix} 1 & 1 \\ 1 & 0 \end{pmatrix}.$$

The Fibonacci Q-matrix has the following remarkable properties:

$$Q^n = \begin{pmatrix} F_{n+1} & F_n \\ F_n & F_{n-1} \end{pmatrix},$$

$$\det Q = -1, \quad \det Q^n = F_{n-1}F_{n+1} - F_n^2 = (-1)^n,$$

n	0	1	2	3	4	5
Q^n	$\begin{pmatrix} 1 & 0 \\ 0 & 1 \end{pmatrix}$	$\begin{pmatrix} 1 & 1 \\ 1 & 0 \end{pmatrix}$	$\begin{pmatrix} 2 & 1 \\ 1 & 1 \end{pmatrix}$	$\begin{pmatrix} 3 & 2 \\ 2 & 1 \end{pmatrix}$	$\begin{pmatrix} 5 & 3 \\ 3 & 2 \end{pmatrix}$	$\begin{pmatrix} 8 & 5 \\ 5 & 3 \end{pmatrix}$
Q^{-n}	$\begin{pmatrix} 1 & 0 \\ 0 & 1 \end{pmatrix}$	$\begin{pmatrix} 0 & 1 \\ 1 & -1 \end{pmatrix}$	$\begin{pmatrix} 1 & -1 \\ -1 & 2 \end{pmatrix}$	$\begin{pmatrix} -1 & 2 \\ 2 & -3 \end{pmatrix}$	$\begin{pmatrix} 2 & -3 \\ -3 & 5 \end{pmatrix}$	$\begin{pmatrix} -3 & 5 \\ 5 & -8 \end{pmatrix}$

Fibonacci numbers in Pascal's triangle.   By studying the so-called diagonal sums of Pascal's triangle, American mathematician George Polya had an unexpected insight as recorded in his book *Mathematical Discovery* [17]. He discovered a fundamental connection between Fibonacci numbers and Pascal's Triangle! It should be noted that this very simple mathematical result, which "lay hidden in plain sight" for several centuries, was considered a recondite "secret" by Blaise Pascal and other mathematicians

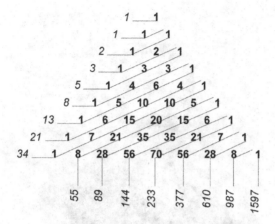

Figure 1.11. Fibonacci numbers in Pascal's triangle.

who studied and understood the relation of Fibonacci numbers to Pascal's triangle.

1.6. Lucas Numbers

1.6.1. *François-Édouard-Anatole Lucas*

Fibonacci did not continue to study the mathematical properties of the numerical sequence (1.23) for which he was to become famous. This was done by other later mathematicians. Since the 19th century mathematical works devoted to Fibonacci numbers "began to reproduce like Fibonacci's rabbits" according to the witty expression of one mathematician.

French mathematician François-Édouard-Anatole Lucas was born in 1842 and became the leader of 19th century Fibonacci studies. Unfortunately he died prematurely in 1891 as the result of a freak accident at a banquet when a dish was smashed and a splinter wounded his cheek. He died from an infection some days later.

1.6.2. *The Lucas numbers*

Lucas introduced the concept of generalized Fibonacci numbers, which can be calculated by following the general recurrence relation:

$$G_n = G_{n-1} + G_{n-2},$$

$$(1.37)$$

Figure 1.12.　François-Édouard-Anatole Lucas (1842–1891).

Source: *Wikipedia, The Free Encyclopedia,*
https://en.wikipedia.org/wiki/Édouard_Lucas

beginning with arbitrary initial terms (seeds) G_1 and G_2.

However, Lucas introduced (1.38) as the main numerical sequence of the recurrence relation (1.37):

$$L_n = L_{n-1} + L_{n-2} \qquad (1.38)$$

with the initial seeds

$$L_1 = 1, \quad L_2 = 3. \qquad (1.39)$$

Then, by using the recurrence relation (1.38) with the (1.39) seeds, we can calculate the following numerical sequence known as *Lucas numbers*:

$$1, 3, 4, 7, 11, 18, 29, 47, 76, 123, 199, \ldots. \qquad (1.40)$$

If we analyze the Lucas numbers (1.40), in a way similar to the Fibonacci numbers (1.23), the following identities for Lucas numbers emerge:

$$L_1 + L_3 + L_5 + \cdots + L_{2n-1} = L_{2n} - 2$$
$$L_2 + L_4 + L_6 + \cdots + L_{2n} = L_{2n+1} - 1$$
$$L_1^2 + L_2^2 + \cdots + L_n^2 = L_n L_{n+1} - 2$$

Table 1.3. The "extended" Lucas numbers.

n	0	1	2	3	4	5	6	7	8	9	10
L_n	2	1	3	4	7	11	18	29	47	76	123
L_{-n}	2	-1	3	-4	7	-11	18	-29	47	-76	123

$$L_n^2 + L_{n+1}^2 = 5F_{2n+1} \tag{1.41}$$

$$\lim_{n \to \infty} \frac{L_n}{L_{n-1}} = \Phi = \frac{1 + \sqrt{5}}{2}. \tag{1.42}$$

The "extended" Lucas numbers. Similar to the Fibonacci numbers (Table 1.2), the Lucas numbers (1.40) can be extended for negative values of the indices n (Table 1.3).

The "extended" Lucas numbers are connected by the following simple relation:

$$L_{-n} = (-1)^n L_n. \tag{1.43}$$

1.7. Binet's Formulas

1.7.1. *Jacques Philippe Marie Binet*

Besides Lucas, Jacques Philippe Marie Binet (1776–1856) was another 19th century French mathematician enthusiastic over Fibonacci numbers and the golden section. Born February 2, 1776 in Renje, he died on May 12, 1856 in Paris.

1.7.2. *Binet's formulas for Fibonacci and Lucas numbers*

Binet's spectacular entrée into Fibonacci number theory was with his now famous Binet's formulas. These formulas link Fibonacci and Lucas numbers directly with the golden ratio and without a doubt are amongst the world's most famous mathematical formulas.

There are a variety of representations of Binet's formulas. In the "theory of Fibonacci numbers" [14, 15] this formula is well-known:

$$\Phi^n = \frac{L_n + F_n\sqrt{5}}{2}. \tag{1.44}$$

It links the power of the golden ratio Φ^n ($n = 0, 1, 2, 3, \ldots$) with the "extended" Fibonacci and Lucas numbers F_n and L_n.

Figure 1.13.   Jacques Philippe Marie Binet (1776–1856).

Source: *Wikipedia, The Free Encyclopedia,*
http://en.wikipedia.org/wiki/Jacques_Philippe_Marie_Binet

By using formula (1.44) we can represent the "extended" Fibonacci and Lucas numbers through the golden ratio. For this purpose it is enough to write the formulas for the sum and the difference of the nth degrees of the golden ratio $\Phi^n + \Phi^{-n}$ and $\Phi^n - \Phi^{-n}$ as follows:

$$\Phi^n + \Phi^{-n} = \frac{(L_n + L_{-n}) + (F_n + F_{-n})\sqrt{5}}{2} \tag{1.45}$$

$$\Phi^n - \Phi^{-n} = \frac{(L_n - L_{-n}) + (F_n - F_{-n})\sqrt{5}}{2}. \tag{1.46}$$

Next consider the expressions of formulas (1.45) and (1.46) for the even values of the indices $n = 2k$. For this we recall the property (1.35) of the "extended" Fibonacci numbers (Table 1.2) and the property (1.43) of the "extended" Lucas numbers (Table 1.3). Then, for the case $n = 2k$ the formulas (1.45) and (1.46) take the following simple forms:

$$\Phi^{2k} + \Phi^{-2k} = L_{2k} \tag{1.47}$$

$$\Phi^{2k} - \Phi^{-2k} = F_{2k}\sqrt{5}. \tag{1.48}$$

For the odd indices $n = 2k+1$, by using the properties (1.35) and (1.43), we can represent the formulas (1.45) and (1.46) as follows:

$$\Phi^{2k+1} + \Phi^{-(2k+1)} = F_{2k+1}\sqrt{5} \tag{1.49}$$

$$\Phi^{2k+1} - \Phi^{-(2k+1)} = L_{2k+1}. \tag{1.50}$$

Considering the formulas (1.48) and (1.49), and (1.47) and (1.50), we can now represent the "extended" Fibonacci and Lucas numbers as follows:

$$F_n = \begin{cases} \dfrac{\Phi^n + \Phi^{-n}}{\sqrt{5}} & \text{for } n = 2k+1 \\[2ex] \dfrac{\Phi^n - \Phi^{-n}}{\sqrt{5}} & \text{for } n = 2k, \end{cases} \tag{1.51}$$

$$L_n = \begin{cases} \Phi^n + \Phi^{-n} & \text{for } n = 2k \\ \Phi^n - \Phi^{-n} & \text{for } n = 2k+1. \end{cases} \tag{1.52}$$

Reflecting on formulas (1.51) and (1.52) may evoke a feeling of "aesthetic pleasure", once again convincing us of the sublime power of mathematics! We know that the "extended" Fibonacci and Lucas numbers are always integers. And yet any degree of the golden ratio is an irrational number. It follows from this that the integer numbers F_n and L_n can be represented by using the formulas (1.51) and (1.52) along with that most significant irrational number, the golden ratio Φ!

Let us consider some examples. According to (1.51) and (1.52) we can represent the Fibonacci number 5 ($n = 5$) and the Lucas number 3 ($n = 2$) as follows:

$$5 = \frac{\left(\frac{1+\sqrt{5}}{2}\right)^5 + \left(\frac{1+\sqrt{5}}{2}\right)^{-5}}{\sqrt{5}}; \tag{1.53}$$

$$3 = \left(\frac{1+\sqrt{5}}{2}\right)^2 + \left(\frac{1+\sqrt{5}}{2}\right)^{-2}. \tag{1.54}$$

It is easy to prove that the identities (1.53) and (1.54) are valid because, according to (1.44), we have:

$$\begin{cases} \left(\dfrac{1+\sqrt{5}}{2}\right)^5 = \dfrac{L_5 + F_5\sqrt{5}}{2} = \dfrac{11 + 5\sqrt{5}}{2}, \\[2ex] \left(\dfrac{1+\sqrt{5}}{2}\right)^{-5} = \dfrac{L_{-5} + F_{-5}\sqrt{5}}{2} = \dfrac{-11 + 5\sqrt{5}}{2}, \end{cases} \tag{1.55}$$

$$\begin{cases} \left(\dfrac{1+\sqrt{5}}{2}\right)^2 = \dfrac{L_2 + F_2\sqrt{5}}{2} = \dfrac{3 + \sqrt{5}}{2}, \\[2ex] \left(\dfrac{1+\sqrt{5}}{2}\right)^{-2} = \dfrac{L_{-2} + F_{-2}\sqrt{5}}{2} = \dfrac{3 - \sqrt{5}}{2}. \end{cases} \tag{1.56}$$

By using (1.55), we can rewrite the identity (1.53) as follows:

$$5 = \frac{11 + 5\sqrt{5}}{\sqrt{5}} + \frac{-11 + 5\sqrt{5}}{\sqrt{5}} = \frac{10\sqrt{5}}{2\sqrt{5}},$$

which confirms the validity of the identity (1.53).

By using (1.56), we can rewrite the identity (1.54) as follows:

$$3 = \frac{3 + \sqrt{5}}{2} + \frac{3 - \sqrt{5}}{2}. \tag{1.57}$$

Note that this reasoning is of a general character, that is, for any "extended" Lucas or Fibonacci numbers, given by the formulas (1.51) and (1.52), all "irrationalities" in the right-hand parts of (1.51) and (1.52) are always mutually canceled out resulting in integer numbers as the outcome!

1.8. The Theory of Fibonacci Numbers in Modern Mathematics

1.8.1. *Fibonacci Association*

Studies by Lucas and Binet stimulated further research in this area of modern mathematics. In 1963 a group of U.S. mathematicians created the Fibonacci Association. In the same year, the Fibonacci Association began publication of *The Fibonacci Quarterly*. In 1984 they began holding regular international conferences focusing upon "Fibonacci numbers and their applications". The Fibonacci Association played a significant role in stimulating future international research.

American mathematician Verner Emil Hoggatt (1921–1981), professor at San Jose State University, was one of the founders of the Fibonacci Association and the magazine *The Fibonacci Quarterly*.

In 1969 Hoggatt published *Fibonacci and Lucas Numbers* [15], considered one of the best books in the field. Hoggatt made timely contributions to the promotion of research in the field of Fibonacci numbers. Scientific supervisor for many master's theses, Hoggat authored numerous articles on Fibonacci numbers.

Learned monk Brother Alfred Brousseau (1907–1988) was another prominent founder of the Fibonacci Association and *The Fibonacci Quarterly*.

The Fibonacci Association has the rather unique and singular purpose of studying only the Fibonacci sequence. This raises some questions:

(1) Why were the members of the Fibonacci Association and many "mathematics lovers" so focused on Fibonacci numbers?

(2) What united these two very different people, mathematician Hoggatt and learned monk Brother Brousseau, in their quest to create the Fibonacci Association and establish *The Fibonacci Quarterly*?

In an attempt to answer these questions regarding Hoggatt's and Brousseau's narrow focus on Fibonacci numbers, we need to examine some of their documents, in particular, their photographs, as well as the books and articles published in *The Fibonacci Quarterly*.

In 1969, TIME magazine published an article titled "The Fibonacci numbers" which was dedicated to the Fibonacci Association. This article contained Brousseau's photo with a cactus in his hands. The cactus is of course one of the most characteristic examples of a Fibonacci botanical object. The article referred to other natural forms involving Fibonacci numbers. For example, Fibonacci numbers are found in the spiral formations of sunflower disks, pine cones, branching patterns of trees, and leaf arrangement (or phyllotaxis) on the branches of trees, etc.

Brousseau recommended that devotees of Fibonacci numbers "pay attention to the search for aesthetic satisfaction in them. There is some kind of mystical connection between these numbers and the Universe."

In the photo in Fig. 1.14, Hoggatt is seen holding a pine cone in his hands The pine cone, of course, is another well-known example of Fibonacci numbers pattern found in Nature. From this comparison it may be reasonable to assume that Hoggatt, like Brousseau, believed in a mystical connection between Fibonacci numbers and the Universe. In our opinion, this belief may have united Hoggatt and Brousseau as a primary motivating factor in their work on Fibonacci numbers.

As previously indicated, Fibonacci numbers are associated with the golden ratio, since the ratio of two adjacent Fibonacci numbers strives to attain the golden ratio in the limit as is shown by the formula (1.20). This means that the Fibonacci numbers, approximating the golden ratio, are expressing the harmony of the Universe, i.e., "there is some kind of mystical connection between these numbers and the Universe" (Brousseau).

Therefore the theory of Fibonacci numbers and their relation to the golden ratio, which began to develop more rapidly following the creation of

Figure 1.14.   Verner Emil Hoggatt (1921–1981).

Source: Museum of Harmony and the Golden Section,
www.goldenmuseum.com/ 1601Mathematics_ engl.html

the Fibonacci Association (1963), was primarily aimed at solving problems of the harmonization of the theoretical natural sciences (physics, chemistry, botany, biology, physiology, medicine and so on), as well as economics, computer science, education and the fine arts. This demonstrates that the "Harmony of Nature" underlies the theory of Fibonacci numbers [14, 15] linking the "Problem of Harmony" to these many diverse fields.

Furthermore, in analyzing the origin and development of Fibonacci number theory in modern mathematics, we are drawn back to the ancient Greek "Doctrine of the Numerical Harmony of the Universe"! Here lie the very roots of the harmonization of mathematics and theoretical natural sciences based upon the golden ratio and Fibonacci numbers [14, 15]. This approach to the Mathematics of Harmony [8], which is a generalization of the theory of Fibonacci numbers [14, 15], emphasizes the interdisciplinary nature of this mathematical discipline.

1.8.2.   *The role of Nikolai Vorobyov in the development of Fibonacci number theory*

In addition to the importance of the creation of the American Fibonacci Association and *The Fibonacci Quarterly*, it should be noted that the

Russian Nikolai Vorobyov (1925–1995) was the first modern mathematician to draw attention to Fibonacci number theory.

In 1961, he published the book *Fibonacci Numbers* [14], which also played a prominent role in the development of Fibonacci number theory. This book became a 20th century bestseller. Passing through several editions, it was translated into multiple languages and became a handbook for Soviet and international scientists.

1.9. The "Golden" Hyperbolic Fibonacci and Lucas Functions

1.9.1. *Classic hyperbolic functions*

The functions:

$$\text{sh}(x) = \frac{e^x - e^{-x}}{2}; \quad \text{ch}(x) = \frac{e^x + e^{-x}}{2} \tag{1.58}$$

are called, respectively, hyperbolic sine and hyperbolic cosine.

These analytical definitions (1.58) can be used to obtain some very important identities of hyperbolic trigonometry. There are several different trigonometric identities, for example, the Pythagorean Theorem for trigonometric functions:

$$\cos^2 \alpha + \sin^2 \alpha = 1. \tag{1.59}$$

We can prove similar identities for the hyperbolic functions:

$$\text{ch}^2 x - \text{sh}^2 x = \left(\frac{e^x + e^{-x}}{2} \right)^2 - \left(\frac{e^x - e^{-x}}{2} \right)^2$$

$$= \frac{e^{2x} + 2 - e^{-2x}}{4} - \frac{e^{2x} - 2 + e^{-2x}}{4} = 1. \tag{1.60}$$

The parity property is another important feature of the hyperbolic functions (1.58):

$$\text{sh}(-x) = -\text{sh}(x); \quad \text{ch}(-x) = \text{ch}(x); \quad \text{th}(-x) = \text{th}(x). \tag{1.61}$$

This means that the hyperbolic sine is an odd function and the hyperbolic cosine is an even function.

Although the French mathematician Johann Heinrich Lambert (1728–1777) is often credited with the creation of hyperbolic functions, it was actually the Italian mathematician Vincenzo Riccati (1707–1775) who accomplished this in the middle of the 18th century. Studying these functions, he used them to obtain solutions to cubic equations. Riccati discovered the standard addition formulas (similar to trigonometric identities) for hyperbolic functions as well as their derivatives. He revealed the relationship between hyperbolic functions and exponential functions. Riccati was the first to use **sh** and **ch** designations for hyperbolic sine and cosine, respectively.

Interest in hyperbolic functions (1.58) increased significantly during the 19th century, when Russian geometer Nikolai Lobachevsky (1792–1856) used them to describe mathematical relationships for non-Euclidean geometry. It was for this reason that Lobachevsky's geometry came to be known as hyperbolic geometry. Due to Lobachevsky's works, the hyperbolic functions (1.58) are now considered a symbol of non-Euclidean geometry.

1.9.2.   *Comparison of Binet's formulas with hyperbolic functions*

The representation of Binet's formulas in the form (1.51) and (1.52) has an important advantage. If we compare the hyperbolic functions (1.58) with the formulas (1.51) and (1.52), we can see that their mathematical structures are similar to one another. This mathematical similarity became the basis for the introduction of a new class of hyperbolic functions called the "golden" hyperbolic Fibonacci and Lucas functions. These functions were first introduced by Ukrainian mathematicians Alexey Stakhov and Ivan Tkachenko. Their article "Hyperbolic Fibonacci trigonometry" [18] is the first publication on the hyperbolic Fibonacci and Lucas functions.

1.9.3.   *The symmetric hyperbolic Fibonacci and Lucas functions*

The symmetric hyperbolic Fibonacci and Lucas sine and cosine, introduced by Stakhov and Rozin in [19], are a very significant step in the development of a general theory of hyperbolic Fibonacci and Lucas functions. They appear as follows:

Symmetric Fibonacci hyperbolic sine.

$$\mathrm{sFs}(x) = \frac{\Phi^x - \Phi^{-x}}{\sqrt{5}} \tag{1.62}$$

Symmetric Fibonacci hyperbolic cosine.

$$cFs(x) = \frac{\Phi^x + \Phi^{-x}}{\sqrt{5}} \tag{1.63}$$

Symmetric Lucas hyperbolic sine.

$$sLs(x) = \Phi^x - \Phi^{-x} \tag{1.64}$$

Symmetric Lucas hyperbolic cosine.

$$cLs(x) = \Phi^x + \Phi^{-x} \tag{1.65}$$

The following simple relations exist between the Fibonacci hyperbolic functions (1.62) and (1.63) and the Lucas hyperbolic functions (1.64) and (1.65):

$$sFs(x) = \frac{sLs(x)}{\sqrt{5}}; \quad cFs(x) = \frac{cLs(x)}{\sqrt{5}}. \tag{1.66}$$

1.9.4. *Parity property*

The main advantage of the symmetric Fibonacci and Lucas hyperbolic functions (1.62)–(1.65), is the preservation of the parity property. The parity properties of the functions (1.62)–(1.65) are demonstrated in [19] as follows:

$$sFs(-x) = -sFs(x); \quad cFs(-x) = cFs(x), \tag{1.67}$$
$$sLs(-x) = -sLs(x); \quad cLs(-x) = cLs(x). \tag{1.68}$$

It follows from (1.67) and (1.68) that the symmetric Fibonacci and Lucas hyperbolic sines (1.62) and (1.64) are odd functions, and the symmetric Fibonacci and Lucas hyperbolic cosines (1.63) and (1.65) are even functions.

1.9.5. *A uniqueness of the Fibonacci and Lucas hyperbolic functions*

Binet's formulas (1.51) and (1.52) set forth the so-called "extended" Fibonacci and Lucas numbers (see Table 1.4).

Comparing Binet's formulas (1.51) and (1.52) with the symmetric Fibonacci and Lucas hyperbolic functions (1.62)–(1.65), it is easy to see that for discrete values of the variable $x(x = 0, \pm1, \pm2, \pm3, \ldots)$ the functions (1.62) and (1.63) coincide with the "extended" Fibonacci numbers

Table 1.4.   The "extended" Fibonacci and Lucas numbers.

n	0	1	2	3	4	5	6	7	8	9	10
F_n	0	1	1	2	3	5	8	13	21	34	55
F_{-n}	0	1	-1	2	-3	5	-8	13	-21	34	-55
L_n	2	1	3	4	7	11	18	29	47	76	123
L_{-n}	2	-1	3	-4	7	-11	18	-29	47	-76	123

calculated according to Binet's formula (1.51), that is:

$$F_n = \begin{cases} sFs(n) & \text{for } n = 2k \\ cFs(n) & \text{for } n = 2k + 1 \end{cases} \tag{1.69}$$

and the functions (1.64) and (1.65) coincide with the "extended" Lucas numbers calculated according to Binet's formula (1.52), that is:

$$L_n = \begin{cases} cLs(n) & \text{for } n = 2k \\ sLs(n) & \text{for } n = 2k + 1 \end{cases} \tag{1.70}$$

where k is any value from the set $k = 0, \pm 1, \pm 2, \pm 3, \ldots$.

Note that the classic hyperbolic functions (1.58) do not possess a recursive property similar to (1.69) and (1.70).

To clearly demonstrate this property, consider the graphs of the symmetric Fibonacci and Lucas hyperbolic functions, represented in Figs. 1.15 and 1.16.

Figure 1.15 represents the graphs of the symmetric Fibonacci hyperbolic sine $y = sFs(x)$ and the symmetric Fibonacci hyperbolic cosine $y = cFs(x)$.

The points on the graph $y = sFs(x)$ correspond to the "extended" Fibonacci numbers with the even indices $2n$:

$$F_{2n} = \{ \ldots, F_{-6} = -8, F_{-4} = -3, F_{-2} = -1,$$
$$F_0 = 0, F_2 = 1, F_4 = 3, F_6 = 8, \ldots \}. \tag{1.71}$$

The points on the graph $y = cFs(x)$ correspond to the "extended" Fibonacci numbers with the odd indices $2n + 1$:

$$F_{2n+1} = \{ \ldots, F_{-7} = 13, F_{-5} = 5, F_{-3} = 2, F_{-1} = 1,$$
$$F_1 = 1, F_3 = 3, F_5 = 5, F_7 = 13, \ldots \}. \tag{1.72}$$

Figure 1.16 represents the graphs of the symmetric Lucas hyperbolic sine $y = sLs(x)$ and the symmetric Lucas hyperbolic cosine $y = cLs(x)$.

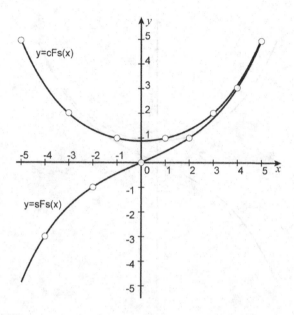

Figure 1.15. The graphs of the symmetric Fibonacci hyperbolic functions.

The points on the graph $y = \text{sLs}(x)$ correspond to the "extended" Lucas numbers with odd indices $2n + 1$:

$$L_{2n+1} = \{\ldots, L_{-7} = -29, L_{-5} = -11, L_{-3} = -4, L_{-1} = -1,$$
$$L_1 = 1, L_3 = 4, L_5 = 11, L_7 = 29, \ldots\}. \tag{1.73}$$

The points on the graph $y = \text{cLs}(x)$ correspond to the "extended" Lucas numbers with even indices $2n$:

$$L_{2n} = \{\ldots, L_{-6} = 18, L_{-4} = 7, L_{-2} = 3, L_0 = 2, L_2 = 3,$$
$$L_4 = 7, L_6 = 18, \ldots\}. \tag{1.74}$$

Here it is necessary to point out that at the point $x = 0$ the symmetric Fibonacci hyperbolic cosine $\text{cFs}(x)$ takes the value $\text{cFs}(0) = \frac{2}{\sqrt{5}}$ (Fig. 1.15), and the symmetric Lucas hyperbolic cosine $\text{cLs}(x)$ takes the value $\text{cLs}(0) = 2$ (Fig. 1.16). It is also important to emphasize that the "extended" Fibonacci numbers F_n with even indices ($n = 0, \pm 2, \pm 4, \pm 6, \ldots$) are "inscribed" into the graph of the symmetric Fibonacci hyperbolic sine $\text{sFs}(x)$ at the discrete points ($x = 0, \pm 2, \pm 4, \pm 6, \ldots$) and the "extended" Fibonacci numbers with odd indices ($n = \pm 1, \pm 3, \pm 5, \ldots$) are "inscribed" into the symmetric Fibonacci hyperbolic cosine $\text{cFs}(x)$ at the discrete points

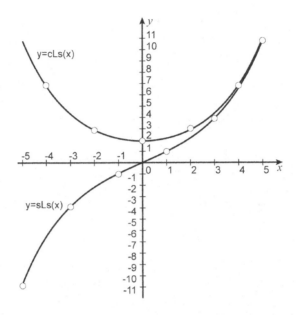

Figure 1.16.   The graphs of the symmetric Lucas hyperbolic functions.

$(x = \pm 1, \pm 3, \pm 5, \ldots)$. On the other hand, the "extended" Lucas numbers with even indices are "inscribed" into the graph of the symmetric Lucas hyperbolic cosine $cLs(x)$ at the discrete points $(x = 0, \pm 2, \pm 4, \pm 6, \ldots)$, and the "extended" Lucas numbers with odd indices are "inscribed" into the graph of the symmetric Lucas hyperbolic cosine $sLs(x)$ at the discrete points $(x = \pm 1, \pm 3, \pm 5, \ldots)$.

Here we can compare the classic hyperbolic functions (1.58) with the symmetric Fibonacci and Lucas hyperbolic functions (1.62)–(1.65). It follows from this comparison that the symmetric Fibonacci and Lucas hyperbolic functions possess all the important properties of the classic hyperbolic functions. Therefore the symmetric Fibonacci and Lucas hyperbolic functions have hyperbolic properties.

On the other hand, a comparison of the "extended" Fibonacci and Lucas numbers, given by Binet's formulas (1.51) and (1.52), with the symmetric Fibonacci and Lucas hyperbolic functions (1.62)–(1.65) demonstrates that according to (1.69) and (1.70) these functions are a generalization of the "extended" Fibonacci and Lucas numbers for a continuous domain. This means that the symmetric Fibonacci and Lucas hyperbolic functions (1.62)–(1.65) possess recursive properties similar to the properties of the "extended" Fibonacci and Lucas numbers (see Table 1.4). Note that the classic hyperbolic functions (1.58), which are the basis of hyperbolic geometry, do not possess recursive properties.

1.9.6. *Hyperbolic properties*

It is demonstrated in [19] that for each identity of the classic hyperbolic functions (1.58) there is an analog in the form of a corresponding identity for the symmetric Fibonacci and Lucas hyperbolic functions (1.62)–(1.65). Consider the following identities for the functions (1.62)–(1.65) in the comparison with the classic hyperbolic functions (1.58).

The following fundamental identity for classic hyperbolic functions (1.58) is well established:

$$[\text{ch}(x)]^2 - [\text{sh}(x)]^2 = 1. \tag{1.75}$$

It is demonstrated in [19] that the symmetric Fibonacci and Lucas hyperbolic functions (1.62)–(1.65) have similar identities, which are as follows:

$$[\text{cFs}(x)]^2 - [\text{sFs}(x)]^2 = \frac{4}{5}; \tag{1.76}$$

$$[\text{cLs}(x)]^2 - [\text{sLs}(x)]^2 = 4. \tag{1.77}$$

Table 1.5 lists the various hyperbolic identities for the symmetric Fibonacci hyperbolic functions (1.62) and (1.63), which are analogs of the corresponding identities for classic hyperbolic functions (1.58). By using the relations (1.66), which link the Fibonacci hyperbolic functions (1.62), (1.63) with the Lucas hyperbolic functions (1.64), (1.65), we can create a table of the hyperbolic identities for the Lucas hyperbolic functions (1.64) and (1.65) similar to Table 1.5.

1.9.7. *The recursive properties of the Fibonacci and Lucas hyperbolic functions*

The recursive properties of the symmetric Fibonacci and Lucas hyperbolic functions (1.62)–(1.65) are another confirmation of the unique nature of this new class of hyperbolic functions, since the classic hyperbolic functions (1.58) do not have similar properties to these.

Let us now consider some recursive properties of the Fibonacci and Lucas hyperbolic functions (1.62)–(1.65) in comparison to similar properties of "extended" Fibonacci and Lucas numbers given by Binet's formulas (1.51) and (1.52).

Table 1.5.   Hyperbolic identities for the Fibonacci hyperbolic functions.

Formulas for the classical hyperbolic functions	Formulas for the Fibonacci hyperbolic functions
$\text{sh}(x) = \dfrac{e^x - e^{-x}}{2}$; $\text{ch}(x) = \dfrac{e^x + e^{-x}}{2}$	$\text{sFs}(x) = \dfrac{\Phi^x - \Phi^{-x}}{\sqrt{5}}$; $\text{cFs}(x) = \dfrac{\Phi^x + \Phi^{-x}}{\sqrt{5}}$
$\text{ch}^2(x) - \text{sh}^2(x) = 1$	$[\text{cFs}(x)]^2 - [\text{sFs}(x)]^2 = \dfrac{4}{5}$
$\text{sh}(x + y) = \text{sh}(x)\text{ch}(y) + \text{ch}(x)\text{sh}(y)$ $\text{sh}(x - y) = \text{sh}(x)\text{ch}(y) - \text{ch}(x)\text{sh}(y)$	$\dfrac{2}{\sqrt{5}}\text{sFs}(x{+}y) = \text{sFs}(x)\text{cFs}(y) + \text{cFs}(x)\text{sFs}(y)$ $\dfrac{2}{\sqrt{5}}\text{sFs}(x{-}y) = \text{sFs}(x)\text{cFs}(y) - \text{cFs}(x)\text{sFs}(y)$
$\text{ch}(x + y) = \text{ch}(x)\text{ch}(y) + \text{sh}(x)\text{sh}(y)$ $\text{ch}(x - y) = \text{ch}(x)\text{ch}(y) - \text{sh}(x)\text{sh}(y)$	$\dfrac{2}{\sqrt{5}}\text{cFs}(x{+}y) = \text{cFs}(x)\text{cFs}(y) + \text{sFs}(x)\text{sFs}(y)$ $\dfrac{2}{\sqrt{5}}\text{cFs}(x{-}y) = \text{cFs}(x)\text{cFs}(y) - \text{sFs}(x)\text{sFs}(y)$
$\text{ch}(2x) = 2\text{sh}(x)\text{ch}(x)$	$\dfrac{1}{\sqrt{5}}\text{cFs}(2x) = \text{sFs}(x)\text{cFs}(x)$
$[\text{ch}(x) \pm \text{sh}(x)]^n = \text{ch}(nx) \pm \text{sh}(nx)$	$[\text{cFs}(x) \pm \text{sFs}(x)]^n$ $= \left(\dfrac{2}{\sqrt{5}}\right)^{n-1}[\text{cFs}(nx) \pm \text{sFs}(nx)]$

Let us start from the simplest recurrence relation for Fibonacci numbers:

$$F_{n+2} = F_{n+1} + F_n. \tag{1.78}$$

It is demonstrated in [19] that the fundamental recurrence relation (1.78) in terms of the Fibonacci hyperbolic functions (1.62) and (1.63) is as follows:

$$\text{sFs}(x + 2) = \text{cFs}(x + 1) + \text{sFs}(x); \tag{1.79}$$

$$\text{cFs}(x + 2) = \text{sFs}(x + 1) + \text{cFs}(x). \tag{1.80}$$

The similar relations we have in the case of recurrence relations for Lucas numbers:

$$L_{n+2} = L_{n+1} + L_n, \tag{1.81}$$

$$\text{sLs}(x + 2) = \text{cLs}(x + 1) + \text{sLs}(x), \tag{1.82}$$

$$\text{cLs}(x + 2) = \text{sLs}(x + 1) + \text{cLs}(x). \tag{1.83}$$

The following relations, similar to Cassini's formula:

$$F_n^2 - F_{n+1}F_{n-1} = (-1)^{n+1}, \tag{1.84}$$

are valid for the symmetric Fibonacci hyperbolic functions (1.62) and (1.63):

$$[sFs(x)]^2 - cFs(x+1)cFs(x-1) = -1; \qquad (1.85)$$
$$[cFs(x)]^2 - sFs(x+1)sFs(x-1) = 1. \qquad (1.86)$$

By using the formulas (1.62)–(1.65), we can prove different identities for the symmetric Fibonacci and Lucas hyperbolic functions. A portion of these identities is presented in Table 1.6.

1.9.8. On the new mathematical term "recursive hyperbolic functions"

The concepts of gnomon, recursion, recursive function, and recurrence relation, which reflect a fundamental property of nature — the principle of self-similarity, are widespread in science, mathematics and computer science. The remarkable book *Gnomon: From Pharaohs to Fractals* by the French mathematician of Egyptian origin Midhat Gazale [20] is devoted to the study of the relationship between these concepts.

In Gazale's book [20] the concept of the gnomon is defined as follows: "Gnomon is the figure, which, when added to another figure, forms a new figure, similar to the original."

This concept is most easily demonstrated by using the so-called Fibonacci rectangles (Fig. 1.17).

Consider the Fibonacci numbers: 1, 1, 2, 3, 5, 8, 13, 21, ... Take two squares with sides equal to 1 (the area of each square will be equal to 1) and put them together. We get a rectangle of size 2×1, called a double square. Then we construct a new square of the size 2×2 on the long side of the double square. Here we get a new Fibonacci rectangle of the size 3×2. Note that the 2×2 square is a gnomon of the 3×2 Fibonacci rectangle. Then we take a new square of the size 3×3 (the gnomon) and place it on the longer side of the preceding Fibonacci rectangle. This results in the new 5×3 Fibonacci rectangle. Continuing this process, we will erect sequential rectangles whose sides are equal to adjacent Fibonacci numbers, that is, the rectangles of the following dimensions: 8×5, 13×8, 21×13, ... (Fig. 1.17). Note that the quadrates of the size 1×1, 2×2, 3×3, 5×5, 8×8 ... (Fibonacci numbers) are gnomons of Fibonacci rectangles.

Note that many objects in Nature are built according to this principle, which is called self-similarity. A self-similar object is that which coincides

Table 1.6.   The recursive identities for Fibonacci and Lucas hyperbolic functions.

The identities for Fibonacci and Lucas numbers	The identities for Fibonacci and Lucas hyperbolic functions
$F_{n+2} = F_{n+1} + F_n$	$sFs(x+2) = cFs(x+1) + sFs(x);$ $cFs(x+2) = sFs(x+1) + cFs(x)$
$L_{n+2} = L_{n+1} + L_n$	$sLs(x+2) = cLs(x+1) + sLs(x);$ $cLs(x+2) = sLs(x+1) + cLs(x)$
$F_n = (-1)^{n+1}F_{-n}$	$sFs(x) = -sFs(-x); cFs(x) = cFs(-x)$
$L_n = (-1)^n L_{-n}$	$sLs(x) = -sLs(-x); cLs(x) = cLs(-x)$
$F_{n+3} + F_n = 2F_{n+2}$	$sFs(x+3) + cFs(x) = 2cFs(x+2);$ $cFs(x+3) + sFs(x) = 2sFs(x+2)$
$F_{n+3} - F_n = 2F_{n+1}$	$sFs(x+3) - cFs(x) = 2sFs(x+1);$ $cFs(x+3) - sFs(x) = 2cFs(x+1)$
$F_{n+6} - F_n = 4F_{n+3}$	$sFs(x+6) - cFs(x) = 4cFs(x+3);$ $cFs(x+6) - sFs(x) = 4cFs(x+3)$
$F_n^2 - F_{n+1}F_{n-1} = (-1)^{n+1}$	$[sFs(x)]^2 - cFs(x+1)cFs(x-1) = -1;$ $[cFs(x)]^2 - sFs(x+1)sFs(x-1) = 1$
$F_{2n+1} = F_{n+1}^2 + F_n^2$	$cFs(2x+1) = [cFs(x+1)]^2 + [sFs(x)]^2;$ $sFs(2x+1) = [sFs(x+1)]^2 + [cFs(x)]^2$
$L_n^2 - 2(-1)^n = L_{2n}$	$[sLs(x)]^2 + 2 = cLs(2x);$ $[cLs(x)]^2 - 2 = cLs(2x)$
$L_n + L_{n+3} = 2L_{n+2}$	$sLs(x) + cLs(x+3) = 2sLs(x+2);$ $cLs(x) + sLs(x+3) = 2cLs(x+2)$
$L_{n+1}L_{n-1} - L_n^2 = -5(-1)^n$	$sLs(x+1)sLs(x-1) - [cLs(x)]^2 = -5;$ $cLs(x+1)cLs(x-1) - [sLs(x)]^2 = 5$
$F_{n+3} - 2F_n = L_n$	$sFs(x+3) - 2cFs(x) = sLs(x);$ $cFs(x+3) - 2sFs(x) = cLs(x)$
$L_{n-1} + L_{n+1} = 5F$	$sLs(x-1) + cLs(x+1) = 5sFs(x);$ $cLs(x-1) + sLs(x+1) = 5cFs(x)$
$L_n + 5F_n = 2L_{n+1}$	$sLs(x) + 5cFs(x) = 2cLs(x+1);$ $cLs(x) + 5sFs(x) = 2sLs(x+1)$
$L_{n+1}^2 + L_n^2 = 5F_{2n+1}$	$[sLs(x+1)]^2 + [cLs(x)]^2 = 5cFs(2x+1);$ $[cLs(x+1)]^2 + [sLs(x)]^2 = 5sFs(2x+1)$

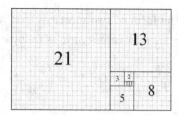

Figure 1.17.   Fibonacci rectangles.

Source: *Wikipedia, the free encyclopedia,*
https://en.wikipedia.org/wiki/Fibonacci_number

exactly or approximately with a part of itself (i.e. the whole has the same form as one or more of its parts). Many real-world objects, such as coastlines, have the property of statistical self-similarity: their parts are statistically homogeneous at different measurement scales. Self-similarity is a characteristic feature of fractals.

In mathematics, the principle of self-similarity is expressed by recursion and recursive function. The Fibonacci and Lucas numerical sequences are the most striking examples of recursive functions in mathematics. They are given by recurrence formulas, in which each element of the numerical series is calculated as a function of the n previous elements. Thus, by using a finite expression (which is a combination of the recurrence formula and the set of values for the first elements of the series) we can obtain an infinite number of elements of the sequence.

This chapter describes a fundamentally novel mathematical result, defining a new class of hyperbolic functions called Fibonacci and Lucas hyperbolic functions. The unique nature of this class of hyperbolic functions is their fundamental connection with Fibonacci and Lucas numbers, which reflect the principle of recursion, recursive functions and ultimately the principle of self-similarity, which is a fundamental principle of Nature.

These considerations allow us to introduce a new mathematical concept of the recursive hyperbolic functions, defined above as Fibonacci and Lucas hyperbolic functions. The word "recursive" in this definition emphasizes the fundamental relationship of a new class of hyperbolic functions with the principle of self-similarity, a fundamental principle of Nature.

1.9.9.   *The golden Q-matrices*

In Stakhov's article [21] the following unusual (2×2)-matrices have been introduced. Let us represent the Fibonacci Q-matrix [15],

$Q^n = \begin{pmatrix} F_{n+1} & F_n \\ F_n & F_{n-1} \end{pmatrix}$ in the form of two matrices that are given for the even ($n = 2k$) and odd ($n = 2k + 1$) values of n as follows:

$$Q^{2k} = \begin{pmatrix} F_{2k+1} & F_{2k} \\ F_{2k} & F_{2k-1} \end{pmatrix}, \quad Q^{2k+1} = \begin{pmatrix} F_{2k+2} & F_{2k+1} \\ F_{2k+1} & F_{2k} \end{pmatrix}. \tag{1.87}$$

By using the relations (1.69)

$$F_n = \begin{cases} \text{sFs}(n) & \text{for } n = 2k \\ \text{cFs}(n) & \text{for } n = 2k + 1 \end{cases}$$

which connect Fibonacci numbers F_n with the Fibonacci hyperbolic functions (1.62) and (1.63), we can represent the matrices (1.87) in terms of the Fibonacci hyperbolic functions (1.62) and (1.63):

$$Q^{2k} = \begin{pmatrix} \text{cFs}(2k + 1) & \text{sFs}(2k) \\ \text{sFs}(2k) & \text{cFs}(2k - 1) \end{pmatrix},$$

$$Q^{2k+1} = \begin{pmatrix} \text{sFs}(2k + 2) & \text{cFs}(2k + 1) \\ \text{cFs}(2k + 1) & \text{sFs}(2k) \end{pmatrix} \tag{1.88}$$

where k is a discrete variable, $k = 0, \pm 1, \pm 2, \pm 3, \dots$.

If we exchange the discrete variable k in the matrices (1.88) with the continuous variable x, we obtain two unusual matrices that are functions of the continuous variable x:

$$Q_1(x) = \begin{pmatrix} \text{cFs}(2x + 1) & \text{sFs}(2x) \\ \text{sFs}(2x) & \text{cFs}(2x - 1) \end{pmatrix},$$

$$Q_2(x) = \begin{pmatrix} \text{sFs}(2x + 2) & \text{cFs}(2x + 1) \\ \text{cFs}(2x + 1) & \text{sFs}(2x) \end{pmatrix}. \tag{1.89}$$

These matrices (1.89) are called the "golden" Q-matrices [21]. They are a generalization of the Fibonacci Q-matrices for continuous domain.

Taking into consideration the properties (1.85) and (1.86), we can write the following remarkable formulas for the determinants of the "golden" Q-matrices (1.89):

$$\det Q_1(x) = \text{cFs}(2x + 1)\text{cFs}(2x - 1) - [\text{sFs}(2x)]^2 = 1 \tag{1.90}$$

$$\det Q_2(x) = \text{cFs}(2x + 2)\text{cFs}(2x) - [\text{sFs}(2x + 1)]^2 = -1. \tag{1.91}$$

1.9.10. The theory of Fibonacci numbers as a degenerate case of the theory of recursive Fibonacci and Lucas hyperbolic functions

It is demonstrated above that the two continuous identities for the recursive Fibonacci and Lucas hyperbolic functions always correspond to every discrete identity for Fibonacci and Lucas numbers. Conversely, one may obtain the discrete identity for Fibonacci and Lucas numbers by using two corresponding continuous identities for the Fibonacci and Lucas hyperbolic functions (1.62)–(1.65). Since the "extended" Fibonacci and Lucas numbers, according to (1.71) and (1.72), are discrete cases of the Fibonacci and Lucas hyperbolic functions (1.62)–(1.65) for the discrete values of the continuous variable x, it follows that with the introduction of the Fibonacci and Lucas hyperbolic functions (1.62)–(1.65), the classic "theory of Fibonacci numbers" [14, 15] degenerates, since this theory is a special limiting (discrete) case of the more general (continuous) theory of the Fibonacci and Lucas hyperbolic functions (1.62)–(1.65). This is an unexpected result following from the theory of the Fibonacci and Lucas hyperbolic functions [19]. And Oleg Bodnar's new geometric theory of phyllotaxis [16, 22–24] is a brilliant confirmation of the unique and fundamental nature of the recursive Fibonacci and Lucas hyperbolic functions (1.62)–(1.65).

1.10. Hyperbolic Geometry of Phyllotaxis (Bodnar's Geometry)

1.10.1. What is a phyllotaxis?

The botanical phenomenon of phyllotaxis is one of the most common features of Nature. It is inherent in many biological systems. An essential feature of phyllotaxis consists in the spiral disposition of leaves on stems of plants and trees, petals in flower heads, seeds in pine cones and sunflower disks, etc. This phenomenon, recognized since the time of Leonardo da Vinci, was a subject of discussion for many scientists, including Kepler, Turing and Weil, amongst others. It is intriguing that more complex concepts of symmetry are used in phyllotaxis, particularly helical symmetry.

Phyllotaxis reveals itself especially in inflorescences and densely-packed botanical structures such as sunflower disks, pine cones, pineapples, cacti, cauliflower heads and many other botanical subjects (see some examples in Fig. 1.18).

(a)   (b)

Figure 1.18.   Examples of phyllotaxis structures: (a) sunflower; (b) pine cone.
*Sources: (a) Wikipedia, the free encyclopedia, https://en.wikipedia.org/wiki/Helianthus;
(b) Wikipedia, the free encyclopedia, https://en.wikipedia.org/wiki/Conifer_ cone*

On the surfaces of such objects, their bio-organs (e.g. seeds on sunflower disks and pine cones) are laid down in left-handed and right-handed spirals. Botanists have demonstrated that the most frequent ratios between these spirals are equal to the ratios of the adjacent Fibonacci numbers, that is,

$$\frac{F_{n+1}}{F_n} : \quad \frac{2}{1}, \frac{3}{2}, \frac{5}{3}, \frac{8}{5}, \frac{13}{8}, \frac{21}{13}, \cdots \to \Phi = \frac{1+\sqrt{5}}{2}. \tag{1.92}$$

These ratios (1.92) are called phyllotaxis orders [16]; they vary from one subject to another. For example, the sunflower disks can have phyllotaxis orders given by Fibonacci's ratios $\frac{34}{21}$, $\frac{55}{34}$, $\frac{89}{55}$, $\frac{144}{89}$ and even $\frac{233}{144}$.

1.10.2.   *The mystery of phyllotaxis*

In observing the well-organized surfaces of phyllotaxis structures, the question arises: how are Fibonacci spirals formed on their surface during growth? It has been shown that many bio-forms change their phyllotaxis orders during their growth. It has been observed, for example, that sunflower disks, located on different levels of the same stalk, have different phyllotaxis orders; moreover, the greater its age, the higher the phyllotaxis order. This means that during their growth, a natural modification of symmetry occurs obeying the following law:

$$\frac{2}{1} \to \frac{3}{2} \to \frac{5}{3} \to \frac{8}{5} \to \frac{13}{8} \to \frac{21}{13} \to \cdots . \tag{1.93}$$

This growth of the phyllotaxis orders according to (1.93) is called dynamic symmetry [16, 22–24]. Many scientists, who have studied the mystery of phyllotaxis, believe that the dynamic symmetry (1.93) is of fundamental interdisciplinary importance. In the opinion of Russian scientist and encyclopaedist Vladimir Vernadsky the problem of biological symmetry is the central problem of biology.

One may assume that this numerical regularity (1.93) reflects some general geometric law, underlying the mechanism of phyllotaxis. Discovering its secret would be of great importance for understanding the phenomenon of phyllotaxis as a whole. The connection of botanic phyllotaxis with Fibonacci numbers, as well as with generalized Fibonacci recurrence sequences, significantly enhances the role of Fibonacci numbers in modern science. In this regard, the Fibonacci and Lucas hyperbolic functions (1.62)–(1.65) are of particular importance as they can be used for the simulation of a variety of biological phenomena. Thus the importance of Bodnar's development of a new geometric theory of phyllotaxis based upon the Fibonacci and Lucas hyperbolic functions [19].

1.10.3. *Significance of Bodnar's geometry for the development of recursive hyperbolic geometry*

We will not delve deeply into the Bodnar's geometry which led to his new geometrical theory of phyllotaxis. Readers who want to become more acquainted with Bodnar's geometry are referred to his book, *The Golden Section and Non-Euclidean Geometry in Nature and Art* [16], and his latest publications on this theme [22–24].

From Bodnar's geometry [16, 22–24] we can derive the following important conclusions:

(1) Bodnar's geometry has revealed a new "hyperbolic world" for science — the world of phyllotaxis and its geometric secrets. The main feature of this world is that its basic mathematical relations are described by the recursive Fibonacci hyperbolic functions (1.62) and (1.63), giving rise to the appearance of Fibonacci numbers on the surface of phyllotaxis objects.

(2) Bodnar's geometry has shown that hyperbolic geometry is much more common in the real world than originally thought. The Fibonacci and Lucas hyperbolic functions (1.62)–(1.65), introduced in [19], appear to be fundamental functions of Nature. They arise in various botanic

structures, including pine cones, pineapples, cacti, sunflower disks and so on.

(3) Bodnar's new hyperbolic geometry applying recursive properties to Nature is of fundamental importance for the future development of the modern life sciences (biology, botany, physiology, medicine, genetics, and so on).

(4) There is a fundamental distinction between Lobachevsky's classic hyperbolic geometry and Bodnar's [16, 22–24] new hyperbolic geometry of phyllotaxis. Lobachevsky's geometry is based on the classic hyperbolic functions (1.58) which use Euler's constant e as the base of these functions. The applications of Lobachevsky's geometry relate, first of all, to the "mineral world" and physical phenomena (Einstein's special theory of relativity, four-dimensional Minkowski's world, etc.). Bodnar's geometry, on the other hand, is a hyperbolic geometry of Living Nature. It is based on the symmetric Fibonacci hyperbolic functions (1.62) and (1.63) which use the golden ratio (Φ) as the base of these functions.

(5) In contrast to the classic hyperbolic functions, the new symmetric Fibonacci and Lucas hyperbolic functions (1.62)–(1.65) have unique mathematical properties. The first property is that they possess recursive properties similar to Fibonacci and Lucas numbers. The second property follows from the fundamental mathematical connection between the symmetric Fibonacci hyperbolic functions and "extended" Fibonacci numbers, giving a simple explanation of why Fibonacci spirals appear on the surface of phyllotaxis objects. It follows from this that Bodnar's geometry has recursive properties.

(6) The previous conclusion suggests the importance of Bodnar's geometry for the future development of hyperbolic geometry. As is well established, Lobachevsky's development of hyperbolic geometry was the result of the replacement of Euclid's Fifth Postulate (the postulate of parallel lines) by a new postulate, Lobachevsky's postulate. Since then, other new non-Euclidean geometries have emerged by way of postulate replacement (e.g. elliptic geometry). In Lobachevsky's time, only one class of hyperbolic functions, determined by the formulas (1.58), was well known. The use of these functions pertains to hyperbolic geometry, known as Lobachevsky's geometry. However, classic hyperbolic functions and classic Lobachevsky's geometry do not have recursive properties. Creating a new class of hyperbolic functions, called recursive Fibonacci and Lucas hyperbolic functions [19], became a prerequisite

for the derivation of a new kind of hyperbolic geometry called Bodnar's geometry [16, 22–24]. Bodnar's geometry resulted from hyperbolic function replacement. This suggests a new way of expressing hyperbolic geometry: the search for new hyperbolic functions can lead to new recursive hyperbolic geometries. This idea will be further developed in Chapters 2 and 3.

(7) Furthermore, the recursive hyperbolic Fibonacci and Lucas functions and Bodnar's geometry are based upon the golden ratio, one of the most beautiful mathematical constants, a symbol of harmony and beauty. This provides an important link to Dirac's Principle of Mathematical Beauty. Bodnar's phyllotaxis geometry is a brilliant embodiment of Dirac's Principle of Mathematical Beauty!

(8) Bodnar's geometry helps reveal the underlying secret of phyllotaxis, one of the most amazing occurrences in the life sciences. If Nature actually operates according to the model suggested by Bodnar, then it is reasonable to suggest that Nature is a great mathematician by employing recursive Fibonacci and Lucas hyperbolic functions throughout its creative efforts.

References

[1] Soroko, E. M. *Structural Harmony of Systems*. Minsk: Nauka i Technika (1984) (Russian).

[2] Shevelev, J. S. *Meta-language of the Living Nature*. Moscow: Publishing House "Voskresenie" (2000) (Russian).

[3] Losev, A. "The history of philosophy as a school of thought", *Communist* (1981) No. 11 (Russian).

[4] Olsen, S. A. *The Golden Section: Nature's Greatest Secret*. New York: Bloomsbury Publishing USA (2006).

[5] Kline, M. *Mathematics. The Loss of Certainty*. New York: Oxford University Press (1980).

[6] Kolmogorov, A. N. *Mathematics in its Historical Development*. Moscow: Nauka (1991) (Russian).

[7] Wenninger, M. *Polyhedron Models*. Cambridge: Cambridge University Press (1971).

[8] Stakhov, A. P., assisted by S. Olsen.*The Mathematics of Harmony. From Euclid to Contemporary Mathematics and Computer Science*. Singapore: World Scientific (2009).

[9] Kann, C. H. *Pythagoras and the Pythagoreans. A Brief History*. Hackett Publishing Co, Inc. (2001).

[10] Zhmud, L. *The Origin of the History of Science in Classical Antiquity*. Walter de Gruyter (2006).

[11] Smorinsky, C. *History of Mathematics. A Supplement.* Springer (2008).

[12] Klein, F. *Lectures on the Icosahedron.* New York: Courier Dover Publications (1956).

[13] Khinchin, A. Ya. *Continued Fractions.* Moscow: Fizmathiz (1960) (Russian).

[14] Vorobyov, N. N. *Fibonacci Numbers.* Moscow: Nauka (1984), (first edition 1961) (Russian).

[15] Hoggat, Jr. V. E. *Fibonacci and Lucas Numbers.* Boston, MA: Houghton Mifflin (1969).

[16] Bodnar, O. Y. *The Golden Section and Non-Euclidean Geometry in Nature and Art.* Lvov: Publishing House "Svit" (1994) (Russian).

[17] Polya, G. *Mathematical Discovery.* John Wiley & Sons (1962).

[18] Stakhov, A. P. and Tkachenko, I. S. "Hyperbolic Fibonacci trigonometry", *Reports of the Ukrainian Academy of Sciences* (1993) Vol. 208, No. 7: 9–14 (Russian).

[19] Stakhov, A. P. and Rozin, B. N. "On a new class of hyperbolic function", *Chaos, Solitons & Fractals* (2004) Vol. 23: 379–389.

[20] Gazale, M. J. *Gnomon. From Pharaohs to Fractals.* Princeton, NJ: Princeton University Press (1999).

[21] Stakhov, A. "The 'golden' matrices and a new kind of cryptography", *Chaos, Solitons & Fractals* (2007), Vol. 32, Issue 3: 1138–1146.

[22] Bodnar, O. Y. "Dynamic symmetry in nature and architecture", *Visual Mathematics* (2010) Vol. 12, No. 4. http://www.mi.sanu.ac.rs/vismath/BOD 2010/index.html (accessed November 4, 2015).

[23] Bodnar, O. Y. "Geometric interpretation and generalization of the non-classical hyperbolic functions", *Visual Mathematics* (2011) Vol. 13, No. 2. http://www.mi.sanu.ac.rs/vismath/bodnarsept2011/SilverF.pdf (accessed November 4, 2015).

[24] Bodnar, O. Y. "Minkovski's geometry in the mathematical modeling of natural phenomena", *Visual Mathematics* (2012) Vol. 14, No. 1. http://www. mi.sanu.ac.rs/vismath/bodnardecembar2011/mink.pdf (accessed November 4, 2015).

Chapter 2

The Mathematics of Harmony and General Theory of Recursive Hyperbolic Functions

2.1. The Mathematics of Harmony: The History, Generalizations and Applications of Fibonacci Numbers and Golden Ratio

2.1.1. The revival of the ancient Greeks' "harmonic ideas" in modern science

Alexey Stakhov's book *The Mathematics of Harmony* [1] was published in 2009. The American philosopher Professor Scott Olsen, author of the international bestseller, *The Golden Section* [2], assisted in the preparation and editing of *The Mathematics of Harmony* [1] for publication.

The publication of these two books is a reflection of an important trend in the development of modern science. This trend involves a return to the "harmonic ideas" of Pythagoras and Plato (the golden ratio and Platonic solids), embodied in Euclid's *Elements* [3], in modern philosophical and scientific endeavors. For example, trends supporting significant discoveries in chemistry and crystallography include, respectively: fullerenes, based on the truncated icosahedron (Nobel Prize — 1996, Kroto, *et al.*), and quasi-crystals, based on the icosahedral or pentagonal symmetry (Nobel Prize — 2011, Shechtman).

Other significant discoveries are the law of structural harmony of systems by Edward Soroko [4] based on the golden *p*-proportions, and the law of spiral biosymmetry transformation by Oleg Bodnar [5] based on the recursive Fibonacci hyperbolic functions. There is also a new theory of the genetic code based on golden genomatrices by Sergey Petoukhov [6]. Publication of these books [1–6] lend support to this important trend in mathematics, which appears to be continuing!

2.1.2.   The phrase "the Mathematics of Harmony"

The notion of the "Mathematics of Harmony" first appeared in the article "Harmony of spheres" in *The Oxford Dictionary of Philosophy* [7, p. 160]:

> *"**Harmony of spheres.** A doctrine often traced to Pythagoras and fusing together mathematics, music, and astronomy. In essence the heavenly bodies being large objects in motion, must produce music. The perfection of the celestial world requires that this music be harmonious, it is hidden from our ears only because it is always present. **The mathematics of harmony was a central discovery of immense significance to the Pythagoreans.**"*

The concept of the Mathematics of Harmony here is closely associated with the harmony of the spheres, also known as harmony of the world (Latin *harmonica mundi*) or world music (Latin *musical mundane*). Harmony of the spheres is the ancient and medieval doctrine of the musical-mathematical construction of the Cosmos, going back to the Pythagorean and Platonic philosophical tradition.

We find another mention of the Mathematics of Harmony, as an ancient Greek discovery, in *A New Kind of Social Science* by Vladimir Dimitrov [8, p. 65]. He writes:

> *"Harmony was a key concept of the Greeks, a conjunction of three strands of meaning. Its root meaning was **aro**, join, so **harmonia** was what was joined. Another meaning was proportion, the balance of things that allowed an easy fit. The quality of joining and proportion then came to be involved in music and the other arts. The precondition for harmony for the Greeks was expressed in the phrase 'nothing too much'. It also had a mysterious positive quality, which became the object of enquiry of their finest minds. Thinkers such as Pythagoras sought to capture the mystery of harmony as something both inexpressible yet also illuminated by mathematics. **The mathematics of harmony** explored by the ancient Greeks is still an inspiring model for contemporary scientists. Crucial to it is their discovery of its quantitative expression in astonishing diversity and complexity of nature through the golden mean (golden ratio), Φ (phi): $\Phi = \frac{1+\sqrt{5}}{2}$, which is approximately equal to 1.618. It is described by Euclid in book five of his* Elements*: 'A straight line is said to have been cut in extreme and mean ratio when, as the whole line is to the greater, so is the greater to the less.' "*

Thus, in the book [8] the concept of the Mathematics of Harmony is directly associated with the golden ratio, the most important ancient mathematical discovery in the field of harmony, which at that time was called the division of a line segment in extreme and mean ratio.

Finally, it is pertinent to mention that this term was used in Stakhov's speech "The golden section and modern harmony mathematics", made at the 7th International Conference on Fibonacci Numbers and Their Applications (Graz, Austria, 1996) [9].

2.1.3. *The most important periods in the development of the Mathematics of Harmony*

As mentioned above, the use of the phrase the "Mathematics of Harmony" was re-introduced to revive an important feature of ancient Greek science (studying the Harmony of the Universe) [8, 9]. Pythagoras, Plato and Euclid focused their researches on the Harmony of the Cosmos. This concept re-emerges in periods of great advances of the human spirit. From this point of view, the study of the Mathematics of Harmony can be divided into the following important periods.

Ancient Greek period. Traditionally, this period is assumed to start with the researches of Pythagoras and Plato (and esoterically with their study with the Egyptian priesthood). Euclid's *Elements* were a culminating event of this period. According to the hypothesis of Proclus [3], Euclid created his *Elements* in order to give a complete geometric theory of the five regular Platonic solids, which had been associated in ancient Greek science with the Harmony of the Cosmos. Euclid placed the fundamentals of the theory of the Platonic solids in Book XIII, the final book of the *Elements*. Euclid simultaneously inserted some of the more advanced achievements of ancient Greek mathematics, including principles of proportion into his *Elements*. In particular, the golden section (Book II) was used by Euclid in the creation of his geometric theory of the Platonic solids.

The Middle Ages. In the Middle Ages, the famous Italian mathematician Leonardo of Pisa (Fibonacci) wrote *Liber Abaci* (1202) where he introduced Hindu–Arabic numerals to the West. And most importantly, he solved the rabbit reproduction problem through the use of this remarkable numerical sequence — the Fibonacci numbers.

The Renaissance. The Renaissance was connected with several prominent figures: Piero della Francesca (1412–1492), Leon Battista Alberti

(1404–1472), Leonardo da Vinci (1452–1519), Luca Pacioli (1445–1517), and Johannes Kepler (1571–1630). Two books were published in that period that best represent the idea of Cosmic harmony. The first was *Divina Proprotione* (*The Divine Proportion*) (1509), written by the outstanding Italian mathematician and learned monk Luca Pacioli. Known as "the monk drunk on beauty", he was under the direct influence of Leonardo da Vinci, who illustrated this wondrous work.

The brilliant 17th century astronomer, Johannes Kepler, made a significant contribution to the development of the harmonic ideas of Pythagoras, Plato and Euclid. In his first book *Mysterium Cosmographicum* (1596) he built the so-called Cosmic Cup — the original model of the Solar system, based on the Platonic solids (see Fig. 1.3). The book *Harmonice Mundi* (or *Harmony of the Worlds*, 1619) is Kepler's primary contribution to the doctrine of Universe Harmony. In *Harmony*, he attempted to explain the proportions of the natural world — particularly the astronomical and astrological aspects — in terms of music. The *Musica Universalis* or *Music of the Spheres*, which had been studied by Pythagoras, Ptolemy and many other early thinkers, was Kepler's central focus in *Harmony*.

The 19th century. This period is connected with the works of the two French mathematicians Jacques Philip Marie Binet (1786–1856) and François Édouard Anatole Lucas (1842–1891), and the two Germans thinkers, the poet and philosopher Adolf Zeising (1810–1876) and the mathematician Felix Klein (1849–1925).

The merit of Binet and Lucas is the fact that their combined research became a launching pad for Fibonacci-type research in the Soviet Union, United States, Great Britain and other countries [10–12].

In 1854 Adolf Zeising published *Neue Lehre von den Proportionen des Menschlichen Körpers...* (*New Teachings on the Proportions of the Human Body ...*). Zeising's primary objective was the formulation of the Law of Proportionality:

> *"A division of the whole into unequal parts is proportional, when the ratio between the parts is the same as the ratio of the bigger part to the whole, this ratio is equal to the golden mean."*

Felix Klein in 1884 published *Lectures on the Icosahedron* [13], dedicated to the geometric theory of the icosahedron and its central role in the general theory of mathematics. Klein treats the icosahedron as a mathematical object, providing a source for the five mathematical theories: geometry, Galois theory, group theory, invariant theory and differential equations.

What is the deep significance of Klein's ideas from the point of view of the Mathematics of Harmony? According to Klein, the Platonic icosahedron, based on the golden ratio, is the main geometric figure of mathematics. It would follow from this that the golden ratio may be the primary geometric object of mathematics, which, according to Klein, can unify all of mathematics.

This idea of Klein's is consistent with the ideas expressed in the article "The generalized golden proportions and a new approach to the geometric definition of a number", published by Stakhov in the *Ukrainian Mathematical Journal* [14]. This article presents the concept of a "golden number theory", which in turn is the basis for "golden mathematics", arising out of the golden ratio and its generalizations.

First half of the 20th century. In the first half of the 20th century, development of the "golden paradigm" of the ancient Greeks may be associated with two Russians, professor of architecture David Grimm (1865–1942), and the religious philosopher Pavel (Paul) Florensky (1882–1937).

The purpose of Grimm's 1935 book, *Proportionality in Architecture* [15], is formulated in its Introduction as follows:

> *"In view of the exceptional significance of the Golden Section in the sense of the proportional division, which establishes a permanent connection between the whole and its parts and gives a constant ratio between them (which is unreachable by any other division), the scheme, based on it, is the main standard and is accepted by us in the future as a basis for checking the proportionality of historical monuments and modern buildings.... Taking this general importance of the Golden Section in all aspects of architectural thought, the theory of proportionality — based on the proportional division of the whole into parts corresponding to the Golden Section, should be recognized as the architectural basis of proportionality...."*

In the 1920s, Pavel Florensky wrote the work *At the Watershed of a Thought*. The third chapter is devoted to the golden ratio. Belorussian philosopher Edward Soroko in his book, *Structural Harmony of Systems* [4], evaluates Florensky's work as follows:

> *"The aim was to derive analytically the stability of the whole object, which is in the field of the effect of oppositely oriented forces. The project was conceived as an attempt to use the 'golden ratio' and its substantial*

basis, which manifests itself not only in a series of experimental obser-
vations of nature, but on the deeper levels of knowledge, for the case of
penetration into the dialectic of movement, into the substance of things."

Second half of the 20th century and the 21st century. In the second half
of the 20th century the interest in this area began increasing in all areas of
science, including mathematics. Several mathematicians became outstanding representatives of this "golden paradigm" trend in mathematics, including the Soviet Nikolai Vorobyov (1925–1995) [10], American Verner Hoggatt
(1921–1981) [11], and Englishman Stefan Vajda [12].

New scientific discoveries in the present century are further reviving the
idea of harmony in modern science. Reliance on the Platonic solids, golden
ratio and Fibonacci numbers throughout the natural sciences (e.g. crystallography, chemistry, astronomy, earth science, quantum physics, botany,
biology, geology, medicine, and genetics), as well as, computer science and
economics have been instrumental in the revival of this ancient idea of
universal harmony in modern science. This has resulted in an exponential
development of the Mathematics of Harmony.

2.2.   Algorithmic Measurement Theory as a Constructive Measurement Theory Based on the Abstraction of Potential Infinity

The algorithmic measurement theory was the first mathematical theory
created by Stakhov at the beginning of his scientific career. Its foundations
are described in his two books [16, 17] published in Russia in 1977 and
1979.

This theory in its original form was of a purely applied character based
upon Stakhov's 1972 doctoral dissertation, "Synthesis of the optimal algorithms for analog-to-digital conversion". However, in creating this theory,
Stakhov ventured deeply into the frameworks of fundamental mathematical
concepts, in particular, the problems of mathematical measurement theory
and mathematical infinity.

Following Aristole's objection to such contradictory concepts, Stakhov
first eliminated from consideration Cantor's abstraction of actual infinity
as being internally contradictory (i.e. the notion of completed infinity).
Thus, from the outset Stakhov had excluded Cantor's axiom (which is the
very basis of classic measurement theory, based on the continuity axioms)

from algorithmic measurement theory. Stakhov discovered a contradiction between Cantor's and Archimedes' axioms [16]. Unfortunately this fundamental contradiction underlies classic measurement theory and mathematics as a whole. Russian mathematician and philosopher Alexander Zenkin further developed this critique of Cantor's theory of infinite sets in the article, "The mistake of Georg Cantor" [18].

By using the so-called constructive approach to algorithmic measurement theory, based upon the abstraction of potential infinity (as opposed to actual infinity), Stakhov solved the problem of the optimal measurement algorithm in his book, *Introduction into Algorithmic Measurement Theory* [16]. This underlying problem had really not been properly addressed in mathematics, much less resolved. Herein we discuss the problem of the synthesis of optimal measurement algorithms, which are a generalization of well-known measurement algorithms: the algorithm of counting, which underlies the Euclidean definition of natural numbers, and the binary algorithm, which underlies the binary system, the basis of modern computers.

Each measurement algorithm corresponds to some positional numeral system. All the well-known positional numeral systems, including the decimal and binary systems, follow from algorithmic measurement theory. It is very important that the new, unknown positional numeral systems, in particular, the Fibonacci and binomial numeral systems, arose within algorithmic measurement theory [16].

We assert that the algorithmic measurement theory is the source for the new positional numeral systems. Note that mathematics was never seriously engaged with numeral systems and advances in this field of mathematics have moved forward very slowly. This is one of the strategic mistakes of mathematics [19]. New positional numeral systems (in particular, Fibonacci p-codes introduced by Stakhov) could become the alternative to the binary system, particularly for mission-critical applications.

2.3. Pascal's Triangle, Fibonacci p-Numbers, and Golden p-Proportions

2.3.1. *Fibonacci p-numbers*

Pascal's triangle is so widely known and well studied that many mathematicians may well ask: what is new that could possibly be found in this triangle? The book *Mathematical Discovery* (Russian translation, 1970) [20] by American mathematician George Polya is well known. In this book

Polya demonstrated an unexpected connection between Pascal's triangle and Fibonacci numbers (Fig. 1.11).

The study of the so-called diagonal sums of Pascal's triangle, carried out by many mathematicians (including Stakhov), has led to a wider generalization of Fibonacci numbers. Here we mean the new recurrent numerical sequences, introduced in [16] and called the Fibonacci p-numbers. These recurrent sequences for the given $p = 0, 1, 2, 3, \ldots$ are generated by the following general recurrence relation:

$$F_p(n) = F_p(n-1) + F_p(n-p-1) \quad \text{given } n > p+1 \qquad (2.1)$$

at the seeds:

$$F_p(1) = F_p(2) = \cdots = F_p(p+1) = 1. \qquad (2.2)$$

Note that for the case $p = 0$ the recurrence relation (2.1) generates the classic binary numbers and for the case $p = 1$ the classic Fibonacci numbers.

2.3.2.   A generalization of the golden ratio

By studying the Fibonacci p-numbers and considering the limit of the ratio of neighboring Fibonacci p-numbers $\lim\limits_{n \to \infty} \frac{F_p(n)}{F_p(n-1)} = x$, Stakhov derived the following algebraic equation, which is the characteristic equation for the recurrence relation (2.1) [16]:

$$x^{p+1} - x^p - 1 = 0 \quad (p = 0, 1, 2, 3, \ldots). \qquad (2.3)$$

The positive roots of equation (2.3) form a set of new mathematical constants Φ_p, which describe some algebraic properties of Pascal's triangle. The classic golden ratio is a special case of the constants Φ_p where $p = 1$. Therefore, the constants Φ_p have been called the golden p-proportions. This discovery led to the admiration of Ukrainian mathematician academician Mitropolsky, who wrote the following [21]:

> "*Let's ponder upon this result. Within several millennia, starting with Pythagoras and Plato, mankind used the widely known classical Golden Proportion as a unique number. And unexpectedly at the end of the 20th century the Ukrainian scientist Stakhov generalized this result and proved the existence of an infinite number of Golden Proportions! And all of them have the same ability to express Harmony, as well as the classical Golden Proportion. Moreover, Stakhov proved that the golden p-proportions Φ_p $(1 \leq \Phi_p \leq 2)$ represented a new class of irrational numbers, which express some otherwise unknown mathematical properties of Pascal triangle. Clearly, such a mathematical result is of fundamental importance for the development of modern science and mathematics.*"

2.4. Fibonacci p-codes

Fibonacci measurement algorithms, based on the Fibonacci p-numbers [16], led to the discovery of new positional numeral systems, called Fibonacci p-codes [16]:

$$N = a_n F_p(n) + a_{n-1} F_p(n-1) + \cdots + a_i F_p(i) + \cdots + a_1 F_p(1), \qquad (2.4)$$

where $p = 0, 1, 2, 3, \ldots$ is a given integer; $a_i \in \{0, 1\}$ is the binary numeral of the ith digit; and the Fibonacci p-number $F_p(i)$ $(i = 1, 2, 3, \ldots, n)$ is the weight of the ith digit.

The formula (2.4) includes an infinite number of different binary representations of natural numbers because every $p = 0, 1, 2, 3, \ldots$ gives its own positional method of number representation. In particular, for the case $p = 0$ the Fibonacci p-code is reduced to the classic binary code:

$$N = a_n 2^{n-1} + a_{n-1} 2^{n-2} + \cdots + a_i 2^{i-1} + \cdots + a_1 2^0, \qquad (2.5)$$

which underlies all modern computers. For the case $p > 0$ all the Fibonacci p-codes are redundant positional binary systems, which can be used for the design of fundamentally new computers (Fibonacci computers), which have excellent informational reliability. The fundamentals of these new Fibonacci arithmetics are described in the books [1] and [16].

2.5. Codes of the Golden Proportion

In 1957, the young 12-year-old wunderkind of American mathematics, George Bergman had described in *Mathematics Magazine* the unique numeral system, called a number system with an irrational base [22]:

$$A = \sum_i a_i \Phi^i, \qquad (2.6)$$

where A is some real number, $a_i \in \{0, 1\}$ is a binary numeral of the ith digit $(i = 0, \pm 1, \pm 2, \pm 3, \ldots)$, Φ^i is the weight of the ith digit, and $\Phi = \frac{1 + \sqrt{5}}{2}$ is a base of Bergman's system (2.6). Unfortunately, 20th century mathematicians did not pay sufficient attention to this brilliant mathematical discovery, despite the fact that Bergman's system (2.6) appears to be the greatest contemporary mathematical discovery in the field of positional numeral systems since the discoveries of the Babylonian positional principle of number representation and the decimal and binary numeral systems.

Stakhov generalized Bergman's number system (2.6) thereby introducing a broad class of positional numeral systems with irrational bases, calling them "codes of the golden p-proportions" in his 1980 article, "The golden ratio in digital technology" [23]:

$$A = \sum_i a_i \Phi_p^i, \qquad (2.7)$$

where $i = 0, \pm 1, \pm 2, \pm 3, \ldots, p = 0, 1, 2, 3, \ldots, a_i \in \{0, 1\}$ is a binary numeral of the ith digit, Φ_p is a base of the numeral system (2.7) (i.e. the golden p-proportion, following from Pascal's triangle), Φ_p^i is the weight of the ith digit, which are thereby connected with the weights of the previous digits by the following identity:

$$\Phi_p^i = \Phi_p^{i-1} + \Phi_p^{i-p-1}. \qquad (2.8)$$

The underlying theory of this generalization of Bergman's number system was set forth in Stakhov's 1984 book, *Codes of the Golden Proportion* [24]. Note that the expression (2.7) includes an infinite number of positional binary numeral systems, because every p ($p = 0, 1, 2, 3, \ldots$) generates its own numeral system. In particular, two special limiting cases result when $p = 0$, and when $p = 1$. For the case $p = 0$ the base $\Phi_p = \Phi_0 = 2$ and the code of the golden p-proportion (2.7) is reduced to the classic binary system; and for the case $p = 1$ (2.7) is reduced to Bergman's system (2.6).

2.6.   The Golden Number Theory

Yuri Mitropolsky, editor-in-chief of the *Ukrainian Mathematical Journal*, invited Stakhov to submit an article on the codes of the golden p-proportions. The article "The generalized golden proportions and a new approach to geometric definition of a number" [14] was published in this prestigious mathematical journal in 2004.

The main idea of the article is the following. The codes of the golden p-proportions, which reflect important algebraic properties of Pascal's triangle, can be considered to be the beginning of a new number theory, the "golden" number theory. Indeed, with the help of the codes of the golden p-proportions we can represent all real numbers, including natural, rational and irrational, through this "golden" number theory. The codes of the golden p-proportions dramatically change our ideas regarding the relationship between rational and irrational numbers, because the special irrational numbers (the golden p-proportions) are the basis of all numbers and, therefore, of all mathematics.

Consider one of the unusual results of the "golden" number theory [14] through the example of Bergman's system (2.6). For this, we represent the natural number N in Bergman's system (2.6) as follows:

$$N = \sum_i a_i \Phi^i. \qquad (2.9)$$

It is proven in [14] that the sum of (2.9) for any natural number N is always finite, that is, any natural number N can be represented as the finite sum of the powers of the golden ratio. As all powers of the golden ratio are irrational numbers (except for $\Phi^0 = 1$), this assertion is far from trivial.

Let us consider the so-called "extended" Fibonacci numbers (see Table 1.2). If we now substitute the extended Fibonacci numbers F_i ($i = 0, \pm 1, \pm 2, \pm 3, \ldots$) instead of the powers Φ^i into the expression (2.9), then, to our surprise, we find in [14] that this sum is equal to 0 for any natural number N, that is,

$$\sum_i a_i F_i = 0. \qquad (2.10)$$

This property has been called the Z-property of natural numbers [14]. Since this property is valid only for natural numbers, it means that in [14] Stakhov found a new property of natural numbers, and this after two and a half thousand years of their theoretical study!

Apart from all of these far-reaching mathematical consequences, a number of other unexpected mathematical results have been obtained in the modern Mathematics of Harmony [1]. The remainder of this chapter is devoted to a presentation of these unexpected results.

2.7. Lucas Sequences

Above we have described a number of important results of the Mathematics of Harmony, in particular, a generalization of Fibonacci numbers and the golden ratio, arising from Pascal's triangle, and their applications in computer science. But it turns out that the very first broad generalization of the Fibonacci recurrence relation had already been made by the French mathematician Édouard Lucas back in the 19th century. His article "The theory of simply periodic numerical functions", first published in 1878 in the *American Journal of Mathematics* and then re-published by the Fibonacci

Association in 1969 [25], doubtless played a significant role in the development of the theory of Fibonacci numbers in the 20th century and, in effect, actually underlies this theory.

The name of Lucas is linked to the well-known Lucas numbers. However, Lucas numbers turn out to be a special limiting case of the wider class of recurrence sequences, which are called Lucas sequences [26].

Lucas sequences are the couples of the linear recurrence sequences $\{U_n(P,Q)\}$ and $\{V_n(P,Q)\}$, which satisfy the same recurrence relation with integer coefficients P and Q:

$$U_{n+2}(P,Q) = PU_{n+1}(P,Q) - QU_n(P,Q); \; U_0(P,Q) = 0; \; U_1(P,Q) = 1$$
$$(2.11)$$

$$V_{n+2}(P,Q) = PV_{n+1}(P,Q) - QV_n(P,Q); \; V_0(P,Q) = 2; \; V_1(P,Q) = P.$$
$$(2.12)$$

Note that many well-known recurrent sequences are partial cases of the Lucas sequences (2.11) and (2.12) and include the following: $\{U_n(1,-1)\}$ are Fibonacci numbers, $\{V_n(1,-1)\}$ are Lucas numbers, $\{U_n(2,-1)\}$ are Pell numbers [27], and $\{V_n(2,-1)\}$ are Pell–Lucas numbers [27].

The characteristic equation of the Lucas sequences $\{U_n(P,Q)\}$ and $\{V_n(P,Q)\}$ are the following:

$$x^2 - Px + Q = 0. \tag{2.13}$$

Its discriminator $D = P^2 - 4Q$ is assumed not to be zero. The roots of the characteristic equation (2.13) are equal:

$$\alpha = \frac{P + \sqrt{D}}{2} \quad \text{and} \quad \beta = \frac{P - \sqrt{D}}{2}$$

and can be used to get the explicit formulas:

$$U_n(P,Q) = \frac{\alpha^n - \beta^n}{\alpha - \beta} = \frac{\alpha^n - \beta^n}{\sqrt{D}} \quad \text{and} \quad V_n(P,Q) = \alpha^n + \beta^n.$$

For a more detailed study of the properties of Lucas sequences, we refer readers to articles [25] and [26].

2.8. The Fibonacci λ-numbers

2.8.1. A brief history

In the late 20th and early 21st centuries, several researchers from different countries, such as Argentinean mathematician Vera W. de Spinadel [28], French mathematician Midhat Gazale [29], American mathematician Jay Kappraff [30], Russian engineer Alexander Tatarenko [31], Armenian philosopher and physicist Hrant Arakelyan [32], Russian researcher Victor Shenyagin [33], Ukrainian physicist Nikolai Kosinov [34], Spanish mathematicians Sergio Falcon and Angel Plaza [35] and others (independent from one another) began to study a new class of recurrence numerical sequences, which are a generalization of the classic Fibonacci numbers. These numerical sequences led to the discovery of a new class of mathematical constants, called the metallic proportions by Vera W. de Spinadel [28].

The interest of the many independent researchers from different countries (USA, Canada, Argentina, France, Spain, Russia, Armenia, Ukraine) in Fibonacci λ-numbers and metallic proportions could not be accidental. We believe this means that the problem of the generalization of Fibonacci numbers and the golden ratio has matured in modern science.

2.8.2. Definition

Let us give a real number $\lambda > 0$ and consider the following recurrence relation:

$$F_\lambda(n+2) = \lambda F_\lambda(n+1) + F_\lambda(n) \qquad (2.14)$$

given for the initial conditions (or seeds):

$$F_\lambda(0) = 0, \quad F_\lambda(1) = 1. \qquad (2.15)$$

The recurrence relation (2.14) at the seeds (2.15) generates an infinite number of new numerical sequences, because every real number $\lambda > 0$ generates its own numerical sequence.

Let us consider the partial cases of the recurrence relation (2.14). For the case $\lambda = 1$ the recurrence relation (2.14) and the seeds (2.15) are reduced to the following:

$$F_1(n+2) = F_1(n+1) + F_1(n). \qquad (2.16)$$

$$F_1(0) = 0, \quad F_1(1) = 1. \qquad (2.17)$$

The recurrence relation (2.14) at the seeds (2.15) generates the classic Fibonacci numbers:

$$0, 1, 1, 2, 3, 5, 8, 13, 21, 34\ldots \tag{2.18}$$

Based on this fact, we will name a general class of the numerical sequences, generated by the recurrence relation (2.14) at the seeds (2.15), the Fibonacci λ-numbers.

For the case $\lambda = 2$ the recurrence relation (2.14) and the seeds (2.15) are reduced to the following:

$$F_2(n+2) = 2F_2(n+1) + F_2(n) \tag{2.19}$$

$$F_2(0) = 0, \quad F_2(1) = 1. \tag{2.20}$$

The recurrence relation (2.19) at the seeds (2.20) generates the so-called Pell numbers [27]:

$$0, 1, 2, 5, 12, 29, 70, 169, 408\ldots \tag{2.21}$$

For the cases $\lambda = 3$, 4 the recurrent relation (2.14) and the seeds (2.15) are reduced to the following:

$$F_2(n+2) = 3F_2(n+1) + F_2(n); \quad F_2(0) = 0, \ F_2(1) = 1 \tag{2.22}$$

$$F_2(n+2) = 4F_2(n+1) + F_2(n); \quad F_2(0) = 0, \ F_2(1) = 1. \tag{2.23}$$

If we compare the recurrence relation (2.14) for the seeds (2.15) with the recurrence relation (2.11), we arrive at the unexpected (if not startling) conclusion that the Fibonacci λ-numbers, given by (2.14) and (2.15), are partial cases of the Lucas sequences, given by (2.11), for the case $P = \lambda$ and $Q = -1$.

2.8.3.   The "extended" Fibonacci λ-numbers

The Fibonacci λ-numbers have many remarkable properties, similar to the properties of classic Fibonacci numbers. It easy to prove that the Fibonacci λ-numbers, as well as the classic Fibonacci numbers, can be "extended" to negative values of the discrete variable n.

Table 2.1 shows the four "extended" Fibonacci λ-sequences, corresponding to the cases of $\lambda = 1, 2, 3, 4$.

Table 2.1. The "extended" Fibonacci λ-numbers ($\lambda = 1, 2, 3, 4$).

n	0	1	2	3	4	5	6	7	8
$F_1(n)$	0	1	1	2	3	5	8	13	21
$F_1(-n)$	0	1	-1	2	-3	5	-8	13	-21
$F_2(n)$	0	1	2	5	12	29	70	169	408
$F_2(-n)$	0	1	-2	5	-12	29	-70	169	-408
$F_3(n)$	0	1	3	10	33	109	360	1189	3927
$F_3(-n)$	0	1	-3	10	-33	109	-360	1199	-3927
$F_4(n)$	0	1	4	17	72	305	1292	5473	23184
$F_4(-n)$	0	1	-4	17	-72	305	-1292	5473	-23184

2.8.4. The generalized Cassini formula

One of the most remarkable identities for the classic Fibonacci numbers is Cassini's formula:

$$F_n^2 - F_{n-1}F_{n+1} = (-1)^{n+1}. \qquad (2.24)$$

In 2012 *Visual Mathematics* published Stakhov's "A generalization of the Cassini formula" [36] as follows:

$$F_\lambda^2(n) - F_\lambda(n-1)F_\lambda(n+1) = (-1)^{n+1}. \qquad (2.25)$$

Numerical examples. Let us consider examples of the validity of the identity (2.25) for various sequences taken from Table 2.1. For example, consider the $F_2(n)$-sequence for the case of $n = 7$. For this case we should consider the following triplet of the Fibonacci 2-numbers (where $\lambda = 2$) of $F_2(n)$:

$$F_2(6) = 70, \quad F_2(7) = 169, \quad F_2(8) = 408.$$

By performing the calculations over them according to (2.25), we get the following result:

$$(169)^2 - 70 \times 408 = 28561 - 28560 = 1,$$

which corresponds to the identity (2.25), because for the case $n = 7$ we have:

$$(-1)^{n+1} = (-1)^8 = 1.$$

Now let us consider the $F_3(n)$-sequence from Table 2.1 for the case where $n = 6$. For this case, we should choose the following triplet of the Fibonacci

3-numbers (where $\lambda = 3$) of $F_3(n)$:

$$F_3(5) = 109, \quad F_3(6) = 360, \quad F_3(7) = 1189.$$

By performing calculations over them according to (2.25) we get the following result:

$$(360)^2 - 109 \times 1189 = 129600 - 129601 = -1,$$

which corresponds to the identity (2.25), because for the case $n = 6$ we have:

$$(-1)^{n+1} = (-1)^7 = -1.$$

Finally, let us consider the $F_4(-n)$ sequence from Table 2.1 for the case where $n = -5$. For this case we should choose the following triplet of the Fibonacci 4-numbers (where $\lambda = 4$) of $F_4(-n)$:

$$F_4(-4) = -72, \quad F_4(-5) = 305, \quad F_4(-6) = -1292.$$

By performing calculations over them according to (2.25), we get the following result:

$$(305)^2 - (-72) \times (-1292) = 93025 - 93024 = 1,$$

which corresponds to the identity (2.25), because for the case $n = -5$ we have:

$$(-1)^{n+1} = (-1)^{-4} = 1.$$

Thus, by studying the generalized Cassini formula (2.25) for the Fibonacci λ-numbers, we have found an infinite number of integer recurrence sequences in the range from $+\infty$ to $-\infty$, with the following unique mathematical property, given by the generalized Cassini formula (2.25):

> *The quadrate of any Fibonacci λ-number $F_\lambda(n)$ is always different from the product of the two adjacent Fibonacci λ-numbers $F_\lambda(n-1)$ and $F_\lambda(n+1)$, which differ from the initial Fibonacci λ-number $F_\lambda(n)$ by 1; herein the sign of the difference of 1 depends on the parity of n: if n is even, then the difference of 1 is taken with the "minus" sign, otherwise, when n is odd, then with the "plus" sign.*

Until now, we have assumed that only the classic Fibonacci numbers have this unusual property, given by Cassini's formula (2.24). However, as is shown above, there are an infinite number of such numerical sequences. All the Fibonacci λ-numbers, generated by the recurrence relation (2.14) at the seeds (2.15), display the same property, given by the generalized Cassini formula (2.25)!

It is well known that the study of integer sequences lies within the area of number theory. The Fibonacci λ-numbers are integers for the cases $\lambda = 1, 2, 3, \ldots$. Therefore, for many mathematicians in the field of number theory, the existence of an infinite number of integer sequences, which satisfy the generalized Cassini formula (2.25), may be a great surprise!

2.9. The Metallic Proportions

2.9.1. *The characteristic equation for the Fibonacci λ-numbers*

Let us represent the recurrence relation (2.14) as follows:

$$\frac{F_\lambda(n+2)}{F_\lambda(n+1)} = \lambda + \frac{1}{\frac{F_\lambda(n+1)}{F_\lambda(n)}}. \tag{2.26}$$

For the case $n \to \infty$ the expression (2.26) is reduced to the following quadratic equation:

$$x^2 - \lambda x - 1 = 0, \tag{2.27}$$

which is the characteristic equation for Fibonacci λ-numbers.

2.9.2. *The metallic means of Vera W. de Spinadel*

The equation (2.27) has two roots, the positive root:

$$x_1 = \frac{\lambda + \sqrt{4 + \lambda^2}}{2} \tag{2.28}$$

and the negative root:

$$x_2 = \frac{\lambda - \sqrt{4 + \lambda^2}}{2}. \tag{2.29}$$

If we sum (2.28) and (2.29), we get:

$$x_1 + x_2 = \lambda. \tag{2.30}$$

If we substitute the root (2.28) in place of x in Eq. (2.27), we get the following identity:

$$x_1^2 = \lambda x_1 + 1. \tag{2.31}$$

If we multiply or divide repeatedly all terms of the identity (2.31) by x_1, we get the following identity:

$$x_1^n = \lambda x_1^{n-1} + x_1^{n-2}, \tag{2.32}$$

where $n = 0, \pm 1, \pm 2, \pm 3, \ldots$.

By using this reasoning for the root x_2, we can get a similar identity for the root x_2:

$$x_2^n = \lambda x_2^{n-1} + x_2^{n-2}. \tag{2.33}$$

Denote the positive root x_1 by Φ_λ, that is,

$$\Phi_\lambda = \frac{\lambda + \sqrt{4 + \lambda^2}}{2}. \tag{2.34}$$

Note that for the case $\lambda = 1$ the formula (2.34) is reduced to the formula for the golden ratio:

$$\Phi_1 = \frac{1 + \sqrt{5}}{2}. \tag{2.35}$$

This means that the formula (2.34) gives a broad class of new mathematical constants, similar to the golden ratio (2.35).

Based on this analogy, de Spinadel [28] named this group of mathematical constants (2.34) the metallic means, also known as the metallic proportions. If we take $\lambda = 1, 2, 3, 4$ in (2.34), we then get the following mathematical constants which, according to de Spinadel, have these special names:

$$\Phi_1 = \frac{1 + \sqrt{5}}{2} \quad \text{(the golden mean, } \lambda = 1\text{);}$$

$$\Phi_2 = 1 + \sqrt{2} \quad \text{(the silver mean, } \lambda = 2\text{);}$$

$$\Phi_3 = \frac{3 + \sqrt{13}}{2} \quad \text{(the bronze mean, } \lambda = 3\text{);}$$

$$\Phi_4 = 2 + \sqrt{5} \quad \text{(the copper mean, } \lambda = 4\text{).}$$

Other metallic means ($\lambda \geq 5$) have not been assigned special names:

$$\Phi_5 = \frac{5 + \sqrt{29}}{2}; \quad \Phi_6 = 3 + 2\sqrt{10}; \quad \Phi_7 = \frac{7 + 2\sqrt{14}}{2}; \quad \Phi_8 = 4 + \sqrt{17}. \tag{2.36}$$

2.9.3.   *Algebraic properties of the metallic proportions*

Let us represent the root x_2, given by (2.29), through the metallic mean (2.34). After simple transformation of (2.29) we can write the root x_2 as follows:

$$x_2 = \frac{\lambda - \sqrt{4 + \lambda^2}}{2} = \frac{(\lambda - \sqrt{4 + \lambda^2})(\lambda + \sqrt{4 + \lambda^2})}{2(\lambda + \sqrt{4 + \lambda^2})}$$

$$= \frac{-4}{2(\lambda + \sqrt{4 + \lambda^2})} = -\frac{1}{\Phi_\lambda}. \tag{2.37}$$

Substituting Φ_λ in place of x_1 and $(-\frac{1}{\Phi_\lambda})$ in place of x_2 in (2.30), we get:

$$\lambda = \Phi_\lambda - \frac{1}{\Phi_\lambda}, \tag{2.38}$$

when Φ_λ is given by (2.34) and $\frac{1}{\Phi_\lambda}$ is given by the formula:

$$\frac{1}{\Phi_\lambda} = \frac{-\lambda + \sqrt{4 + \lambda^2}}{2}. \tag{2.39}$$

Using the formulas (2.34) and (2.39), we can write the following identity:

$$\Phi_\lambda + \frac{1}{\Phi_\lambda} = \sqrt{4 + \lambda^2}. \tag{2.40}$$

Also it is easy to prove the following identity:

$$\Phi_\lambda^n = \lambda \Phi_\lambda^{n-1} + \Phi_\lambda^{n-2}, \tag{2.41}$$

where $n = 0, \pm 1, \pm 2, \pm 3, \ldots$.

2.9.4.   *Two surprising representations of the metallic means*

For the case where $n = 2$ the identity (2.41) can be represented in the form:

$$\Phi_\lambda^2 = 1 + \lambda \Phi_\lambda. \tag{2.42}$$

If we take the square root of the left and right parts of (2.42), we obtain the following representation of the metallic proportion Φ_λ:

$$\Phi_\lambda = \sqrt{1 + \lambda \Phi_\lambda}. \tag{2.43}$$

Substituting $\sqrt{1 + \lambda \Phi_\lambda}$ in place of Φ_λ in the right-hand part of (2.43), we get:

$$\Phi_\lambda = \sqrt{1 + \lambda \sqrt{1 + \Phi_\lambda}}. \tag{2.44}$$

Continuing this process *ad infinitum*, that is, repeatedly substituting $\sqrt{1 + \lambda \Phi_\lambda}$ in place of Φ_λ in the right-hand part of (2.44), we get the following surprising representation of the metallic mean Φ_λ in the form of a nested radical:

$$\Phi_\lambda = \sqrt{1 + \lambda\sqrt{1 + \lambda\sqrt{1 + \lambda\sqrt{1 + \cdots}}}}\,. \qquad (2.45)$$

Next represent the identity (2.42) in the form:

$$\Phi_\lambda = \lambda + \frac{1}{\Phi_\lambda}. \qquad (2.46)$$

Substituting $\lambda + \frac{1}{\Phi_\lambda}$ in place of Φ_λ in the right-hand part of (2.46), we get:

$$\Phi_\lambda = \lambda + \frac{1}{\lambda + \frac{1}{\Phi_\lambda}}. \qquad (2.47)$$

Continuing this process *ad infinitum*, that is, substituting $\lambda + \frac{1}{\Phi_\lambda}$ repeatedly in place of Φ_λ in the right-hand part of (2.47), we get the following surprising representation of the metallic proportion Φ_λ in the form of a continued fraction:

$$\Phi_\lambda = \lambda + \frac{1}{\lambda + \frac{1}{\lambda + \frac{1}{\lambda + \cdots}}}. \qquad (2.48)$$

Note that for the case $\lambda = 1$ the representations (2.45) and (2.48) are similar in their mathematical structure to the following well-known nested radical and continued fraction representations of the classic golden ratio [2, p. 52]:

$$\Phi = \sqrt{1 + \sqrt{1 + \sqrt{1 + \sqrt{1 + \cdots}}}}; \quad \Phi = 1 + \frac{1}{1 + \frac{1}{1 + \frac{1}{1 + \cdots}}}. \qquad (2.49)$$

The representations of the metallic means in the forms of a nested radical (2.45) and continued fraction (2.48), similar to the golden ratio's surprising representations (2.49), are an additional confirmation of the fact that the metallic proportions Φ_λ are new and striking constants of mathematics!

2.10.   Gazale's Formulas

2.10.1.   *Gazale's formula for the Fibonacci λ-numbers*

The formulas (2.14) and (2.15) define the Fibonacci λ-numbers $F_\lambda(n)$ recursively. However, we can express the numbers $F_\lambda(n)$ in explicit form through the metallic proportion Φ_λ. We can represent the Fibonacci λ-numbers $F_\lambda(n)$ by using the roots x_1 and x_2 as follows:

$$F_\lambda(n) = k_1 x_1^n + k_2 x_2^n, \tag{2.50}$$

where k_1 and k_2 are constant coefficients, which are the solutions of the following system of the algebraic equations:

$$\begin{cases} F_\lambda(0) = k_1 x_1^0 + k_2 x_2^0 = k_1 + k_2 \\ F_\lambda(1) = k_1 x_1^1 + k_2 x_2^1 = k_1 \Phi_\lambda - k_2 \dfrac{1}{\Phi_\lambda} \end{cases} \tag{2.51}$$

Consider that $F_\lambda(0) = 0$ and $F_\lambda(1) = 1$, we can rewrite the system (2.51) as follows:

$$k_1 = -k_2 \tag{2.52}$$

$$k_1 \Phi_\lambda + k_1 \frac{1}{\Phi_\lambda} = k_1 \left(\Phi_\lambda + \frac{1}{\Phi_\lambda} \right) = 1. \tag{2.53}$$

Given equations (2.52) and (2.53), we can find the following formulas for the coefficients k_1 and k_2:

$$k_1 = \frac{1}{\sqrt{4 + \lambda^2}}; \quad k_2 = -\frac{1}{\sqrt{4 + \lambda^2}}. \tag{2.54}$$

Using (2.54), we can rewrite the formula (2.50) as follows:

$$F_\lambda(n) = \frac{1}{\sqrt{4 + \lambda^2}} x_1^n - \frac{1}{\sqrt{4 + \lambda^2}} x_2^n = \frac{1}{\sqrt{4 + \lambda^2}} (x_1^n - x_2^n). \tag{2.55}$$

Furthermore given that $x_1 = \Phi_\lambda$ and $x_2 = -\frac{1}{\Phi_\lambda}$, we can rewrite the formula (2.55) as follows:

$$F_\lambda(n) = \frac{\Phi_\lambda^n - \left(-\frac{1}{\Phi_\lambda} \right)^n}{\sqrt{4 + \lambda^2}} \tag{2.56}$$

or

$$F_\lambda(n) = \frac{1}{\sqrt{4 + \lambda^2}} \left[\left(\frac{\lambda + \sqrt{4 + \lambda^2}}{2} \right)^n - \left(\frac{\lambda - \sqrt{4 + \lambda^2}}{2} \right)^n \right]. \tag{2.57}$$

For the partial case $\lambda = 1$, the formula (2.57) is reduced to the formula:

$$F_1(n) = \frac{1}{\sqrt{5}} \left[\left(\frac{1 + \sqrt{5}}{2} \right)^n - \left(\frac{1 - \sqrt{5}}{2} \right)^n \right]. \tag{2.58}$$

This is called Binet's formula derived by French mathematician Jacques Philip Marie Binet in 1843, although the result was known to Leonhard Euler, Daniel Bernoulli, and Abraham de Moivre more than a century earlier. In particular, de Moivre apparently derived this formula in 1718.

Note that for the first time this formula (2.56) was deduced by the French mathematician Midhat J. Gazale in his 1999 book, *Gnomon: From Pharaohs to Fractals* [29]. This formula sets forth in analytical form an infinite number of Fibonacci λ-numbers, which for the case $\lambda = 1$ are reduced to Binet's formula (2.58) for the classic Fibonacci numbers. In Stakhov's article [37] this unique formula (2.56) was named Gazale's formula for the Fibonacci λ-numbers or simply Gazale's formula.

2.10.2.   *Surprising properties of the Fibonacci λ-numbers*

Let us explore some surprising properties of the Fibonacci λ-numbers by using Gazale's formula (2.56). First of all, compare the Fibonacci λ-numbers $F_\lambda(n)$ and $F_\lambda(-n)$. We can rewrite the formula (2.56) as follows:

$$F_\lambda(n) = \frac{\Phi_\lambda^n - (-1)^n \Phi_\lambda^{-n}}{\sqrt{4 + \lambda^2}}. \qquad (2.59)$$

Let us first consider the formula (2.59) for the negative values of n, that is:

$$F_\lambda(-n) = \frac{\Phi_\lambda^{-n} - (-1)^{-n} \Phi_\lambda^n}{\sqrt{4 + \lambda^2}}. \qquad (2.60)$$

By comparing the formulas (2.59) and (2.60) for the even ($n = 2k$) and odd ($n = 2k + 1$) values of n, we find:

$$\begin{cases} F_\lambda(2k) = -F_\lambda(-2k) \\ F_\lambda(2k + 1) = F_\lambda(-2k - 1) \end{cases}. \qquad (2.61)$$

This means that the sequences of Fibonacci λ-numbers $F_\lambda(n)$ and $F_\lambda(-n)$ in the range $n = 0, \pm 1, \pm 2, \pm 3, \ldots$ are symmetrical sequences relative to $F_\lambda(0) = 0$. If we take into consideration that, according to (2.61), the Fibonacci λ-numbers $F_\lambda(2k)$ and $F_\lambda(-2k)$ are opposite in sign, then the following simple relation follows from (2.61):

$$F_\lambda(-n) = (-1)^{n+1} F_\lambda(n). \qquad (2.62)$$

This property of the Fibonacci λ-numbers $F_\lambda(n)$ and $F_\lambda(-n)$ is represented in Table 2.1.

2.10.3. Gazale's formula for the Lucas λ-numbers

In his Gazale formula article [37], Stakhov obtained the following important result: the Gazale formula also relates to Lucas numbers. By analogy to the classic Lucas numbers, we can consider the formula:

$$L_\lambda(n) = x_1^n + x_2^n, \tag{2.63}$$

where x_1 and x_2, respectively, are the roots (2.28) and (2.29) of algebraic equation (2.27).

It is clear that for the case $\lambda = 1$ this formula sets forth the classic Lucas numbers: 2, 1, 3, 4, 7, 11, 18, ...

Therefore, for the general case of λ the formula (2.63) sets forth the Lucas λ-numbers. For the given λ, we can find some peculiarities of the Lucas λ-numbers. First of all, by using (2.63), we can calculate the seeds of the Lucas λ-numbers. In fact, for the cases $n = 0$ and $n = 1$ we have, respectively:

$$L_\lambda(0) = x_1^0 + x_2^0 = 1 + 1 = 2, \tag{2.64}$$

$$L_\lambda(1) = x_1^1 + x_2^1 = \lambda. \tag{2.65}$$

By using (2.32) and (2.33), we can represent the formula (2.63) as follows:

$$\begin{aligned} L_\lambda(n) = x_1^n + x_2^n &= \lambda x_1^{n-1} + x_1^{n-2} + \lambda x_2^{n-1} + x_2^{n-2} \\ &= \lambda(x_1^{n-1} + x_2^{n-1}) + (x_1^{n-2} + x_2^{n-2}). \end{aligned} \tag{2.66}$$

Taking into consideration the definition (2.63), we can rewrite (2.66) as the following recurrence relation:

$$L_\lambda(n) = \lambda L_\lambda(n-1) + L_\lambda(n-2). \tag{2.67}$$

It is clear that this recurrence relation (2.67) at the seeds (2.64) and (2.65) recursively sets forth the Lucas λ-numbers.

If we make the substitution $x_1 = \Phi_\lambda$ and $x_2 = -\frac{1}{\Phi_\lambda}$ into the formula (2.63), we can express the Lucas λ-numbers through the metallic mean Φ_λ as follows:

$$L_\lambda(n) = \left[\Phi_\lambda^n + \left(\frac{-1}{\Phi_\lambda} \right)^n \right]. \tag{2.68}$$

Although this formula is absent in Gazale's *Gnomon* book [29], according to Stakhov's suggestion [37] the formula (2.68) was named Gazale's formula for the Lucas λ-numbers after Midhat Gazale, who first introduced the formula (2.56).

Table 2.2.   The "extended" Lucas λ-numbers.

n	0	1	2	3	4	5	6	7	8
$L_1(n)$	2	1	3	4	7	11	18	29	47
$L_1(-n)$	2	1	-3	4	-7	11	-18	29	-47
$L_2(n)$	2	2	6	14	34	82	198	478	1154
$L_2(-n)$	2	-2	6	-14	34	-82	198	-478	1154
$L_3(n)$	2	3	11	36	119	393	1298	4287	14159
$L_3(-n)$	2	-3	11	-36	119	-393	1298	-4287	14159
$L_4(n)$	2	4	18	76	322	1364	5778	24476	103682
$L_4(-n)$	2	-4	18	-76	322	-1364	5778	-24476	103682

We can rewrite the formula (2.68) as follows:

$$L_\lambda(n) = \Phi_\lambda^n + (-1)^n \Phi_\lambda^{-n}. \tag{2.69}$$

Let us consider the formula (2.69) for the negative values of n, that is,

$$L_\lambda(-n) = \Phi_\lambda^{-n} + (-1)^{-n} \Phi_\lambda^n. \tag{2.70}$$

By comparing the formulas (2.69) and (2.70) for the even ($n = 2k$) and odd ($n = 2k + 1$) values of n, we can write:

$$\begin{cases} L_\lambda(2k) = L_\lambda(-2k) \\ L_\lambda(2k + 1) = -L_\lambda(-2k - 1) \end{cases}. \tag{2.71}$$

This means that the sequences of Lucas λ-numbers in the range $n = 0, \pm 1, \pm 2, \pm 3, \dots$ are symmetrical sequences relative to $L_\lambda(0) = 2$. Considering that the Lucas λ-numbers $L_\lambda(2k + 1)$ and $L_\lambda(-2k - 1)$ are opposite in sign, then the λ-numbers $L_\lambda(2k + 1)$ and $L_\lambda(-2k - 1)$ are connected by the following simple relation:

$$L_\lambda(-n) = (-1)^n L_\lambda(n). \tag{2.72}$$

This property of the Lucas λ-numbers $L_\lambda(n)$ and $L_\lambda(-n)$ is demonstrated in Table 2.2.

Note that for the case $\lambda = 1$ the Lucas λ-numbers coincide with the classic Lucas numbers, and for the case $\lambda = 2$ they coincide with the Pell–Lucas numbers [27].

2.11. Hyperbolic Fibonacci and Lucas λ-functions

2.11.1. *Definition*

Gazale's formulas (2.56) and (2.69) are the source for the introduction of a new class of hyperbolic functions. In order to determine these functions, we use the identities (2.62) and (2.72) to represent Gazale's formulas (2.56) and (2.69) in the following forms [37, 38]:

$$
F_\lambda(n) = \begin{cases} \dfrac{\Phi_\lambda^n - \Phi_\lambda^{-n}}{\sqrt{4+\lambda^2}} & \text{for } n = 2k \\[4mm] \dfrac{\Phi_\lambda^n + \Phi_\lambda^{-n}}{\sqrt{4+\lambda^2}} & \text{for } n = 2k+1 \end{cases} \tag{2.73}
$$

$$
L_\lambda(n) = \begin{cases} \Phi_\lambda^n - \Phi_\lambda^{-n} & \text{for } n = 2k+1 \\[2mm] \Phi_\lambda^n + \Phi_\lambda^{-n} & \text{for } n = 2k \end{cases}. \tag{2.74}
$$

Comparing Gazale's formulas (2.73) and (2.74) with the classic hyperbolic functions:

$$
\text{sh}(x) = \frac{e^x - e^{-x}}{2}; \quad \text{ch}(x) = \frac{e^x + e^{-x}}{2}, \tag{2.75}
$$

we can see that the formulas (2.73) and (2.74) are similar to the formulas (2.75) in terms of their mathematical structure. This similarity became the reason for introducing a general class of hyperbolic functions called the λ-Fibonacci and λ-Lucas hyperbolic functions [36, 38]:

λ-Fibonacci hyperbolic sine and cosine.

$$
\begin{aligned}
sF_\lambda(x) &= \frac{\Phi_\lambda^x - \Phi_\lambda^{-x}}{\sqrt{4+\lambda^2}} \\[3mm]
&= \frac{1}{\sqrt{4+\lambda^2}} \left[\left(\frac{\lambda + \sqrt{4+\lambda^2}}{2} \right)^x - \left(\frac{\lambda + \sqrt{4+\lambda^2}}{2} \right)^{-x} \right] \tag{2.76}
\end{aligned}
$$

$$
\begin{aligned}
cF_\lambda(x) &= \frac{\Phi_\lambda^x + \Phi_\lambda^{-x}}{\sqrt{4+\lambda^2}} \\[3mm]
&= \frac{1}{\sqrt{4+\lambda^2}} \left[\left(\frac{\lambda + \sqrt{4+\lambda^2}}{2} \right)^x + \left(\frac{\lambda + \sqrt{4+\lambda^2}}{2} \right)^{-x} \right]. \tag{2.77}
\end{aligned}
$$

λ-Lucas hyperbolic sine and cosine.

$$\mathrm{sL}_\lambda(x) = \Phi_\lambda^x - \Phi_\lambda^{-x} = \left(\frac{\lambda + \sqrt{4 + \lambda^2}}{2}\right)^x - \left(\frac{\lambda + \sqrt{4 + \lambda^2}}{2}\right)^{-x}$$

(2.78)

$$\mathrm{cL}_\lambda(x) = \Phi_\lambda^x + \Phi_\lambda^{-x} = \left(\frac{\lambda + \sqrt{4 + \lambda^2}}{2}\right)^x + \left(\frac{\lambda + \sqrt{4 + \lambda^2}}{2}\right)^{-x},$$

(2.79)

where x is a continuous variable and $\lambda > 0$ is a given positive real number.

It is easy to see that the functions (2.76), (2.77) and (2.78), (2.79) are connected by the simple relations:

$$\mathrm{sF}_\lambda(x) = \frac{\mathrm{sL}_\lambda(x)}{\sqrt{4 + \lambda^2}}; \quad \mathrm{cF}_\lambda(x) = \frac{\mathrm{cL}_\lambda(x)}{\sqrt{4 + \lambda^2}}.$$

(2.80)

This means that the λ-Lucas hyperbolic functions (2.78) and (2.79) coincide with the λ-Fibonacci hyperbolic functions (2.76) and (2.77) through the constant coefficient $\frac{1}{\sqrt{4+\lambda^2}}$.

Comparing the formula (2.73) with the formulas (2.76) and (2.77) and then the formula (2.74) with the formulas (2.78) and (2.79), it is easy to prove the following fundamental relationship of the λ-Fibonacci and λ-Lucas hyperbolic functions with λ-Fibonacci and λ-Lucas numbers, given by the Gazale formulas (2.73) and (2.74):

$$F_\lambda(n) = \begin{cases} \mathrm{sF}_\lambda(n) & \text{for } n = 2k \\ \mathrm{cF}_\lambda(n) & \text{for } n = 2k+1 \end{cases}$$

(2.81)

$$L_\lambda(n) = \begin{cases} \mathrm{cL}_\lambda(n) & \text{for } n = 2k \\ \mathrm{sL}_\lambda(n) & \text{for } n = 2k+1 \end{cases}$$

(2.82)

where k takes the values from the set $k = 0, \pm 1, \pm 2, \pm 3, \ldots$.

2.11.2.   The graphs of the λ-Fibonacci and λ-Lucas hyperbolic functions

The graphs of the λ-Fibonacci and λ-Lucas hyperbolic functions (Fig. 2.1) are similar to the graphs of the symmetric Fibonacci and Lucas hyperbolic functions (see Figs. 1.17 and 1.18).

It is important to note that at the point $x = 0$, the λ-Fibonacci hyperbolic cosine $\mathrm{cF}_\lambda(x)$ (2.77) takes the value $\mathrm{cF}_\lambda(0) = \frac{2}{\sqrt{4+\lambda^2}}$, and the λ-Lucas hyperbolic cosine $\mathrm{cL}_\lambda(x)$ (2.79) takes the value $\mathrm{cL}_\lambda(0) = 2$. It is also

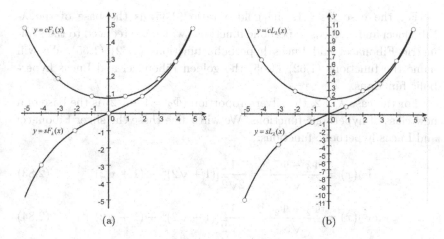

(a)   (b)

Figure 2.1.   Graphs of the (a) λ-Fibonacci, and (b) λ-Lucas hyperbolic functions.

important to note that the λ-Fibonacci numbers $F_\lambda(n)$ with even values of $n = 0, \pm 2, \pm 4, \pm 6, \ldots$ are "inscribed" into the graph of the λ-Fibonacci hyperbolic sine $sF_\lambda(x)$ at the even discrete points $x = 0, \pm 2, \pm 4, \pm 6, \ldots$ and the λ-Fibonacci numbers $F_\lambda(n)$ with odd values of $n = 0, \pm 1, \pm 3, \pm 5, \ldots$ are "inscribed" into the λ-Fibonacci hyperbolic cosine $cF_\lambda(x)$ at the odd discrete points $x = \pm 1, \pm 3, \pm 5, \ldots$.

On the other hand, the λ-Lucas numbers $L_\lambda(n)$ with even values of n are "inscribed" into the graph of the λ-Lucas hyperbolic cosine $cL_\lambda(x)$ at the even discrete points $x = 0, \pm 2, \pm 4, \pm 6, \ldots$ and the λ-Lucas numbers $L_\lambda(n)$ with odd values of n are "inscribed" into the graph of the λ-Lucas hyperbolic sine $sL_\lambda(x)$ at the odd discrete points $x = \pm 1, \pm 3, \pm 5, \ldots$.

By analogy with the symmetric Fibonacci and Lucas hyperbolic functions (1.62)–(1.65), we can introduce other kinds of these λ-Fibonacci and λ-Lucas functions, in particular, λ-Fibonacci and λ-Lucas hyperbolic tangents and cotangents, secants and cosecants and so on.

2.12.   The Partial Cases of the λ-Fibonacci and λ-Lucas Hyperbolic Functions

2.12.1.   *The golden, silver, bronze, and copper λ-hyperbolic functions*

Let us consider the partial cases of the λ-Fibonacci and λ-Lucas hyperbolic functions (2.76) through (2.79) for different values of λ.

For the case $\lambda = 1$, the golden ratio (2.35) is the base of the λ-Fibonacci and λ-Lucas hyperbolic functions, which are reduced to the symmetric Fibonacci and Lucas hyperbolic functions (1.62)–(1.65). We will name the functions (1.62)–(1.65) the golden Fibonacci and Lucas hyperbolic functions.

For the case $\lambda = 2$, the silver proportion ($\Phi_2 = 1 + \sqrt{2}$) is the base of a new class of hyperbolic functions. We will name these the silver Fibonacci and Lucas hyperbolic functions:

$$sF_2(x) = \frac{\Phi_2^x - \Phi_2^{-x}}{\sqrt{8}} = \frac{1}{2\sqrt{2}}[(1+\sqrt{2})^x - (1+\sqrt{2})^{-x}] \qquad (2.83)$$

$$cF_2(x) = \frac{\Phi_2^x + \Phi_2^{-x}}{\sqrt{8}} = \frac{1}{2\sqrt{2}}[(1+\sqrt{2})^x + (1+\sqrt{2})^{-x}] \qquad (2.84)$$

$$sL_2(x) = \Phi_2^x - \Phi_2^{-x} = (1+\sqrt{2})^x - (1+\sqrt{2})^{-x} \qquad (2.85)$$

$$cL_2(x) = \Phi_2^x + \Phi_2^{-x} = (1+\sqrt{2})^x + (1+\sqrt{2})^{-x}. \qquad (2.86)$$

For the case $\lambda = 3$, the bronze proportion ($\Phi_3 = \frac{3+\sqrt{13}}{2}$) is the base of a new class of hyperbolic functions. We will name these the bronze Fibonacci and Lucas hyperbolic functions:

$$sF_3(x) = \frac{\Phi_3^x - \Phi_3^{-x}}{\sqrt{13}} = \frac{1}{\sqrt{13}}\left(\frac{3+\sqrt{13}}{2}\right)^x - \left(\frac{3+\sqrt{13}}{2}\right)^{-x} \qquad (2.87)$$

$$cF_3(x) = \frac{\Phi_3^x + \Phi_3^{-x}}{\sqrt{13}} = \frac{1}{\sqrt{13}}\left(\frac{3+\sqrt{13}}{2}\right)^x + \left(\frac{3+\sqrt{13}}{2}\right)^{-x} \qquad (2.88)$$

$$sL_3(x) = \Phi_3^x - \Phi_3^{-x} = \left(\frac{3+\sqrt{13}}{2}\right)^x - \left(\frac{3+\sqrt{13}}{2}\right)^{-x} \qquad (2.89)$$

$$cL_3(x) = \Phi_3^x + \Phi_3^{-x} = \left(\frac{3+\sqrt{13}}{2}\right)^x + \left(\frac{3+\sqrt{13}}{2}\right)^{-x}. \qquad (2.90)$$

For the case $\lambda = 4$, the copper proportion ($\Phi_4 = 2 + \sqrt{5}$) is the base of a new class of hyperbolic functions. We will name these the copper Fibonacci and Lucas hyperbolic functions:

$$sF_4(x) = \frac{\Phi_4^x - \Phi_4^{-x}}{2\sqrt{5}} = \frac{1}{2\sqrt{5}}[(2+\sqrt{5})^x - (2+\sqrt{5})^{-x}] \qquad (2.91)$$

$$cF_4(x) = \frac{\Phi_4^x + \Phi_4^{-x}}{2\sqrt{5}} = \frac{1}{2\sqrt{5}}[(2 + \sqrt{5})^x + (2 + \sqrt{5})^{-x}] \qquad (2.92)$$

$$sL_4(x) = \Phi_4^x - \Phi_4^{-x} = (2 + \sqrt{5})^x - (2 + \sqrt{5})^{-x} \qquad (2.93)$$

$$cL_4(x) = \Phi_4^x + \Phi_4^{-x} = (2 + \sqrt{5})^x + (2 + \sqrt{5})^{-x}. \qquad (2.94)$$

Note that the list of these functions can be continued *ad infinitum*. Furthermore, note that since $\lambda > 0$ is a positive real number, the number of the λ-Fibonacci and Lucas hyperbolic functions is equal to the number of positive real numbers.

2.12.2. *Comparison of the classic hyperbolic functions with the λ-Lucas hyperbolic functions*

Let us compare the λ-Lucas hyperbolic functions (2.78) and (2.79) with the classic hyperbolic functions (2.75). It is easy to prove [37, 38] that for the case:

$$\Phi_\lambda = \frac{\lambda + \sqrt{4 + \lambda^2}}{2} = e \qquad (2.95)$$

the λ-Lucas hyperbolic functions (2.78) and (2.79) coincide with the classic hyperbolic functions (2.75) through the constant coefficient $\frac{1}{2}$, that is:

$$sh(x) = \frac{sL_\lambda(x)}{2} \quad \text{and} \quad ch(x) = \frac{cL_\lambda(x)}{2}. \qquad (2.96)$$

By using (2.96), after simple transformations we can calculate the value of λ_e, for which the formula (2.95) is valid:

$$\lambda_e = e - \frac{1}{e} = 2sh(1) \approx 2.35040238. \qquad (2.97)$$

Thus, according to (2.96) the classic hyperbolic functions (2.75) are a partial case of the λ-Lucas hyperbolic functions for the case (2.97).

2.13. The Most Important Formulas for the λ-Fibonacci and λ-Lucas Hyperbolic Functions

2.13.1. *The relationships between the metallic proportions and the golden proportion*

We emphasize once again that the number of the λ-Fibonacci and λ-Lucas hyperbolic functions, given by (2.76)–(2.79), can be extended to infinity. It is important to emphasize that they are related to some well-known numerical sequences, in particular, to the Fibonacci, Lucas and Pell numbers. These functions maintain all the important properties of the classic hyperbolic functions and symmetric Fibonacci and Lucas hyperbolic functions (1.62)–(1.65); this means that, on the one hand, they have recurrence properties, similar to λ-Fibonacci and λ-Lucas numbers and, on the other hand, hyperbolic properties, similar to the classic hyperbolic functions, which are a special limiting case of the λ-Lucas hyperbolic functions.

The metallic proportions (2.34), which are a generalization of the classic golden ratio (2.35), are the bases of these functions. Let us start with the relations between the metallic proportions and the golden proportion (Table 2.3).

Table 2.3.   Comparative table for the golden and metallic proportions.

The golden proportion ($\lambda = 1$)	The metallic proportions ($\lambda > 0$)
$\Phi = \dfrac{1 + \sqrt{5}}{2}$	$\Phi_\lambda = \dfrac{\lambda + \sqrt{4 + \lambda^2}}{2}$
$\Phi = \sqrt{1 + \sqrt{1 + \sqrt{1 + \sqrt{\cdots}}}}$	$\Phi_\lambda = \sqrt{1 + \lambda\sqrt{1 + \lambda\sqrt{1 + \lambda\sqrt{\cdots}}}}$
$\Phi = 1 + \dfrac{1}{1 + \frac{1}{1 + \frac{1}{1+\cdots}}}$	$\Phi_\lambda = \lambda + \dfrac{1}{\lambda + \frac{1}{\lambda + \frac{1}{\lambda+\cdots}}}$
$\Phi^n = \Phi^{n-1} + \Phi^{n-2} = \Phi \times \Phi^{n-1}$	$\Phi_\lambda^n = \lambda\Phi_\lambda^{n-1} + \Phi_\lambda^{n-2} = \Phi_\lambda \times \Phi_\lambda^{n-1}$
$F(n) = \dfrac{\Phi^n - (-1)^n \Phi^{-n}}{\sqrt{5}}$	$F_\lambda(n) = \dfrac{\Phi_\lambda^n - (-1)^n \Phi_\lambda^{-n}}{\sqrt{4 + \lambda^2}}$
$L(n) = \Phi^n + (-1)^n \Phi^{-n}$	$L_\lambda(n) = \Phi_\lambda^n + (-1)^n \Phi_\lambda^{-n}$
$\mathrm{sFs}(x) = \dfrac{\Phi^x - \Phi^{-x}}{\sqrt{5}}$ $\mathrm{cFs}(x) = \dfrac{\Phi^x + \Phi^{-x}}{\sqrt{5}}$	$\mathrm{sF}_\lambda(x) = \dfrac{\Phi_\lambda^x - \Phi_\lambda^{-x}}{\sqrt{4 + \lambda^2}}$ $\mathrm{cF}_\lambda(x) = \dfrac{\Phi_\lambda^x + \Phi_\lambda^{-x}}{\sqrt{4 + \lambda^2}}$
$\mathrm{sLs}(x) = \Phi^x - \Phi^{-x}$ $\mathrm{cLs}(x) = \Phi^x + \Phi^{-x}$	$\mathrm{sL}_\lambda(x) = \Phi_\lambda^x - \Phi_\lambda^{-x}$ $\mathrm{cL}_\lambda(x) = \Phi_\lambda^x + \Phi_\lambda^{-x}$

2.13.2. The recursive properties

Let us now prove some recurrence properties for the λ-Fibonacci and λ-Lucas hyperbolic functions.

Theorem 2.1. *The following relations, which are similar to the recurrence relation for the λ-Fibonacci numbers $F_\lambda(n+2) = \lambda F_\lambda(n+1) + F_\lambda(n)$, are valid for the λ-Fibonacci and Lucas hyperbolic functions*:

$$sF_\lambda(x+2) = \lambda cF_\lambda(x+1) + sF_\lambda(x) \tag{2.98}$$

$$cF_\lambda(x+2) = \lambda sF_\lambda(x+1) + cF_\lambda(x). \tag{2.99}$$

Proof.

$$sF_\lambda(x+2) = \lambda cF_\lambda(x+1) + sF_\lambda(x) = \lambda \frac{\Phi_\lambda^{x+1} + \Phi_\lambda^{-x-1}}{\sqrt{4+\lambda^2}} + \frac{\Phi_\lambda^x - \Phi_\lambda^{-x}}{\sqrt{4+\lambda^2}}$$

$$= \frac{\Phi_\lambda^x(\lambda\Phi+1) - \Phi_\lambda^{-x}(1 - \lambda\Phi_\lambda^{-1})}{\sqrt{4+\lambda^2}}. \tag{2.100}$$

As $\lambda\Phi_\lambda + 1 = \Phi_\lambda^2$ and $1 - \Phi_\lambda^{-1} = \Phi_\lambda^{-2}$, we can represent (2.100) as follows:

$$\lambda cF_\lambda(x+1) + sF_\lambda(x) = \frac{\Phi_\lambda^{x+2} - \Phi_\lambda^{-x-2}}{\sqrt{4+\lambda^2}} sF_\lambda(x+2),$$

which proves the identity (2.98). By analogy we can prove the identity (2.99).

Above we demonstrated the generalized Cassini formula for the λ-Fibonacci numbers, given by (2.25). This formula can be generalized for the continuous domain for the case of the λ-Fibonacci hyperbolic functions as follows.

Theorem 2.2. *(A generalization of the Cassini formula for the continuous domain). The following relations, which are similar to the generalized Cassini formula for the λ-Fibonacci numbers $F_\lambda^2(n) - F_\lambda(n-1)F_\lambda(n+1) = (-1)^{n+1}$ are valid for the λ-Fibonacci hyperbolic functions*:

$$[sF_\lambda(x)]^2 - cF_\lambda(x+1)cF_\lambda(x-1) = -1 \tag{2.101}$$

$$[cF_\lambda(x)]^2 - sF_\lambda(x+1)sF_\lambda(x-1) = 1. \tag{2.102}$$

Note that Theorems 2.1 and 2.2 are examples of the so-called recursive properties of the λ-Fibonacci hyperbolic functions.

Table 2.4.   Hyperbolic properties of the λ-Fibonacci hyperbolic functions.

Formulas for the classic hyperbolic functions	Formulas for the λ-Fibonacci hyperbolic functions
$\text{sh}(x) = \dfrac{e^x - e^{-x}}{2};$ $\text{ch}(x) = \dfrac{e^x + e^{-x}}{2}$	$\text{sF}_\lambda(x) = \dfrac{\Phi_\lambda^x - \Phi_\lambda^{-x}}{\sqrt{4+\lambda^2}};\ \ \text{cF}_\lambda(x) = \dfrac{\Phi_\lambda^x + \Phi_\lambda^{-x}}{\sqrt{4+\lambda^2}};$
$\text{sh}(x+2) = 2\text{sh}(1)\text{ch}(x+1) + \text{sh}(x)$ $\text{ch}(x+2) = 2\text{sh}(1)\text{sh}(x+1) + \text{ch}(x)$	$\text{sF}_\lambda(x+2) = \lambda \text{cF}_\lambda(x+1) + \text{sF}_\lambda(x)$ $\text{cF}_\lambda(x+2) = \lambda \text{sF}_\lambda(x+1) + \text{cF}_\lambda(x)$
$\text{sh}^2(x) - \text{ch}(x+1)\text{ch}(x-1) = -\text{ch}^2(1)$ $\text{ch}^2(x) - \text{sh}(x+1)\text{sh}(x-1) = \text{ch}^2(1)$	$[\text{sF}_\lambda(x)]^2 - \text{cF}_\lambda(x+1)\text{cF}_\lambda(x-1) = -1$ $[\text{cF}_\lambda(x)]^2 - \text{sF}_\lambda(x+1)\text{sF}_\lambda(x-1) = 1$
$\text{ch}^2(x) - \text{sh}^2(x) = 1$	$[\text{cF}_\lambda(x)]^2 - [\text{sF}_\lambda(x)]^2 = \dfrac{4}{4+\lambda^2}$
$\text{sh}(x+y) = \text{sh}(x)\text{ch}(y) + \text{ch}(x)\text{sh}(y)$ $\text{sh}(x-y) = \text{sh}(x)\text{ch}(y) - \text{ch}(x)\text{sh}(y)$	$\dfrac{2}{\sqrt{4+\lambda^2}}\text{sF}_\lambda(x+y)$ $\qquad = \text{sF}_\lambda(x)\text{cF}_\lambda(y) + \text{cF}_\lambda(x)\text{sF}_\lambda(y)$ $\dfrac{2}{\sqrt{4+\lambda^2}}\text{sF}_\lambda(x-y)$ $\qquad = \text{sF}_\lambda(x)\text{cF}_\lambda(y) - \text{cF}_\lambda(x)\text{sF}_\lambda(y)$
$\text{ch}(x+y) = \text{ch}(x)\text{ch}(y) + \text{sh}(x)\text{sh}(y)$ $\text{ch}(x-y) = \text{ch}(x)\text{ch}(y) - \text{sh}(x)\text{sh}(y)$	$\dfrac{2}{\sqrt{4+\lambda^2}}\text{cF}_\lambda(x+y)$ $\qquad = \text{cF}_\lambda(x)\text{cF}_\lambda(y) + \text{sF}_\lambda(x)\text{sF}_\lambda(y)$ $\dfrac{2}{\sqrt{4+\lambda^2}}\text{cF}_\lambda(x-y)$ $\qquad = \text{cF}_\lambda(x)\text{cF}_\lambda(y) - \text{sF}_\lambda(x)\text{sF}_\lambda(y)$
$\text{ch}(2x) = 2\text{sh}(x)\text{ch}(x)$	$\dfrac{1}{\sqrt{4+\lambda^2}}\text{cF}_\lambda(2x) = \text{sF}_\lambda(x)\text{cF}_\lambda(x)$
$[\text{ch}(x) \pm \text{sh}(x)]^n = \text{ch}(nx) \pm \text{sh}(nx)$	$[\text{cF}_\lambda(x) \pm \text{sF}_\lambda(x)]^n$ $\qquad = \left(\dfrac{2}{\sqrt{4+\lambda^2}}\right)^{n-1} [\text{cF}_\lambda(nx) \pm \text{sF}_\lambda(nx)]$

2.13.3. The hyperbolic properties

Let us formulate some hyperbolic properties for the functions (2.76)–(2.79). We start with the parity properties, which are valid for the classic hyperbolic functions (2.95). Of course the hyperbolic cosine is an even function, and the hyperbolic sine is an odd function. We accept without proof the parity properties of the functions (2.76)–(2.79).

Theorem 2.3. *The λ-Fibonacci and λ-Lucas hyperbolic sines are odd functions, and the λ-Fibonacci and λ-Lucas hyperbolic cosines are even functions, that is:*

$$sF_\lambda(-x) = -sF_\lambda(x), \quad cF_\lambda(-x) = cF_\lambda(x) \tag{2.103}$$

$$sL_\lambda(-x) = -sL_\lambda(x), \quad cL_\lambda(-x) = cL_\lambda(x). \tag{2.104}$$

2.14. A General Theory of the Recursive Hyperbolic Functions

In Chapter 1 we introduced a fundamentally new class of hyperbolic functions, namely the symmetric Fibonacci and Lucas hyperbolic functions:

$$sFs(x) = \frac{\Phi^x - \Phi^{-x}}{\sqrt{5}}, \quad cFs(x) = \frac{\Phi^x + \Phi^{-x}}{\sqrt{5}}; \tag{2.105-a}$$

$$sLs(x) = \Phi^x - \Phi^{-x}, \quad cLs(x) = \Phi^x + \Phi^{-x}. \tag{2.105-b}$$

These functions have also been called recursive hyperbolic functions. This name emphasizes their important properties, i.e. the existence of the so-called recursive properties, which are confirmed by their deep mathematical connections with "extended" Fibonacci and Lucas numbers, as defined by Binet's formulas:

$$F_n = \begin{cases} \dfrac{\Phi^n + \Phi^{-n}}{\sqrt{5}} & \text{for } n = 2k+1 \\[2mm] \dfrac{\Phi^n - \Phi^{-n}}{\sqrt{5}} & \text{for } n = 2k \end{cases}; \tag{2.106-a}$$

$$L_n = \begin{cases} \Phi^n + \Phi^{-n} & \text{for } n = 2k \\ \Phi^n - \Phi^{-n} & \text{for } n = 2k+1 \end{cases}, \tag{2.106-b}$$

where k takes the values from the set $k = 0, \pm1, \pm2, \pm3, \ldots$.

Note that at the discrete points of the continuous variable $x = 0, \pm1, \pm2, \pm3, \ldots$ the recursive hyperbolic functions (2.105-a) and (2.105-b) coincide with the "extended" Fibonacci and Lucas sequences for the infinite interval, from $-\infty$ to $+\infty$. This relationship is given by very simple mathematical relations:

$$F_n = \begin{cases} \mathrm{sFs}(n) & \text{for } n = 2k \\ \mathrm{cFs}(n) & \text{for } n = 2k + 1 \end{cases} ; \qquad (2.107\text{-a})$$

$$L_n = \begin{cases} \mathrm{cLs}(n) & \text{for } n = 2k \\ \mathrm{sLs}(n) & \text{for } n = 2k + 1 \end{cases} , \qquad (2.107\text{-b})$$

where k takes the values from the set $k = 0, \pm 1, \pm 2, \pm 3, \ldots$.

Let us compare the formulas (2.105-a), (2.105-b), (2.106-a), (2.106-b), (2.107-a), and (2.107-b) with the above derived formulas for the λ-Fibonacci and λ-Lucas hyperbolic functions, Gazale's formulas, and other formulas defined by the recursive properties:

$$\mathrm{sF}_\lambda(x) = \frac{\Phi_\lambda^x - \Phi_\lambda^{-x}}{\sqrt{4 + \lambda^2}}, \quad \mathrm{cF}_\lambda(x) = \frac{\Phi_\lambda^x + \Phi_\lambda^{-x}}{\sqrt{4 + \lambda^2}} \qquad (2.108\text{-a})$$

$$\mathrm{sL}_\lambda(x) = \Phi_\lambda^x - \Phi_\lambda^{-x}, \quad \mathrm{cL}_\lambda(x) = \Phi_\lambda^x + \Phi_\lambda^{-x} \qquad (2.108\text{-b})$$

$$F_\lambda(n) = \begin{cases} \dfrac{\Phi_\lambda^n - \Phi_\lambda^{-n}}{\sqrt{4 + \lambda^2}} & \text{for } n = 2k \\[2ex] \dfrac{\Phi_\lambda^n + \Phi_\lambda^{-n}}{\sqrt{4 + \lambda^2}} & \text{for } n = 2k + 1 \end{cases} \qquad (2.109\text{-a})$$

$$L_\lambda(n) = \begin{cases} \Phi_\lambda^n - \Phi_\lambda^{-n} & \text{for } n = 2k + 1 \\ \Phi_\lambda^n + \Phi_\lambda^{-n} & \text{for } n = 2k \end{cases} \qquad (2.109\text{-b})$$

$$F_\lambda(n) = \begin{cases} \mathrm{sF}_\lambda(n) & \text{for } n = 2k \\ \mathrm{cF}_\lambda(n) & \text{for } n = 2k + 1 \end{cases} \qquad (2.110\text{-a})$$

$$L_\lambda(n) = \begin{cases} \mathrm{cL}_\lambda(n) & \text{for } n = 2k \\ \mathrm{sL}_\lambda(n) & \text{for } n = 2k + 1 \end{cases} \qquad (2.110\text{-b})$$

where k takes the values from the set $k = 0, \pm 1, \pm 2, \pm 3, \ldots$. We see that these formulas are all similar in their mathematical structure.

The following conclusions concerning the properties of the λ-Fibonacci and λ-Lucas hyperbolic functions result from the above observations:

(1) The λ-Fibonacci and λ-Lucas hyperbolic functions are, on the one hand, a generalization of the classic hyperbolic functions (2.75) and they preserve all hyperbolic properties of the classic hyperbolic functions (2.75);

on the other hand, they are a generalization of the recursive Fibonacci and Lucas hyperbolic functions (2.105-a) and (2.105-b), which are special limiting cases of the functions (2.108-a) and (2.108-b), where $\lambda = 1$.

(2) The next unique feature of the functions (2.108-a) and (2.108-b) is the fact that they define a theoretically infinite number of new hyperbolic functions, because every positive real number $\lambda > 0$ generates a new, previously unknown class of hyperbolic functions.

(3) The cases $\lambda = 1, 2, 3, \ldots$ select a special class of the λ-Fibonacci and λ-Lucas hyperbolic functions, which have an additional unique property. The uniqueness of this class of the λ-Fibonacci and λ-Lucas hyperbolic functions (2.108-a) and (2.108-b) when compared with the classic hyperbolic functions (2.75) is the fact that they possess all of the recursive properties inherent in recursive Fibonacci and Lucas hyperbolic functions (2.105-a) and (2.105-b). They have fundamental relationships with "extended" Fibonacci and Lucas λ-numbers, as defined by Gazale's formulas (2.110-a) and (2.110-b). This unexpected relationship, defined by (2.105-a) and (2.105-b), allows us to introduce a special new class of hyperbolic functions, the generalized recursive hyperbolic functions, based on metallic proportions. Note that the recursive Fibonacci and Lucas hyperbolic functions (2.105-a) and (2.105-b) are special limiting cases of the generalized recursive hyperbolic functions (2.109-a) and (2.109-b), as introduced in [37, 38].

2.15. Conclusions for Chapter 2

(1) The Mathematics of Harmony, developed in 2009 Stakhov's book of the same name [1], creates new vistas for the development of modern mathematics and computer science. However, from the mathematical point of view, λ-Fibonacci and λ-Lucas hyperbolic functions are of greatest interest for mathematics, particularly for non-Euclidean geometry.

(2) The beauty of the basic formulas for the λ-Fibonacci and λ-Lucas hyperbolic functions, presented in Tables 2.3 and 2.4, is astonishing. It suggests that Dirac's Principle of Mathematical Beauty may be fully applicable to the metallic proportions (2.34) and the λ-Fibonacci and λ-Lucas hyperbolic functions (2.108-a) and (2.108-b). This, in turn, inspires the hope that these results can be employed as effective models for natural phenomena and their corresponding sciences.

(3) The primary result of Stakhov's articles on Gazale's formulas [37] and the λ-Fibonacci and λ-Lucas hyperbolic functions [38] has been the

introduction of a new class of hyperbolic functions, the λ-Fibonacci and λ-Lucas hyperbolic functions (where $\lambda > 0$ is a given real number), based upon the metallic proportions [28–35]. This new class of hyperbolic functions is similar to classic hyperbolic functions and retains all of their useful hyperbolic properties. In addition, they are a generalization of the λ-Fibonacci and λ-Lucas numbers, which coincide with the λ-Fibonacci and λ-Lucas hyperbolic functions for discrete values of the continuous variable $x = 0, \pm 1, \pm 2, \pm 3, \ldots$, and retains all of their useful recurrence properties.

(4) In general, we are able to conclude that the formulas (2.105-a), (2.105-b), (2.106-a), (2.106-b), (2.107-a), (2.107-b), (2.108-a), (2.108-b), (2.109-a), (2.109-b), (2.110-a), and (2.110-b) provide the mathematical foundation for a general theory of recursive hyperbolic functions.

References

[1] Stakhov, A. P., assisted by S. Olsen. *The Mathematics of Harmony. From Euclid to Contemporary Mathematics and Computer Science*. Singapore: World Scientific (2009).

[2] Olsen, S. *The Golden Section: Nature's Greatest Secret*. New York: Walker Publishing Company (2006).

[3] Stakhov, A. P. "The mathematics of harmony: clarifying the origins and development of mathematics", *Congressus Numerantium* (2008) Vol. CXCIII: 5–48.

[4] Soroko, E. M. *Structural Harmony of Systems*. Minsk: Nauka i Technika (1984) (Russian).

[5] Bodnar, O. Y. *The Golden Section and Non-Euclidean Geometry in Nature and Art*. Lvov: Publishing House "Svit" (1994) (Russian).

[6] Petoukhov, S. V. "Metaphysical aspects of the matrix analysis of genetic code and the golden section", in *Metaphysics: Century XXI*. Moscow: BINOM (2006): 216–250 (Russian).

[7] "Harmony of spheres", in *The Oxford Dictionary of Philosophy*. Oxford University Press (1994, 1996, 2005).

[8] Dimitrov, V. *A New Kind of Social Science. Study of Self-organization of Human Dynamics*. Morrisville: Lulu Press (2005).

[9] Stakhov, A. P. "The golden section and modern harmony mathematics", *Applications of Fibonacci Numbers*. Kluwer Academic Publishers (1998) Vol. 7: 393–399.

[10] Vorobyov, N. N. *Fibonacci Numbers*. Moscow: Nauka (1961) (Russian).

[11] Hoggat, Jr. V. E. *Fibonacci and Lucas Numbers*. Boston, MA: Houghton Mifflin (1969).

[12] Vajda, S. *Fibonacci & Lucas Numbers, and the Golden Section. Theory and Applications*. Ellis Horwood Limited (1989).

[13] Klein, F. *Lectures on the Icosahedron.* New York: Courier Dover Publications (1956).

[14] Stakhov, A. P. "The generalized golden proportions and a new approach to geometric definition of a number", *Ukrainian Mathematical Journal* (2004) Vol. 56: 1143–1150 (Russian).

[15] Grimm, G. D. *Proportionality in Architecture.* Leningrad-Moscow: ONTI (1935) (Russian).

[16] Stakhov, A. P. *Introduction into Algorithmic Measurement Theory.* Moscow: Soviet Radio (1977) (Russian).

[17] Stakhov, A. P. *Algorithmic Measurement Theory.* Moscow: Nauka (1979) (Russian).

[18] Zenkin, A. A. "The mistake of Georg Cantor", *Philosophy Problems* (2000) No. 2: 163–168 (Russian).

[19] Stakhov, A. P. "The 'strategic mistakes' in the mathematics development and the role of the harmony mathematics for their overcoming", *Visual Mathematics* (2008) Vol. 10, No. 2.

[20] Polya, G. *Mathematical Discovery.* Moscow: Nauka (1970) (Russian).

[21] Stakhov, A. P. "Mitropolsky's commentary", in *The Mathematics of Harmony. From Euclid to Contemporary Mathematics and Computer Science.* Singapore: World Scientific (2009).

[22] Bergman, G. "A number system with an irrational base", *Mathematics Magazine* (1957) No. 31: 98–119.

[23] Stakhov, A. P. "The golden ratio in digital technology", *Automation and Computer Technology* (1980) No. 1: 27–33 (Russian).

[24] Stakhov, A. P. *Codes of the Golden Proportion.* Moscow: Radio and Communication (1984) (Russian).

[25] Lucas, E. "The theory of simply periodic numerical functions", *American Journal of Mathematics* (1878) (reprinted by Fibonacci Association in 1969).

[26] "Lucas sequence", *Wikipedia, The Free Encyclopedia,* http://en.wikipedia. org/wiki/Lucas_sequence (accessed November 4, 2015).

[27] "Pell number", *Wikipedia, the free encyclopaedia,* http://en.wikipedia.org/ wiki/Pell_number (accessed November 4, 2015).

[28] de Spinadel, V. W. *From the Golden Mean to Chaos.* Nueva Libreria (first edition 1998); Nobuko (second edition 2004).

[29] Gazale, M. J. *Gnomon. From Pharaohs to Fractals.* Princeton, NJ: Princeton University Press (1999).

[30] Kappraff, J. *Connections. The Geometric Bridge Between Art and Science. Second Edition.* Singapore: World Scientific (2001).

[31] Tatarenko, A. "The golden T_m-harmonies and D_m-fractals is an essence of soliton-similar m-structure of the world", Academy of Trinitarism, Moscow: Electronic number 77-6567, publication 12691 (2005) (Russian). http:// www. trinitas.ru/rus/doc/0232/009a/02320010.htm (accessed November 4, 2015).

[32] Arakelyan, H. *The Numbers and Magnitudes in Modern Physics.* Yerevan: Armenian Academy of Sciences (1989) (Russian).

[33] Shenyagin, V. P. "Pythagoras, or how every man creates its own myth. The fourteen years after the first publication of the quadratic mantissa s-proportions", Academy of Trinitarism, Moscow: Electronic number 77-6567, publication 17031 (2011) (Russian). http://www.trinitas.ru/rus/doc/0232/013a/02322050.htm (accessed November 4, 2015).

[34] Kosinov, N. V. "The golden ratio, golden constants, and golden theorems", Academy of Trinitarism, Moscow: Electronic number 77-6567, publication 14379 (2007) (Russian). http://www.trinitas.ru/rus/doc/0232/009a/02321049.htm (accessed November 4, 2015).

[35] Falcon, S. and Plaza, A. "On the Fibonacci k-numbers", *Chaos, Solitons & Fractals* (2007) Vol. 32, Issue 5: 1615–1624.

[36] Stakhov, A. P. "A generalization of the Cassini formula", *Visual Mathematics* (2012) Vol. 14, No. 2. http://www.mi.sanu.ac.rs/vismath/stakhovsept2012/cassini.pdf (accessed November 4, 2015).

[37] Stakhov, A. P. "Gazale's formulas, a new class of hyperbolic Fibonacci and Lucas Functions and the improved method of the 'golden' cryptography", Academy of Trinitarism, Moscow: Electronic number 77-6567, publication 14098 (2006) (Russian). http://www.trinitas.ru/rus/doc/0232/004a/02321063.htm (accessed November 4, 2015).

[38] Stakhov, A. P. "On the general theory of hyperbolic functions based on the hyperbolic Fibonacci and Lucas functions and on Hilbert's Fourth Problem", *Visual Mathematics* (2013) Vol. 15, No. 1. http://www.mi.sanu.ac.rs/vismath/pap.htm (accessed November 4, 2015).

Chapter 3

Hyperbolic and Spherical Solutions of Hilbert's Fourth Problem: The Way to the Recursive Non-Euclidean Geometries

3.1. Non-Euclidean Geometry

On February 23, 1826 at a meeting of the Mathematics and Physics Faculty of Kazan University, Russian mathematician Nikolai Lobachevsky (1792–1856) proclaimed that he had discovered a new geometry called imaginary geometry. It was based upon Euclid's traditional postulates, with the exception of the Fifth Parallel Postulate. Lobachevsky formulated a new Fifth Postulate as follows:

> *"At the plane through a point outside a given straight line, we can construct two and only two straight lines parallel to this line, as well as an endless set of straight lines, which do not overlap with this line and are not parallel to this line, and another endless set of straight lines, intersecting the given straight line."*

This new geometry was outlined for the first time by Lobachevsky in an 1829 article, "About the foundations of geometry" in the magazine *Kazan Bulletin.*

Independently of Lobachevsky, the Hungarian mathematician János Bolyai (1802–1860) arrived at similar ideas. He published his work "Appendix" in 1832 three years after Lobachevsky. In addition the prominent German mathematician Carl Friedrich Gauss (1777–1855) came to the same ideas. After his death some of his unpublished sketches on non-Euclidean geometry were discovered.

Twelve years following his death, Lobachevsky's geometry received full recognition and wide distribution. It was at that time that it became clear that an axiomatic scientific theory was considered fully complete only when

Figure 3.1.   Nikolai Lobachevsky (1792–1856).

Source: *Wikipedia, The Free Encyclopedia,*
http://en.wikipedia.org/wiki/Nikolai_Lobachevsky

Figure 3.2.   János Bolyai (1802–1860).

Source: *Wikipedia, The Free Encyclopedia,*
https://en.wikipedia.org/wiki/János_Bolyai

Figure 3.3.   Carl Friedrich Gauss (1777–1855).
Source: *Wikipedia, The Free Encyclopedia,*
http://en.wikipedia.org/wiki/Carl_Friedrich_Gauss

its system of axioms meets three conditions: independence, consistency and completeness. Lobachevsky's geometry had satisfied all three conditions.

This finally became clear in 1868 when the Italian mathematician Eugenio Beltrami (1835–1899) wrote in his memoirs, *The Experience of the Non-Euclidean Geometry Interpretation,* that in Euclidean space at pseudo-spherical surfaces, if geodesic lines are taken to be straight lines, the geometry of Lobachevsky's plane arises.

Later the German mathematician Felix Klein (1849–1925) and the French mathematician Henri Poincaré (1854–1912) proved the consistency of non-Euclidean geometry, by means of the creation of models corresponding to Lobachevsky's plane. The interpretation of Lobachevsky's geometry on the surfaces of Euclidean space led to a general recognition of Lobachevsky's ideas.

The creation of Riemannian geometry by Georg Riemann (1826–1866), became a further major consequence of this non-Euclidean approach. Riemannian geometry developed a mathematical doctrine about geometric space, a notion of differential and distance between elements of diversity, and a doctrine about curvature.

The introduction of the generalized Riemannian spaces, whose particular cases are Euclidean space and Lobachevsky's space, and the so-called Riemannian geometry, had opened new avenues in the development of

Figure 3.4.   Eugenio Beltrami (1835–1899).

Source: Wikipedia, The Free Encyclopedia,
http://en.wikipedia.org/wiki/Eugenio_Beltrami

Figure 3.5.   Henri Poincaré (1854–1912).

Source: Wikipedia, The Free Encyclopedia,
https://en.wikipedia.org/wiki/Henri_Poincaré

Figure 3.6.   Georg Riemann (1826–1866).

Source: Wikipedia, The Free Encyclopedia,
http://en.wikipedia.org/wiki/Bernhard_Riemann

Figure 3.7.   Vincenzo Riccati (1707–1775).

Source: Wikipedia, The Free Encyclopedia,
http://en.wikipedia.org/wiki/Vincenzo_Riccati

geometry. They found for example applications in physics (e.g. theory of relativity) and other branches of the theoretical natural sciences.

Lobachevsky's geometry is also called hyperbolic geometry because it is based on hyperbolic functions, introduced in the 18th century by the Italian mathematician Vincenzo Riccati (1707–1775).

The most famous classical interpretations of Lobachevsky's plane with Gaussian curvature $K < 0$ (which is a measure of the curvature of a surface in a neighborhood of some point), are the following:

— Lobachevsky's interpretation on a plane;
— Beltrami's interpretation on a disk;
— Poincare's interpretation on a disk;
— Klein's interpretation on a half-plane and others.

We will use the following Lobachevsky's metric form on a half-plane (see, for example, [1]):

$$(ds)^2 = R^2[(du)^2 + \text{sh}^2(u)\,(dv)^2] \tag{3.1}$$

with the Gaussian curvature $K = -\frac{1}{R^2} < 0$, where the variables (u, v) belong to the half-plane:

$$\Pi^+ : (u, v), 0 < u < +\infty, -\infty < v < +\infty. \tag{3.2}$$

Here ds is called an arc length, $\text{sh}(u) = \frac{e^u - e^{-u}}{2}$ is a hyperbolic sine, and $R > 0$ is a radius of curvature for this metric.

Lobachevsky's geometry has remarkable applications in several fields of modern natural sciences. This concerns not only applied aspects (cosmology, electrodynamics, plasma theory), but it primarily concerns the fundamental areas of mathematics (number theory, theory of automorphic functions created by Poincaré, geometry of surfaces, and so on).

Since on the closed surfaces of negative Gaussian curvature, Lobachevsky's geometry is fulfilled and Lobachevsky's plane is a universal covering for these surfaces, it is very fruitful to study various objects (dynamical systems with continuous and discrete time, layers, fabrics, and so on), defined on these surfaces. By developing this idea, we can raise these objects to the level of universal covering, which is replenished by the absolute ("infinity"), and further we can study smooth topological properties of these objects by using the notion of the absolute.

Samuil Aranson studied this problem for about four decades. Aranson's works [2–8], also written with co-authors, present these results and research methods. Aranson's doctoral dissertation "Global problems of qualitative theory of dynamic systems on surfaces" (1990) is devoted to this theme.

3.2. Hilbert's Problems and Hilbert's Philosophy

In the lecture "Mathematical problems" [9], presented at the Second International Congress of Mathematicians (Paris, 1900), David Hilbert (1862–1943) formulated his famous twenty-three mathematical problems. These problems were a considerable factor in the development of 20th century mathematics. His lecture was a unique phenomenon in the history of mathematics and literature.

Hilbert's work [9] is of interest from both a philosophical and methodological point of view. His arguments, concerning various aspects of solving mathematical problems, are of great importance. In the preamble to Hilbert's "Mathematical problems" [9] he writes:

> *"History teaches the continuity of the development of science. We know that every age has its own problems, which the following age either solves or casts aside as profitless and replaces by new ones. If we would obtain an idea of the probable development of mathematical knowledge in the immediate future, we must let the unsettled questions pass before our minds and look over the problems which the science of today sets and whose solution we expect from the future. To such a review of problems the present day, lying at the meeting of the centuries, seems to me well*

Figure 3.8.   David Hilbert (1862–1943).

Source: *Wikipedia, The Free Encyclopedia,*
http://en.wikipedia.org/wiki/David_Hilbert

*adapted. For the close of a great epoch not only invites us to look back
into the past but also directs our thoughts to the unknown future.*

*The deep significance of certain problems for the advance of mathe-
matical science in general and the important role which they play in the
work of the individual investigator are not to be denied. As long as a
branch of science offers an abundance of problems, so long is it alive; a
lack of problems foreshadows extinction or the cessation of independent
development. Just as every human undertaking pursues certain objects,
so also mathematical research requires its problems. It is by the solution
of problems that the investigator tests the temper of his steel; he finds
new methods and new outlooks, and gains a wider and freer horizon."*

Hilbert warns about the dangers of modern isolation of Pure Mathematics from experience:

*"In the meantime, while the creative power of pure reason is at work,
the outer world again comes into play, forces upon us new questions
from actual experience, opens up new branches of mathematics, and while
we seek to conquer these new fields of knowledge for the realm of pure
thought, we often find the answers to old unsolved problems and thus at
the same time advance most successfully the old theories. And it seems
to me that the numerous and surprising analogies and that apparently
pre-arranged harmony which the mathematician so often perceives in the
questions, methods and ideas of the various branches of his science, have
their origin in this ever-recurring interplay between thought and experience."*

This wonderful quotation contains some important thoughts. The first
is that *"while the creative power of pure reason is at work, the outer world
again comes into play."* From experience, *"we often find the answers to old
unsolved problems and thus at the same time advance most successfully the
old theories."* The second important thought is that *"the numerous and surprising analogies and that **apparently pre-arranged harmony ... have
their origin in this ever-recurring interplay between thought and
experience."***

In this quote Hilbert refers to the idea of a "pre-arranged harmony"
(from Leibniz's Doctrine on Pre-established Harmony), which has its origin
in the "ever-recurring interplay between thought and experience."

Hilbert puts forth rigor and simplicity as general requirements for the
solution of mathematical problems, while emphasizing:

*"Besides it is an error to believe that rigor in the proof is the enemy of
simplicity. On the contrary we find it confirmed by numerous examples
that the rigorous method is at the same time the simpler and the more*

easily comprehended. The very effort for rigor forces us to find out simpler methods of proof. It also frequently leads the way to methods which are more capable of development than the old methods of less rigor."

Then Hilbert expresses the idea that a strict simple proof can be implemented in the language of mathematical formulas and geometric forms, since "the arithmetical symbols are written diagrams and the geometrical figures are graphic formulas." [9].

3.3. Klein's Icosahedral Idea

3.3.1. The "Icosahedral Idea" by Felix Klein and Proclus' hypothesis

Next we turn to an important quote from Hilbert's "Mathematical problems" lecture [9] regarding Felix Klein's book, *Lectures on the Icosahedron* [10]:

"And how convincingly has F. Klein, in his work on the icosahedron, pictured the significance which attaches to the problem of the regular polyhedra in elementary geometry, in group theory, in the theory of equations and in that of linear differential equations."

The German mathematician Felix Klein (1849–1925) is well known in mathematics. In the 19th century Klein tried to use the regular icosahedron (dual to the dodecahedron) to unite all of the branches of mathematics [10]. It is important to emphasize that Hilbert drew attention to an unexpected parallel between his "Mathematical problems" [9] and Klein's "Icosahedral Idea". Like Hilbert's problems, Klein's "Icosahedral Idea" is very important for the development of mathematics. Klein interprets the regular icosahedron, based on the golden ratio, as the geometric object connecting the five mathematical theories: geometry, Galois theory, group theory, invariant theory, and differential equations. Klein's main idea is extremely simple: "Each unique geometric object is connected one way or another with the properties of the regular icosahedron." Unfortunately, this remarkable idea was left undeveloped in contemporary mathematics, which is one of its "strategic mistakes" [11].

As pointed out by Hilbert, Klein's "Icosahedral Idea" originates in the regular polyhedra or Platonic solids, in ancient Greece associated with the Harmony of the Universe [12].

Here one may recall another fundamental mathematical idea concerning Euclid's *Elements*. It is well known that Book XIII, the final book of Euclid's *Elements*, is devoted to the geometric theory of Platonic solids. The Greek philosopher and mathematician Proclus Diadochus (412–485 AD) was the first to draw attention to this fact. Proclus derived the surprising hypothesis from this that the main purpose of Euclid's creation of the *Elements* was to provide a complete geometric theory of the Platonic solids including their harmonic relationships. This means that Euclid's *Elements* can historically be considered a first version of the Mathematics of Harmony which, according to Proclus, was embodied in this outstanding work of Greek mathematics.

This approach leads to a novel and unexpected conclusion for many mathematicians. It is found in [11], that in parallel with Classical Mathematics, another mathematical theme — Harmony Mathematics — was studied in Greek science. Similar to Classical Mathematics, Harmony Mathematics has its origin in Euclid's *Elements*. However, Classical Mathematics focuses on the axiomatic approach of Euclid, while Harmony Mathematics is based upon the golden ratio (Proposition II.11) and Platonic solids described in Book XIII of Euclid's *Elements*. Thus, Euclid's *Elements* is the source of these two independent directions in the development of mathematics — Classical Mathematics and Harmony Mathematics.

It should be noted that, by referring to Klein's book *Lectures on the Icosahedron* [10], Hilbert had actually returned to Pythagoras and Plato's "harmonic ideas", embodied in Euclid's *Elements*.

Unfortunately this important conclusion coming from the analysis of Hilbert's "Mathematical problems" [9] was ignored by many mathematicians who had read and analyzed it. The basic purpose of the present chapter is to consider non-Euclidean geometry and relate it to Hilbert's Fourth Problem through the Mathematics of Harmony [13].

3.3.2.   Classical mathematics and harmony mathematics

For several centuries, the primary focus of mathematicians has been on Classical Mathematics, which became the Tsarina of the Natural Sciences. However, many prominent thinkers and mathematicians, starting with Pythagoras, Plato, Euclid, Pacioli, and Kepler up to Lucas, Binet, Coxeter, Vorobyov, Hoggatt, and others, were developing the basic concepts and applications of Harmony Mathematics.

Unfortunately, Classical Mathematics and Harmony Mathematics developed separately from one another. Moreover, Pythagoras and Plato's "harmonic ideas", embodied in Euclid's *Elements,* were essentially ignored in Classical Mathematics [11]. The time has come to unite them. This unique union may lead to novel discoveries in mathematics and the natural sciences. Some important discoveries in natural sciences are suggestive of this union, in particular, fullerenes (Nobel Prize of 1996), based on the Archimedean truncated icosahedron, and Shechtman's quasi-crystals (Nobel Prize of 2011), based on Plato's icosahedron. These fundamental mathematical theories should be joined for the sole purpose of discovering and explaining the Laws of Nature.

The modern Mathematics of Harmony [11, 13] is a reflection and development of the ancient "Harmony Idea", embodied in Euclid's *Elements.* Here, new and original mathematical discoveries have been made with Fibonacci hyperbolic functions [14–17], based on the Fibonacci numbers and the golden proportion, and their generalizations, the hyperbolic Fibonacci λ-functions [18, 19], in turn based on the λ-Fibonacci numbers and Spinadel's metallic proportions [20]. We used these classes of hyperbolic functions to obtain a solution to Hilbert's Fourth Problem, as set out in [21–24]. According to this solution, there is an infinite number of hyperbolic geometries, which are close to Lobachevsky's geometry.

3.4. Hilbert's Fourth Problem

By applying the aforementioned analysis of Hilbert's "Mathematical problems" [9], specifically to Hilbert's Fourth Problem, which has direct relation to non-Euclidean geometry, our solution led to the following unexpected conclusions.

Upon first inspection of Hilbert's Fourth Problem, like a "game between thought and experience", it appears vaguely formulated [25, 26]. This refers to the game between the axiomatic approach and the language of the strict and simple formulas of the Mathematics of Harmony [13], which has both theoretical and practical importance for science. Stakhov's 2013 paper [27] presents a similar idea, that of a game between postulates and hyperbolic functions.

It should be noted that the axiomatic approach to solving Hilbert's Fourth Problem was used in the second half of the 20th century by the Russian and Ukrainian mathematician Pogorelov (1919–2002) [28].

However, at present the majority of mathematicians recognize Hilbert's Fourth Problem as being too vague to understand or solve [25, 26]. Therefore, it is not correct to insist that Pogorelov had completely solved this problem. As pointed out by Aranson [29] we are apparently talking about a partial solution to Hilbert's Fourth Problem, based on the axiomatic approach.

There is also another approach to solving this problem through the use of the new classes of hyperbolic functions, which arose in the Mathematics of Harmony [13]. Here again we cannot claim that this complex and important mathematical problem is completely solved.

The Russian translation of Hilbert's "Mathematical problems" [9] lecture and its comments are given in the book *Hilbert's Problems* [30]. In Hilbert's original work [9], his Fourth Problem is known as the "Problem of the straight line as the shortest distance between two points". It has been formulated in [9] as follows:

> "*Another problem relating to the foundations of geometry is this: If from among the axioms necessary to establish ordinary Euclidean geometry, we exclude the axiom of parallels, or assume it as not satisfied, but retain all other axioms, we obtain, as is well known, the geometry of Lobachevsky's (hyperbolic geometry). We may therefore say that this is a geometry standing next to Euclidean geometry . . . .*
>
> *The more general question now arises: Whether from other suggestive standpoints geometries may not be devised which, with equal right, stand next to Euclidean geometry . . . .*
>
> *The theorem of the straight line as the shortest distance between two points and the essentially equivalent theorem of Euclid about the sides of a triangle, play an important part not only in number theory but also in the theory of surfaces and in the calculus of variations. For this reason, and because I believe that the thorough investigation of the conditions for the validity of this theorem will throw a new light upon the idea of distance, as well as upon other elementary ideas, e.g., upon the idea of the plane, and the possibility of its definition by means of the idea of the straight line, the construction and systematic treatment of the geometries here possible seem to me desirable.*"

Hilbert's Fourth Problem is very important to mathematics as it touches upon the very foundations of geometry, number theory, the theory of surfaces and the calculus of variations. It is of fundamental importance not only for mathematics, but also for all theoretical natural sciences. Are there non-Euclidean geometries, which are next to Euclidean geometry and are interesting from "other suggestive standpoints"?

If we consider this problem in the context of the theoretical natural sciences, then Hilbert's Fourth Problem aims at searching for **new**

hyperbolic worlds of nature, which reflect some new and novel properties of Nature's structures and phenomena and is close to Euclidean geometry.

Hilbert considers Lobachevsky's geometry and spherical geometry as those nearest to Euclidean geometry. As is stated in Wikipedia [26],

> *"In mathematics, Hilbert's Fourth Problem in the 1900 'Hilbert problems' was a foundational question in geometry. In one statement derived from the original, it was to find geometries whose axioms are closest to those of Euclidean geometry if the ordering and incidence axioms are retained, the congruence axioms weakened, and the equivalent of the parallel postulate omitted."*

In mathematical literature Hilbert's Fourth Problem is sometimes considered as very vaguely formulated making its final solution difficult. As noted in Wikipedia [26], *"The original statement of Hilbert, however, has also been judged too vague to admit a definitive answer."*

American geometer Herbert Busemann analyzes the whole range of issues related to Hilbert's Fourth Problem in [31], concluding that the question posed by this problem is unnecessarily broad.

Unfortunately, attempts made at solving Hilbert's Fourth Problem by German mathematician Hamel (1901) and later by Pogorelov [28] (1974) have not led to significant progress, as pointed out in the Wikipedia articles [25, 26]. In [25], the status of the problem is formulated as being "too vague to be stated resolved or not" and Pogorelov's solution [28] is not even mentioned. More about Pogorelov's axiomatic approach to solving Hilbert's Fourth Problem can be found in Aranson's article [29].

A more cautious approach regarding Pogorelov's solution to Hilbert's Fourth Problem is presented in the book, *The Honors Class — Hilbert's Problems and Their Solvers* [32]. The modern mathematical community blames Hilbert for not clearly formulating the Fourth Problem which is the main reason why it was not solved until now. In spite of this critical attitude, we should emphasize the deep importance of this problem for mathematics and the theoretical natural sciences. Without a doubt, Hilbert's intuition led him to the conclusion that Lobachevsky's geometry, spherical geometry and Minkovski's geometry do not exhaust all possible variants of non-Euclidean geometries. Hilbert's Fourth Problem directs researchers to search for new non-Euclidean geometries, which are close to the traditional Euclidean geometry.

According to [27], one cause of the difficulties associated with the solution of Hilbert's Fourth Problem lies elsewhere. All known attempts to solve this problem (Hamel and Pogorelov) were in the traditional framework of the so-called "game of postulates" [27].

This "game" started with the works of Lobachevsky and Bolyai, when Euclid's Fifth Postulate was replaced with another. This was the major step in the development of non-Euclidean geometry, leading to Lobachevsky's geometry. Known as *hyperbolic geometry*, it changed the traditional geometric ideas. Its name highlights the fact that it is based on the classical hyperbolic functions (3.1) and (3.2), a key idea of Lobachevsky's geometry.

It is important to emphasize that the very name of hyperbolic geometry contains another approach to the solution of Hilbert's Fourth Problem: **searching for new classes of hyperbolic functions**, which can be the basis for new hyperbolic geometries. Every new class of the hyperbolic functions generates a new variant of the hyperbolic geometry. By analogy with the "game of postulates", this approach to solving Hilbert's Fourth Problem can be named the "game of functions" [27].

In connection with this introduction of a new class of hyperbolic functions based on the golden ratio [14–17], a new geometric theory of phyllotaxis (Bodnar's geometry) [33] was formulated relating the new hyperbolic geometries to the world of Nature surrounding us.

Stakhov [18, 19] in making a broad generalization of the symmetric Fibonacci and Lucas hyperbolic functions [15–17] developed the so-called λ-Fibonacci and λ-Lucas *hyperbolic functions*. The existence of infinite variants of hyperbolic functions, the basis for new hyperbolic geometries, is demonstrated in [18, 19].

Developing this idea is the main purpose of this chapter, that is, to create new hyperbolic geometries, based on the λ-Fibonacci and λ-Lucas hyperbolic functions [18, 19], and to obtain a simple and original solution to Hilbert's Fourth Problem. This study can be considered to be an original solution, the "pseudo-spherical solution", of Hilbert's Fourth Problem based on the metallic proportions [20].

3.5.   Hyperbolic Solution of Hilbert's Fourth Problem

3.5.1.   *Preliminary information*

Lobachevsky's classic metric (3.1) can be obtained, when considering the upper half of the two-sheeted hyperboloid (pseudo-sphere of radius R):

$$M^2 : Z^2 - X^2 - Y^2 = R^2; \quad Z \geq R > 0, \tag{3.3}$$

embedded in the three-dimensional pseudo-Euclidean space (X, Y, Z) with Minkowski's metric:

$$(dl)^2 = (dZ)^2 - (dX)^2 - (dY)^2.$$

Here dl is an arc element in the space (X, Y, Z), with the next parametrization of the surface (3.3) in the form:

$$M^2\colon X = R\operatorname{sh}(u)\cos(v), \quad Y = R\operatorname{sh}(u)\sin(v), \quad Z = R\operatorname{ch}(u), \qquad (3.4)$$

where the half-plane (3.2) $\Pi^+\colon (u, v)$, $0 < u < +\infty$, $-\infty < v < +\infty$ is the domain of existence of curvilinear coordinates (u, v). With this parametrization of M^2 for the metric (3.1) we have $(ds)^2 = -(dl)^2$.

In the special theory of relativity (STR) we use the following coordinate system: the spatial coordinates X, Y, and the time coordinate $Z = c_0 t > 0$, where c_0 is the speed of light in a vacuum, and t is time.

Figure 3.9 gives a visual representation of the surface M^2. The surface M^2 belongs to the so-called time-homothetic domain, bounded by the upper half of the upper half of the isotropic cone (or light cone) $K^2\colon Z^2 - X^2 - Y^2 = 0$, $Z > 0$.

Note that the surface M^2 is considered to be the upper half of a two-sheeted hyperboloid, embedded in Minkowski's three-dimensional space.

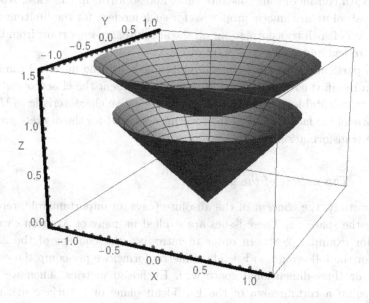

Figure 3.9. The surface M^2.

From this point of view, the surface M^2 is considered to be an open two-dimensional manifold. All parametric representations, concerning this surface in those or other curvilinear coordinates, are nothing as a covering of the manifold by the different sets of the cards with recalculation from one card to another card, while in this situation, each such card completely covers all the two-dimensional manifolds M^2.

Furthermore, each card will have an independent life isolated from the surface M^2 as interpreted in these or other Lobachevsky models with specifically stated metrics (e.g. Lobachevsky's plane with Lobachevsky's metric, Poincaré's model on the disk, Klein's model on the half-plane), and also in our infinite set of models, which preserve invariance on the half-plane.

Each model with its specifically introduced metric has its own unique geometric and differential properties, and each requires a separate investigation.

At the same time, when we convert one model into another, these properties are interpreted differently. In such a case, questions like the following arise: does the Gaussian curvature remain constant; how do geodesic lines, angles, squares of figures in the particular model change; what are movements, compressions, conversions; which are not movements; how should we study dynamic systems, as the one-parameter groups of transformations with continuous and discrete time; and so forth. In this case, studies of the absolute arithmetic properties for such models, for the limiting continuation of such transformations on the absolute, are important from both a theoretical and a practical point of view.

Of particular interest is the study of models like the universal ramified or non-ramified covering spaces $\overline{M}$, where we present the closed orientable and non-orientable surface M with Euler's negative characteristic $\chi(M)$ in the form of the factor $\overline{M}/G$ of the covering spaces $\overline{M}$ for the discrete groups of the transformations G.

3.5.2. The notion of the absolute

In our study, the concept of the absolute plays an important role, replenishing the space $\overline{M}$. These issues are studied in many of Aranson's works (see for example [2–8]). In order to introduce the concept of the absolute for the following non-Euclidean metric forms, we pre-equip the entire plane or three-dimensional space with Euclidean metrics. Then we further equip a certain area of the Euclidean plane or a surface in three-dimensional Euclidean space with the non-Euclidean metric form. We will call the absolute of the non-Euclidean metric form the boundary ∂D of the domain of definition D of this form, such that approaching ∂D "inside" of the domain D, the non-Euclidean form degenerates.

For Lobachevsky's metric form (3.1) with Gaussian curvature $K = -\frac{1}{R^2} < 0$, where the variables (u, v) belong to the half-plane (3.2), the coefficients of the form (3.1) are the following:

$$E = R^2 > 0, \quad F = 0, \quad G = R^2 \text{sh}^2(u) > 0. \tag{3.5}$$

Then, we have:

$$D = \Pi^+ : 0 < u < +\infty, -\infty < v < +\infty,$$

$$\partial D = (u = 0, -\infty < v < +\infty) \cup (u = +\infty, -\infty < v < +\infty).$$

When $u = 0$, we have $E = R^2 > 0$, $F = 0$, $G = 0$, $EG - F^2 = 0$, and when $u = +\infty$, we have $E = R^2 > 0$, $F = 0$, $G = +\infty$, $EG - F^2 = +\infty$. Hence, we get the line $(u = 0, -\infty < v < +\infty)$, $(u = +\infty, -\infty < v < +\infty)$ which is an "infinitely" distant line.

3.5.3. *Metric properties of surfaces*

First let us recall some well-known facts of differential geometry of surfaces [1, 34]. Let the surface M^2 be given in parametric form:

$$M^2 : x = x(u, v), \quad y = y(u, v), \quad z = z(u, v), \tag{3.6}$$

where (u, v) belongs to any domain D of surface parameters.

The first quadratic form. The first quadratic form (i.e. the differential of arc length) in this case is as follows:

$$(ds)^2 = E(du)^2 + 2F du dv + G(dv)^2, \tag{3.7}$$

where $E = E(u, v) > 0$, $F = F(u, v)$, $G = G(u, v) > 0$, $EG - F^2 > 0$.

Metric forms of the surface.

(1) A length of the arc s: $x = x(u(t), v(t)), y = y(u(t), v(t)), z = z(u(t), v(t))$ between the points $M_1(x_1, y_1, z_1)$, $M_2(x_2, y_2, z_2)$ on the surface M^2, corresponding to the values of the parameter t_1 and t_2,

equal:

$$s = \int_{t_1}^{t_2} ds = \int_{t_1}^{t_2} \sqrt{E\left(\frac{du}{dt}\right)^2 + 2F\frac{du}{dt}\frac{dv}{dt} + G\left(\frac{dv}{dt}\right)^2} \, dt. \qquad (3.8)$$

Here are

$$M_1 : x_1 = x[u(t_1), v(t_1)], \quad y_1 = y[u(t_1), v(t_1)], \quad z_1 = z[u(t_1), v(t_1)],$$
$$M_2 : x_2 = x[u(t_2), v(t_2)], \quad y_2 = y[u(t_2), v(t_2)], \quad z_2 = z[u(t_2), v(t_2)],$$
$$E = E[u(t), v(t)], \quad F = F[u(t), v(t)], \quad G = G[u(t), v(t)].$$

(2) The area of an arbitrary piece D of the surface M^2 equals:

$$\iint_D \sqrt{EG - F^2} du \wedge dv. \qquad (3.9)$$

Here $du \wedge dv$ is the external (vector) product of the differentials and

$$E = E(u, v) > 0, \quad F = F(u, v), \quad G = G(u, v) > 0, \quad EG - F^2 > 0.$$

(3) The angle between two curves on the surface M^2 is determined as follows. If the curves

$$L_1: x = x(u_1(t), v_1(t)), \quad y = y(u_1(t), v_1(t)), \quad z = z(u_1(t), v_1(t)),$$
$$L_2: x = x(u_2(t), v_2(t)), \quad y = y(u_2(t), v_2(t)), \quad z = z(u_2(t), v_2(t)),$$

belonging to the surface M^2, intersect at the point M_0 (x_0, y_0, z_0), then the intersection angle of θ (the angle between the positive directions of the tangents at the point M_0) is determined by the formula:

$$\cos\theta$$
$$= \frac{E\frac{du_1}{dt}\frac{du_2}{dt} + F\left(\frac{du_1}{dt}\frac{dv_2}{dt} + \frac{dv_1}{dt}\frac{du_2}{dt}\right) + G\frac{dv_1}{dt}\frac{dv_2}{dt}}{\sqrt{E(\frac{du_1}{dt})^2 + 2F\frac{du_1}{dt}\frac{dv_1}{dt} + G\left(\frac{dv_1}{dt}\right)^2} \sqrt{E\left(\frac{du_2}{dt}\right)^2 + 2F\frac{du_2}{dt}\frac{dv_2}{dt} + G\left(\frac{dv_2}{dt}\right)^2}}.$$
$$\qquad (3.10)$$

(4) Gaussian curvature K of a surface is calculated by the formula:

$$K = \frac{\Delta_1 - \Delta_2}{(EG - F^2)^2}, \quad \Delta_1 = \det(a_{ij}), \quad \Delta_2 = \det(b_{ij}), \quad i, j = 1, 2, 3,$$
$$\qquad (3.11)$$

where

$$a_{11} = -\frac{1}{2}\frac{\partial^2 G}{(\partial u)^2} + \frac{\partial^2 F}{\partial u \partial v} - \frac{1}{2}\frac{\partial^2 E}{(\partial v)^2},$$

$$a_{12} = \frac{\partial E}{\partial u}, \quad a_{13} = \frac{\partial F}{\partial u} - \frac{1}{2}\frac{\partial E}{\partial v},$$

$$a_{21} = \frac{\partial F}{\partial v} - \frac{1}{2}\frac{\partial G}{\partial u}, \quad a_{22} = E, \quad a_{23} = F,$$

$$a_{31} = \frac{1}{2}\frac{\partial G}{\partial u}, \quad a_{32} = F, \quad a_{33} = G,$$

$$b_{11} = 0, \quad b_{12} = \frac{\partial E}{\partial v}, \quad b_{13} = \frac{\partial G}{\partial u},$$

$$b_{21} = \frac{1}{2}\frac{\partial E}{\partial v}, \quad b_{22} = E, \quad b_{23} = F,$$

$$b_{31} = \frac{1}{2}\frac{\partial G}{\partial u}, \quad b_{32} = F, \quad b_{33} = G.$$

In our situation, we consider the metric forms, for which we have $F = F(u,v) \equiv 0$, and then, according to this remark, we get from (3.11) the following formula for the Gaussian curvature:

$$K = K(u,v) = -\frac{1}{AB}\left[\frac{\partial}{\partial u}\left(\frac{\frac{\partial B}{\partial u}}{A}\right) + \frac{\partial}{\partial v}\left(\frac{\frac{\partial A}{\partial v}}{B}\right)\right],$$

$$A = \sqrt{E(u,v)}, \quad B = \sqrt{G(u,v)}. \tag{3.12}$$

If we put the point $M_1(x_1, y_1, z_1) \in M^2$ in with the point $M_2(x_2, y_2, z_2) \in M^2$, then we get the one-to-one mapping $f\colon M^2 \mapsto M^2$ of the surface the surface M^2 on itself. This mapping $f\colon M^2 \mapsto M^2$ is called:

(1) preserving lengths (isometric) if its length for an arbitrary curve remains unchanged;
(2) preserving angles (conformal) if the angles between any two intersecting curves remain unchanged;
(3) preserving areas (equiareal) if areas of an arbitrary piece of the surface remain unchanged.

We denote by E_1, F_1, G_1 and E_2, F_2, G_2 the coefficients of the metric form (3.7), corresponding to the points M_1 and M_2 on the surface M^2. Table 3.1 presents the necessary and sufficient conditions for the quadratic form (3.7), when the above metric elements at the mapping $f\colon M^2 \mapsto M^2$ remain unchanged.

Table 3.1.

Mapping	The necessary and sufficient conditions for the quadratic form
Preserving length (isometric)	$E_2 = E_1,\ F_2 = F_1,\ G_2 = G_1$
Preserving angle (conformal)	$E_2 = \lambda_0 E_1,\ F_2 = \lambda_0 F_1,\ G_2 = \lambda_0 G_1,\ \lambda_0(u,v) > 0$
Preserving area (equiareal)	$E_2 G_2 - (F_2)^2 = E_1 G_1 - (F_1)^2$

These relations must be satisfied at each point on the surface. Each mapping preserving length is conformal and equiareal. If the conformal mapping preserves the area, then this mapping preserves the length.

3.5.4.   *Examples*

First we demonstrate the conservation or lack of conservation of lengths, angles and areas with linear transformations, and then we do so with nonlinear transformations of the pseudo-sphere and sphere. Under such transformations of the plane the constant zero Gaussian curvature $K = 0$ remains unchanged; for the pseudo-sphere the constant negative Gaussian curvature $K = -\frac{1}{R^2} < 0$ remains unchanged; and for the sphere the constant positive Gaussian curvature $K = \frac{1}{R^2} > 0$ remains unchanged.

Example 3.1 (Transformations on the plane which change lengths, angles and areas, but the Gaussian curvature $K = 0$ remains unchanged). Given the plane Π^2: $x + y + z = 0$ in the Euclidean space $R^3(x, y, z)$, consider the two parametrizations of the plane Π^2:

(1) The first parametrization:

$$x = x_1(u,v) = u, \quad y = y_1(u,v) = v, \quad z = z_1(u,v) = -u - v. \quad (3.13)$$

(2) The second parametrization:

$$x = x_2(u,v) = \alpha u, \quad y = y_2(u,v) = v, \quad z = z_2(u,v) = -\alpha u - v. \quad (3.14)$$

We assume that the parameters (u, v) in (3.13) and (3.14) satisfy the conditions that $-\infty < u < +\infty, -\infty < v < +\infty$, and the coefficient α is any real number with the additional condition:

$$\alpha \neq 0, \quad \alpha \neq \pm 1. \quad (3.15)$$

Consider the system of equations:

$$\begin{cases} x_1 = u, & y_1 = v, & z_1 = -u - v \\ x_2 = \alpha u, & y_2 = v, & z_2 = -\alpha u - v \end{cases}.$$

If we exclude (u, v) from the system of equations, we then obtain the point mapping $f_\alpha \colon \Pi^2 \mapsto \Pi^2$ of the plane $\Pi^2 \colon x + y + z = 0$ of the form:

$$f_\alpha \colon x_2 = \alpha x_1, \quad y_2 = y_1, \quad z_2 = -\alpha x_1 - y_1. \tag{3.16}$$

Here we have:

$$\forall (x_1, y_1, z_1) \in \Pi^2, \quad (x_2, y_2, z_2) = f_\alpha(x_1, y_1, z_1),$$

where the symbol $\forall$ means "any".

The inverse point mapping $f_\alpha^{-1} \colon \Pi^2 \mapsto \Pi^2$ has the form:

$$f_\alpha^{-1} \colon x_1 = \frac{1}{\alpha} x_2, \quad y_1 = y_2, z_1 = -\frac{1}{\alpha} x_2 - y_2. \tag{3.17}$$

Find the metric forms on the plane $\Pi^2 \colon x + y + z = 0$, induced by the Euclidean metric:

$$(dl)^2 = (dx)^2 + (dy)^2 + (dz)^2 \tag{3.18}$$

of the space $R^3(x, y, z)$.

At the first parametrization (3.13), we get:

$$dx_1 = \frac{\partial x_1}{\partial u} du + \frac{\partial x_1}{\partial v} dv = 1 \times du + 0 \times dv = du,$$

$$dy_1 = \frac{\partial y_1}{\partial u} du + \frac{\partial y_1}{\partial v} dv = 0 \times du + 1 \times dv = dv,$$

$$dz_1 = \frac{\partial z_1}{\partial u} du + \frac{\partial z_1}{\partial v} dv = -1 \times du - 1 \times dv = -du - dv.$$

Hence on the plane Π^2 we get the following metric form:

$$(dl_1)^2 = (ds_1)^2 = (dx_1)^2 + (dy_1)^2 + (dz_1)^2$$
$$= (du)^2 + (dv)^2 + (-du - dv)^2 = 2(du)^2 + 2dudv + 2(dv)^2.$$

Rewrite this in the standard form:

$$(ds_1)^2 = E_1(du)^2 + 2F_1 dudv + G_1(dv)^2, \tag{3.19}$$

where

$$E_1 = 2, \quad F_1 = 1, \quad G_1 = 2. \tag{3.20}$$

We can do the same with the second parametrization (3.14):

$$dx_2 = \frac{\partial x_2}{\partial u}du + \frac{\partial x_2}{\partial v}dv = \alpha \times du + 0 \times dv = \alpha du,$$

$$dy_2 = \frac{\partial y_2}{\partial u}du + \frac{\partial y_2}{\partial v}dv = 0 \times du + 1 \times dv = dv,$$

$$dz_2 = \frac{\partial z_2}{\partial u}du + \frac{\partial z_2}{\partial v}dv = -\alpha \times du - 1 \times dv = -\alpha du - dv.$$

Hence on the plane Π^2 we get the following metric form:

$$(dl_2)^2 = (ds_2)^2 = (dx_1)^2 + (dy_1)^2 + (dz_1)^2$$
$$= 2\alpha^2(du)^2 + 2\alpha du dv + 2(dv)^2.$$

Rewrite this in the standard form:

$$(ds_2)^2 = E_2(du)^2 + 2F_2 du dv + G_2(dv)^2, \qquad (3.21)$$

where

$$E_2 = 2\alpha^2, \quad F_2 = \alpha, \quad G_2 = 2. \qquad (3.22)$$

We now show that the point mapping $f_\alpha^{-1}: \Pi^2 \mapsto \Pi^2$ of the form (3.16) is **not isometric**, that is, it does not preserve length. First we assume the opposite that length is preserved. Then, according to the necessary and sufficient conditions for the preservation of lengths (see Table 3.1), the following relations must be satisfied:

$$E_2 = E_1, \quad F_2 = F_1, \quad G_2 = G_1. \qquad (3.23)$$

But in our situation, by virtue of (3.22) and (3.20), we get:

$$2\alpha^2 = 2, \quad \alpha = 1, \quad 2 = 2 \Rightarrow \alpha = 1,$$

which is impossible, since, under the conditions of (3.15), $\alpha \neq 0$, $\alpha \neq \pm 1$.

Now we show that the point mapping $f_\alpha^{-1} : \Pi^2 \mapsto \Pi^2$ of the form (3.16) is **not conformal**, that is, it does not preserve angles. First we assume the opposite, that it does in fact preserve angles. Then, according to the necessary and sufficient conditions for the preservation of angles (see Table 3.1), the following relations must be satisfied:

$$E_2 = \lambda_0 E_1, \quad F_2 = \lambda_0 F_1, \quad G_2 = \lambda_0 G_1, \quad \lambda_0(u,v) > 0. \qquad (3.24)$$

But in our situation, by virtue of (3.22) and (3.20), we get:

$$2\alpha^2 = \lambda_0 \times 2, \quad \alpha = \lambda_0 \times 1, \quad 2 = \lambda_0 \times 2 \Rightarrow \alpha = \lambda_0 = 1,$$

which is impossible, since, under the conditions of (3.15), $\alpha \neq 0$, $\alpha \neq \pm 1$.

We now show that the point mapping $f_\alpha^{-1}\colon \Pi^2 \mapsto \Pi^2$ of the form (3.16) is **not equiareal**, that is, it does not preserve areas. First we assume the opposite, that it does in fact preserve areas. Then, according to the necessary and sufficient conditions for the preservation of angles (see Table 3.1), the following relations must be satisfied:

$$E_2 G_2 - (F_2)^2 = E_1 G_1 - (F_1)^2. \tag{3.25}$$

But in our situation, by virtue of (3.22) and (3.20), we get:

$$E_2 = 2\alpha^2, \quad F_2 = \alpha, \quad G_2 = 2,$$
$$E_2 G_2 - (F_2)^2 = (2\alpha^2 \times 2) - \alpha^2 = 4\alpha^2 - \alpha^2 = 3\alpha^2,$$
$$E_1 = 2, F_1 = 1, G_1 = 2,$$
$$E_1 G_2 - (F_2)^2 = (2 \times 2 - 1^2) = 4 - 1 = 3,$$
$$E_2 G_2 - (F_2)^2 = E_1 G_1 - (F_1)^2,$$
$$2\alpha^2 \times 2 - \alpha^2 = 2 \times 2 - 1^2 \Rightarrow 3\alpha^2 = 3 \Rightarrow \alpha = \pm 1$$

which is impossible, since, under the conditions of (3.15), $\alpha \neq 0$, $\alpha \neq \pm 1$.

On the half-plane $D^2(U, V), -\infty < U < +\infty, -\infty < V < +\infty$, the following transformation corresponds to the point mapping $f_\alpha^{-1}\colon \Pi^2 \mapsto \Pi^2$ of the form (3.16):

$$\bar{f}_\alpha\colon (u, v) \mapsto (\alpha u, v). \tag{3.26}$$

Hence we obtain the following relation between the differentials:

$$\begin{pmatrix} d(\alpha u) \\ dv \end{pmatrix} = \begin{pmatrix} \alpha & 0 \\ 0 & 1 \end{pmatrix} \begin{pmatrix} du \\ dv \end{pmatrix}. \tag{3.27}$$

Therefore, under the action of the mapping (3.26), the metric form (3.19):

$$(ds_1)^2 = 2(du)^2 + 2dudv + 2(dv)^2$$

is converted into the metric form (3.21):

$$(ds_2)^2 = 2\alpha^2 (du)^2 + 2\alpha dudv + 2(dv)^2.$$

In this situation, the metric forms (3.21) for the conditions (3.22) at the changing of the coefficient $\alpha \neq 0$, $\alpha \neq \pm 1$ define an infinite set of new geometries in the plane $\Pi^2\colon x + y + z = 0$ in the Euclidean space $R^3(x, y, z)$.

Note that in this case, the metric form:

$$(ds_2)^2 = 2\alpha^2 (du)^2 + 2\alpha dudv + 2(dv)^2$$

for any $\alpha \neq 0$, according to (3.11), remains unchanged under the constant Gaussian curvature $K = 0$.

Note also that the geodesic lines L in the plane Π^2: $x + y + z = 0$ are obtained as the lines of the intersection of this plane with the planes Π_*^2: $ax + by + cz + d = 0$, where

$$a^2 + b^2 + c^2 > 0, \quad (a - b)^2 + (a - c)^2 + (b - c)^2 > 0.$$

Example 3.2 (Transformations on the pseudo-sphere, which preserve lengths, angles, areas and Gaussian curvature). In the space $Q^3(X, Y, Z)$ with Minkowski's metric $(dl)^2 = (dZ)^2 - (dX)^2 - (dY)^2$ let the following pseudo-sphere M^2 of the form (3.3) be defined as:

$$Z^2 - X^2 - Y^2 - R^2 = 0; \quad Z \geq R > 0.$$

Consider the two parametrizations of the pseudo-sphere M^2:

(1) The first parametrization (see also (3.4))

$$\begin{cases} X = X_1(u, v) = R\mathrm{sh}(u)\cos(v), \\ Y = Y_1(u, v) = R\mathrm{sh}(u)\sin(v), \\ Z = Z_1(u, v) = R\mathrm{ch}(u) \end{cases} \qquad (3.28)$$

induces the metric form:

$$(dl_1)^2 = (dZ_1)^2 - (dX_1)^2 - (dY_1)^2 = -R^2[(du)^2 + \mathrm{sh}^2(u)\,(dv)^2]. \quad (3.29)$$

The metric form

$$(ds_1)^2 = -(dl_1)^2 = R^2[(du)^2 + \mathrm{sh}^2(u)(dv)^2] \qquad (3.30)$$

is called Lobachevsky's metric form in [5] (see (3.4)).

(2) The second parametrization

$$\begin{cases} X = X_2(u, v) = R\mathrm{sh}(u)\cos(v - \alpha), \\ Y = Y_2(u, v) = R\mathrm{sh}(u)\sin(v - \alpha), \\ Z = Z_2(u, v) = R\mathrm{ch}(u) \end{cases} \qquad (3.31)$$

induces the metric form:

$$\begin{aligned} (dl_2)^2 &= (dZ_2)^2 - (dX_2)^2 - (dY_2)^2 \\ &= -R^2[(du)^2 + \mathrm{sh}^2(u)\,(d(v - \alpha))^2] \\ &= -R^2[(du)^2 + \mathrm{sh}^2(u)\,(dv)^2], \end{aligned}$$

that is, we again get the same Lobachevsky's metric form:

$$(ds_2)^2 = -(dl_2)^2 = R^2[(du)^2 + \mathrm{sh}^2(u)\,(d(v - \alpha))^2]. \qquad (3.32)$$

For the metric forms (3.30) and (3.32), we assume that the parameters (u, v) satisfy the conditions $0 < u < +\infty, -\infty < v < +\infty$, and the coefficient α is any real number.

Consider the system of equations:

$$\begin{cases} X_1 = R\operatorname{sh}(u)\cos(v), & Y_1 = R\operatorname{sh}(u)\sin(v), & Z_1 = R\operatorname{ch}(u), \\ X_2 = R\operatorname{sh}(u)\cos(v - \alpha), & Y_2 = R\operatorname{sh}(u)\sin(v - \alpha), & Z_2 = R\operatorname{ch}(u). \end{cases}$$

$$(3.33)$$

We exclude the variables (u, v) from the equations (3.33), and then on the pseudo-sphere M^2 for each $\alpha \in (-\infty, +\infty)$ we get the one-to-one point mapping $f_\alpha \colon M^2 \mapsto M^2$ of the form:

$$\begin{pmatrix} X_2 \\ Y_2 \\ Z_2 \end{pmatrix} = \begin{pmatrix} \cos(\alpha) & \sin(\alpha) & 0 \\ -\sin(\alpha) & \cos(\alpha) & 0 \\ 0 & 0 & 1 \end{pmatrix} \begin{pmatrix} X_1 \\ Y_1 \\ Z_1 \end{pmatrix}. \tag{3.34}$$

Indeed, if we consider the second relation in (3.33), we obtain:

$$\begin{aligned} X_2 &= R\operatorname{sh}(u)\cos(v - \alpha) \\ &= R\operatorname{sh}(u)[\cos(v)\cos(\alpha) + \sin(v)\sin(\alpha)] \\ &= \cos(\alpha)[R\sin(u)\cos(v)] + \sin(\alpha)[R\sin(u)\sin(v)], \\ Y_2 &= R\operatorname{sh}(u)\sin(v - \alpha) \\ &= R\operatorname{sh}(u)[\sin(v)\cos(\alpha) - \cos(v)\sin(\alpha)] \\ &= -\sin(\alpha)[R\operatorname{sh}(u)\cos(v)] + \cos(\alpha)[R\operatorname{sh}(u)\sin(v)]. \end{aligned}$$

Substituting here the first relation from (3.33), we obtain the mapping f_α of the form (3.34).

Note that the point mapping (3.34) $f_\alpha \colon M^2 \mapsto M^2$ on the pseudo-sphere M^2 is a rotation around the Z axis by the angle α.

From the point of view of differential dynamics, any one-to-one point mapping is a dynamic system with discrete time (also known as a cascade).

When the metric forms (3.30) and (3.32) have their coefficients at any $\alpha \neq 0$ they coincide and have the form:

$$E_2 = E_1 = R^2, \quad F_1 = F_2 = 0, \quad G_2 = G_1 = R^2\operatorname{sh}^2(u). \tag{3.35}$$

But then the point mapping $f_\alpha \colon M^2 \mapsto M^2$ with a changing α preserves the M^2 lengths, angles and areas (see Table 3.1).

Note that, when $u = 0$, the metric forms (3.30) and (3.32) degenerate, because in the case where $u = 0$ their coefficients G_1 and G_2 vanish. Therefore, $u = 0$ is an absolute for the metric forms (3.30) and (3.32).

It is interesting to note that the absolute $u = 0$ of the metric forms (3.30) and (3.32) corresponds on the pseudo-sphere M^2 to the fixed point $(X^* = 0, Y^* = 0, Z^* = R)$ of the point mapping $f_\alpha : M^2 \mapsto M^2$, and vice versa.

The point mapping $f_\alpha: M^2 \mapsto M^2$ of the form (3.34) on the parameter plane $D^2(u, v): 0 < u < +\infty, -\infty < v < +\infty$, corresponds to the transformation:

$$f_\alpha: (u, v) \mapsto (u, v - \alpha). \tag{3.36}$$

Hence, we get the following connection between differentials:

$$\begin{pmatrix} du \\ d(v - \alpha) \end{pmatrix} = \begin{pmatrix} 1 & 0 \\ 0 & 1 \end{pmatrix} \begin{pmatrix} du \\ dv \end{pmatrix}. \tag{3.37}$$

Therefore, under the action of the mapping (3.36) $f_\alpha: (u, v) \mapsto (u, v - \alpha)$, the metric form (3.30) $(ds_1)^2 = (du)^2 + \text{sh}^2(u)(dv)^2$ is converted into the metric form (3.32) $(ds_2)^2 = (du)^2 + \text{sh}^2(u)(dv)^2$, which coincides with the metric form (3.30).

In this sense, the metric form (3.32), when changing the coefficient $\alpha \neq 0$, preserves all metric properties of the pseudo-sphere of the form (3.3), which is defined in the space $Q^3(x, y, z)$, having Minkowski's metric.

3.5.5. Transformations on the pseudo-sphere, which alter lengths, angles and areas, but preserve Gaussian curvature $K = -\frac{1}{R^2} < 0$

Let pseudo-sphere M^2 of the form (3.3) $Z^2 - X^2 - Y^2 - R^2 = 0; Z \geq R > 0$ be placed in the space $R^3(X, Y, Z)$ with Minkovski's metric $(dl)^2 = (dZ)^2 - (dX)^2 - (dY)^2$.

Consider the two parametrizations of the pseudo-sphere M^2:

(1) The first parametrization (see (3.28))

$$\begin{cases} X = X_1(u, v) = R\text{sh}(u)\cos(v), \\ Y = Y_1(u, v) = R\text{sh}(u)\sin(v), \\ Z = Z_1(u, v) = R\text{ch}(u) \end{cases}$$

induces the metric form (3.29):

$$(dl_1)^2 = (dZ_1)^2 - (dX_1)^2 - (dY_1)^2 = -R^2[(du)^2 + \text{sh}^2(u)(dv)^2].$$

The metric form (3.30) $(ds_1)^2 = -(dl_1)^2 = R^2[(du)^2 + \text{sh}^2(u)(dv)^2]$ is called Lobachevsky's metric form (see (3.4)).

(2) The second parametrization

$$\begin{cases} X = X_2(u,v) = R\mathrm{sh}(\alpha u)\cos(v), \\ Y = Y_2(u,v) = R\mathrm{sh}(\alpha u)\sin(v), \\ Z = Z_2(u,v) = R\mathrm{ch}(\alpha u) \end{cases} \tag{3.38}$$

induces the metric form:

$$\begin{aligned} (dl_2)^2 &= (dZ_2)^2 - (dX_2)^2 - (dY_2)^2 \\ &= -R^2[(\alpha u)^2 + \mathrm{sh}^2(\alpha u)(dv)^2] \\ &= R^2[\alpha^2(du)^2 + \mathrm{sh}^2(\alpha u)(dv)^2]. \end{aligned}$$

Hence we get the metric form:

$$(ds_2)^2 = -(dl_2)^2 = R^2[\alpha^2(du)^2 + \mathrm{sh}^2(\alpha u)(dv)^2]. \tag{3.39}$$

In the metric forms (3.30) and (3.39) we assume that the parameters (u,v) satisfy the conditions $0 < u < +\infty, -\infty < v < +\infty$, and the coefficient α is any real number with the additional condition $\alpha > 0$.

Consider the system of equations:

$$\begin{cases} X_1 = R\mathrm{sh}(u)\cos(v), & Y_1 = R\mathrm{sh}(u)\sin(v), & Z_1 = R\mathrm{ch}(u), \\ X_2 = R\mathrm{sh}(\alpha u)\cos(v), & Y_2 = R\mathrm{sh}(\alpha u)\sin(v), & Z_2 = R\mathrm{ch}(\alpha u). \end{cases} \tag{3.40}$$

Excluding the parameters (u,v) from the equations (3.40), then for every $\alpha > 0$ we get on the pseudo-sphere M^2 the nonlinear one-to-one point mapping $f_\alpha: M^2 \mapsto M^2$ of the form:

$$f_\alpha: X_2 = \frac{\mathrm{sh}(\alpha u)}{\mathrm{sh}(u)}X_1, \quad Y_2 = \frac{\mathrm{sh}(\alpha u)}{\mathrm{sh}(u)}Y_1, \quad Z_2 = \frac{\mathrm{ch}(\alpha u)}{\mathrm{ch}(u)}Z_1, \tag{3.41}$$

where

$$u = \mathrm{Arch}\left(\frac{Z_1}{R}\right) = \ln\left(\frac{Z_1 + \sqrt{(Z_1)^2 - R^2}}{R}\right), Z_1 \geq R > 0.$$

For the case $\alpha = 1$, by virtue of (3.41), we have $X_2 = X_1$, $Y_2 = Y_1$, $Z_2 = Z_1$, that is, every point of the pseudo-sphere M^2 remains unchanged. For the case where $\alpha > 0$ and $\alpha \neq 1$ we have:

$$M_2(X_2, Y_2, Z_2) \neq M_1(X_1, Y_1, Z_1).$$

We now show that for any value of $\alpha > 0$ for the metric form (3.39) $(ds_2)^2 = R^2[\alpha^2(du)^2 + \mathrm{sh}^2(\alpha u)(dv)^2]$ the Gaussian curvature $K = -\frac{1}{R^2} < 0$ is preserved.

The coefficients of the metric form (3.39) have the following form:

$$E_2 = R^2\alpha^2, \quad F_2 = 0, \quad G_2 = R^2\text{sh}^2(\alpha u),$$

$$\alpha > 0, \quad u > 0, \quad -\infty < v < +\infty. \tag{3.42}$$

Hence we get:

$$A = \sqrt{E_2(u, v)} = R\alpha > 0, \quad B = \sqrt{G_2(u, v)} = R\text{sh}(\alpha u) > 0,$$

$$\frac{\partial A}{\partial v} = 0, \quad \frac{\partial^2 B}{\partial u^2} = R\alpha^2\text{sh}(\alpha u), \quad \frac{\partial}{\partial u}\left(\frac{\frac{\partial B}{\partial u}}{A}\right) = \frac{R\alpha^2\text{sh}(\alpha u)}{R\alpha} = \alpha\text{sh}(\alpha u)$$

$$AB = R^2\alpha\text{sh}(\alpha u)$$

$$K = K(u, v) = -\frac{1}{AB}\left[\frac{\partial}{\partial u}\left(\frac{\frac{\partial B}{\partial u}}{A}\right) + \frac{\partial}{\partial v}\left(\frac{\frac{\partial A}{\partial v}}{B}\right)\right]$$

$$= -\frac{\alpha\text{sh}(\alpha u)}{R^2\alpha\text{sh}(\alpha u)} = -\frac{1}{R^2}.$$

As for the other metric properties (lengths, angles and areas), with any value of $\alpha > 0$, $\alpha \neq 1$, the other metric properties, when compared with the case of $\alpha = 1$, are **not preserved**.

We now show that the point mapping $f_\alpha: M^2 \mapsto M^2$ of the form (3.41) for the pseudo-sphere M^2 is **not isometric**, that is, it does not preserve lengths. First we assume the opposite, that it is isometric. Then, according to the necessary and sufficient conditions for the preservation of lengths (see Table 3.1), the relations (3.23) $E_2 = E_1$, $F_2 = F_1$, $G_2 = G_1$ must be satisfied.

But then in our situation for the case where $\alpha > 0$, $\alpha \neq 1$ for the metric forms:

$$(ds_1)^2 = R^2[(du)^2 + \text{sh}^2(u)(dv)^2],$$

$$(ds_2)^2 = R^2[\alpha^2(du)^2 + \text{sh}^2(\alpha u)(dv)^2],$$

with the coefficients $E_1 = R^2$, $F_1 = 0$, $G_1 = R^2\text{sh}^2(u)$, $E_2 = R^2\alpha^2$, $F_2 = 0$, $G_2 = R^2\text{sh}^2(\alpha u)$, the (3.23) relations must be satisfied. Then, from the first relation from (3.23) we get for the given case $E_2 = R^2\alpha^2 = E_1 = R^2 \Rightarrow \alpha = 1$, which is impossible.

We now show that the point mapping $f_\alpha: M^2 \mapsto M^2$ of the form (3.41) for the pseudo-sphere M^2 is **not conformal**, that is, it does not preserve angles. First assume the opposite, that it is conformal. Then, according to the necessary and sufficient conditions for the preservation of angles (see Table 3.1), the following relations must be satisfied:

$$E_2 = \lambda_0 E_1, \quad F_2 = \lambda_0 F_1, \quad G_2 = \lambda_0 G_1, \quad \lambda_0(u, v) > 0.$$

As $E_1 = R^2$, $F_1 = 0$, $G_1 = R^2\text{sh}^2(u)$, $E_2 = R^2\alpha^2$, $F_2 = 0$, $G_2 = R^2\text{sh}^2(\alpha u)$, then in the given case we get:

$$\begin{cases} E_2 = R^2\alpha^2 = \lambda_0 E_1 = \lambda_0 R^2 \Rightarrow \alpha = \sqrt{\lambda_0}, \quad \lambda_0 = \alpha^2, \\ F_2 = 0 = \lambda_0 F_1 = \lambda_0 \times 0 = 0 \Rightarrow 0 = 0, \\ G_2 = R^2\text{sh}^2(\alpha u) = \lambda_0 G_1 = \lambda_0 R^2\text{sh}^2(u) \Rightarrow \text{sh}(\alpha u) = \sqrt{\lambda_0}\text{sh}(u). \end{cases} \quad (3.43)$$

Now we substitute $\alpha = \sqrt{\lambda_0}$ into the relation $\text{sh}(\alpha u) = \sqrt{\lambda_0}\text{sh}(u)$ As α ($\alpha > 0$, $\alpha \neq 1$) is any constant, and $u(0 < u < +\infty)$ is a variable, then we get the following identity:

$$\text{sh}(\alpha u) \equiv \alpha\text{sh}(u). \quad (3.44)$$

The identity (3.44) must also be satisfied for the case where $\alpha = 2$, $u = 1$. As $\text{sh}(2 \times 1) = 3.626$, $\text{sh}(1) = 0.841$, $2 \times 0.841 = 1.682$, then for this case we get from (3.44): $3.626 = 1.682$, which is impossible.

We now show that the point mapping $f_\alpha\colon M^2 \mapsto M^2$ of the form (3.41) for the pseudo-sphere M^2 is **not equiareal**, that is, it does not preserve areas. First assume the opposite, that it is equiareal. Then, according to the necessary and sufficient conditions for the preservation of areas (see Table 3.1), the relations (3.25) $E_2 G_2 - (F_2)^2 = E_1 G_1 - (F_1)^2$ must be satisfied.

As $E_1 = R^2$, $F_1 = 0$, $G_1 = R^2\text{sh}^2(u)$, $E_2 = R^2\alpha^2$, $F_2 = 0$, $G_2 = R^2\text{sh}^2(\alpha u)$, then, taking into consideration the conditions $\alpha > 0$, $\alpha \neq 1$, $0 < u < +\infty$, we get the following identities:

$$R^4\alpha^2\text{sh}^2(\alpha u) \equiv R^4\text{sh}^2(u) \Rightarrow \alpha\text{sh}(\alpha u) \equiv \text{sh}(u). \quad (3.45)$$

The identity (3.45) must also be satisfied for the case where $\alpha = 2$, $u = 1$. As $\text{sh}(2 \times 1) = 3.626$, $\text{sh}(1) = 0.841$, $2 \times 3.626 = 7.252$, then for this case we get from (3.45) that $7.252 = 0.841$, which is impossible.

By performing a direct verification with formula (3.12), we find that for all $\alpha > 0$, $u > 0$ in the metric form (3.39) the Gaussian curvature is preserved and is equal to $K = -\frac{1}{R^2} < 0$. Note that when $\alpha = 1$, the metric form (3.39) coincides with Lobachevsky's metric form (3.30).

The point mapping $f_\alpha\colon M^2 \mapsto M^2$ of the form (3.41) on the half-plane $D^2(u,v)$, $0 < u < +\infty, -\infty < v < +\infty$ corresponds to the transformation $\bar{f}_\alpha\colon (u,v) \mapsto (\alpha u, v)$. Hence, we get the following connection between differentials:

$$\begin{pmatrix} d(\alpha u) \\ dv \end{pmatrix} = \begin{pmatrix} \alpha & 0 \\ 0 & 1 \end{pmatrix} \begin{pmatrix} du \\ dv \end{pmatrix}$$

For the case $u = 0$, the metric forms (3.30) and (3.39) degenerate, because for the case where $u = 0$ their coefficients G_1 and G_2 vanish. Therefore the case $u = 0$ is the absolute for the metric forms (3.30) and (3.39).

The absolute $u = 0$ of the metric forms (3.30) and (3.32) corresponds on the pseudo-sphere M^2 to the fixed points $(X^* = 0, Y^* = 0, Z^* = R)$ of the point mapping $f_\alpha : M^2 \mapsto M^2$, and vice versa.

The geodesic lines L on the pseudo-sphere M^2: $Z^2 - X^2 - Y^2 - R^2 = 0$; $Z \geq R > 0$ are the intersection lines of this pseudo-sphere with the following planes Π^2_*: $aX + bY + cZ = 0$, where $a^2 + b^2 + c^2 > 0$.

3.5.6.   Normalized distance between Lobachevsky's metric form (3.30) and the metric form (3.39)

Since the Gaussian curvature $K = -\frac{1}{R^2} < 0$ is preserved for any value of $\alpha > 0$, we can then introduce the concept of normalized distance $\bar{\rho}_{12}$ between Lobachevsky's metric form (3.30) and the metric form (3.39), which does not depend on the particular values of the Gaussian curvature $K = -\frac{1}{R^2} < 0$. We define the normalized distance as follows:

$$\bar{\rho}_{12} = |\alpha - 1| \geq 0, \quad \alpha > 0. \tag{3.46}$$

The normalized distance $\bar{\rho}_{12}$ has the following properties:

(1) For the case $\bar{\rho}_{12} = 0$, the parameter $\alpha = 1$, that is, the metric form (3.39) coincides with Lobachevsky's metric form (3.30).
(2) For the case $\bar{\rho}_{12} > 0$, the parameter $\alpha \in (0, 1) \cup (1, +\infty)$, and then the metric form (3.39) does not coincide with Lobachevsky's metric form (3.30).
(3) For the case $\bar{\rho}_{12} \to 0$, the parameter $\alpha \to 1$, that is, the metric forms (3.39) are reduced in the limit to Lobachevsky's metric form (3.30).

Write the distance $\bar{\rho}_{12}$ (3.46) by using the metallic proportions:

$$\Phi_\lambda = \frac{\lambda + \sqrt{4 + \lambda^2}}{2}, \quad \lambda > 0.$$

To do this, we set the parameter $\alpha = \ln(\Phi_\lambda)$ and introduce the normalized distance $\bar{\rho}_{12} = |\ln(\Phi_\lambda) - 1|$ (Fig. 3.10).

Table 3.2 gives the numerical values of the metallic proportions Φ_λ and the normalized distances $\bar{\rho}_{12}$ for the natural values of $\lambda = 1, 2, 3, 4, \ldots$. We can see that the metric form (3.39), based on the silver proportion $\Phi_{\lambda=2} = 1 + \sqrt{2} \approx 2.414$, is closest to Lobachevsky's metric form (3.30); for

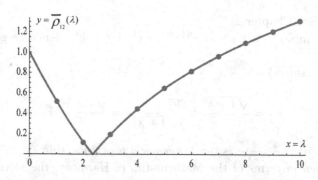

Figure 3.10.   The graph of the function $\bar\rho_{12}$.

Table 3.2.   Metallic proportions and normalized distances.

λ	Name of metallic proportions	Φ_λ	Approximate values of Φ_λ	The normalized distance $\bar\rho_{12}$
1	Golden	$\dfrac{1+\sqrt5}{2}$	1.618	0.518
2	**Silver**	$\mathbf{1+\sqrt2}$	**2.414**	**0.118**
3	Bronze	$\dfrac{3+\sqrt{13}}{2}$	3.303	0.194
4	Copper	$2+\sqrt5$	4.236	0.443

the case of the silver proportion ($\lambda = 2$) the distance of the metric form (3.39) to Lobachevsky's metric form (3.30) is $\bar\rho_{12} = 0.118$ (see Table 3.2).

Note that if $\bar\rho_{12} = 0$, then $\lambda = 2\operatorname{sh}(1) \approx 2.3504$, $\Phi_\lambda = e \approx 2.7182$, $\alpha = \ln(\Phi_\lambda = e) = 1$. In this case, the metric form (3.39) coincides with Lobachevsky's metric form (3.30).

3.5.7. *Hyperbolic solution of Hilbert's Fourth Problem in the language of metallic proportions and λ-Fibonacci hyperbolic functions*

We represent the metric form (3.39) $(ds_2)^2 = R^2[\alpha^2(du)^2 + \operatorname{sh}^2(\alpha u)(dv)^2]$ by using the notions of the metallic proportions and λ-*Fibonacci* hyperbolic functions (see Chapter 2).

As $\alpha = \ln(\Phi_\lambda)$, then $e^{\alpha u} = e^{u \ln \Phi_\lambda} = e^{\ln(\Phi_\lambda^u)} = \Phi_\lambda^u$. Hence we get:

$$\mathrm{sh}(\alpha u) = \frac{e^{\alpha u} - e^{-\alpha u}}{2} = \frac{\Phi_\lambda^u - \Phi_\lambda^{-u}}{2}$$

$$= \frac{\sqrt{4 + \lambda^2}}{2} \times \frac{\Phi_\lambda^u - \Phi_\lambda^{-u}}{\sqrt{4 + \lambda^2}} = \frac{\sqrt{4 + \lambda^2}}{2} \mathrm{sF}_\lambda(u) \qquad (3.47)$$

where $\mathrm{sF}_\lambda(u) = \frac{\Phi_\lambda^u - \Phi_\lambda^{-u}}{\sqrt{4 + \lambda^2}}$ is the λ-*Fibonacci hyperbolic sine*.

But then in terms of the Mathematics of Harmony, the metric forms (3.39) with $\bar{\rho}_{12} \geq 0$ take the following form:

$$(ds_2)^2 = R^2[\alpha^2 (du)^2 + \mathrm{sh}^2(\alpha u)\, (dv)^2]$$

$$= R^2 \left[\ln^2(\Phi_\lambda)(du)^2 + \frac{4 + \lambda^2}{4}(\mathrm{sF}_\lambda(u))^2 (dv)^2 \right]. \qquad (3.48)$$

Substitute $\alpha = \ln(\Phi_\lambda)$ into (3.40), then for every $\lambda > 0$ on the pseudo-sphere M^2 there is a nonlinear one-to-one point mapping of the form (3.41)

$$f_\lambda^* = f_{\alpha = \ln(\Phi_\lambda)}\colon M^2 \mapsto M^2.$$

When we change the parameter $\lambda > 0$, the metric form (3.40) induces an infinite number of new geometries on the pseudo-sphere M^2: $Z^2 - X^2 - Y^2 - R^2 = 0$; $Z \geq R > 0$. These geometries preserve the Gaussian curvature $K = -\frac{1}{R^2} < 0$, but do not preserve the other metric values as found in Lobachevsky's metric (3.30):

$$(ds_1)^2 = R^2[(du)^2 + \mathrm{sh}^2(u)(dv)^2]$$

for the case where $\bar{\rho}_{12} = 0$ ($\lambda = 2\mathrm{sh}(1)$).

In this sense, for the case where $\bar{\rho}_{12} = |\ln(\Phi_\lambda) - 1| > 0$ ($\lambda > 0, \lambda \neq 2\mathrm{sh}(1)$) the metric forms (3.48):

$$(ds_2)^2 = R^2 \left[\ln^2(\Phi_\lambda)\, (du)^2 + \frac{4 + \lambda^2}{4}(\mathrm{sF}_\lambda(u))^2 (dv)^2 \right]$$

are a hyperbolic pseudo-spherical solution to Hilbert's Fourth Problem.

3.5.8.   Recursive hyperbolic functions and geometry

The above solution to Hilbert's Fourth Problem selects from among an infinite number of hyperbolic geometries (defined by (3.39)) a special class of hyperbolic geometries (defined by (3.48)) named "harmonic" hyperbolic geometries. The peculiarity of these hyperbolic geometries is that their bases are the "harmonic" proportions, in particular, the famous golden ratio $\Phi_1 = \frac{1+\sqrt{5}}{2}$ (widely known in the ancient world) and the metallic proportions $\Phi_\lambda = \frac{\lambda+\sqrt{4+\lambda^2}}{2}$, $\lambda > 0$, which are generalizations of the golden ratio and are reduced to it for the case $\lambda = 1$. The subset of such hyperbolic geometries, corresponding to the discrete values of $\lambda = 1, 2, 3, 4, \ldots$, is of particular interest because it reveals the fundamental relationship of these hyperbolic geometries with a new class of recurrent integer sequences, known as λ-Fibonacci numbers, which are reduced to the classic Fibonacci numbers for the case $\lambda = 1$. Fibonacci numbers are, of course, extremely widespread throughout nature; in particular, they underlie the botanical phenomenon of phyllotaxis [33].

In contrast to classic hyperbolic geometry, the "harmonic" hyperbolic geometry, corresponding to the metric form (3.48):

$$(ds_2)^2 = R^2 \left[\ln^2(\Phi_\lambda)(du)^2 + \frac{4+\lambda^2}{4}(\mathrm{sF}_\lambda(u))^2 \right],$$

has recursive properties, which brings the "harmonic" hyperbolic geometry in line with mathematical, computer and natural structures. Therefore, the above "harmonic" hyperbolic functions and geometries can rightfully be called recursive hyperbolic functions and geometries.

As is known, recursion underlies fractals. Here are some objects of animate and inanimate nature having fractal properties. Examples from animate nature include corals, starfish and sea urchins, sea shells, flowers and plants (broccoli, cabbage), fruits (pineapple), trees and leaves, the circulatory system and bronchi of humans and animals. Examples from inanimate nature include the boundaries of geographic objects (countries, regions, cities), coastlines, mountain ranges, snowflakes, clouds, lightning, crystals, stalactites, stalagmites. In physics, fractals naturally arise in the modeling of nonlinear processes such as the turbulent flow of fluids, the complex processes of diffusion-absorption, flames, clouds, and so on. Fractals are used in the modeling of porous materials, such as petrochemicals. In biology, they are used to simulate the population and to describe systems of internal organs (e.g. the vascular system). In computer science, there are algorithms for image compression using fractals. They are based upon the idea that the image can be stored by a contractive mapping for which this image is the fixed point. This list, far from complete, includes areas of mathematics, science and nature, in which the recursive functions and hyperbolic geometries can be employed as models. In essence, the above recursive hyperbolic geometries are a breakthrough in the field of hyperbolic geometry, derived from the Mathematics of Harmony.

In conclusion, we can only express our admiration for the brilliant intuition of David Hilbert, who in his famous lecture "Mathematical problems", made at the 2nd Congress of Mathematicians in Paris (1900), addressed the "Idea of a Pre-established Harmony" and drew a parallel between his "Mathematical problems" and Klein's "Icosahedral Idea" [10]. Hearkening back to the ancient golden ratio, Platonic solids and Euclid's *Elements*, he thereby foreshadowed the creation of recursive hyperbolic geometries, as described in this chapter.

3.6.   Spherical Fibonacci Functions

3.6.1.   Preliminaries

In this section from the point of view of the Mathematics of Harmony, we will study the geometric and metric properties of the two-dimensional sphere not only as a surface in three-dimensional Euclidean space, but also as a two-dimensional manifold associated with its various parametrizations. The result of comparing these properties with Lobachevsky's geometry, induced by a similar study of the pseudo-sphere in Minkowski's three-dimensional space, allows us to talk about the new solution to Hilbert's Fourth Problem for spherical geometry.

General concepts regarding the history of spherical geometry and a modern view of this geometry as non-Euclidean geometry with a positive Gaussian curvature are presented in the book *Modern Geometry. Methods and Applications* [1] and the article "Spherical geometry" [35]. Spherical geometry is not only of theoretical significance but also has many practical applications in various branches of the natural sciences including cartography, navigation and astronomy.

3.6.2.   Standard and perturbed spherical metric forms

A sphere S^2 of the radius $R > 0$ with the center $(0, 0, 0)$ in the Euclidean space $R^3 = (X, Y, Z)$ is defined by the equation:

$$S^2:\ X^2 + Y^2 + Z^2 = R^2. \tag{3.49}$$

Euclidean metric in the space $R^3 = (X, Y, Z)$ has the following form:

$$(dl)^2 = (dX)^2 + (dY)^2 + (dZ)^2, \tag{3.50}$$

in which dl is an element of length in this space.

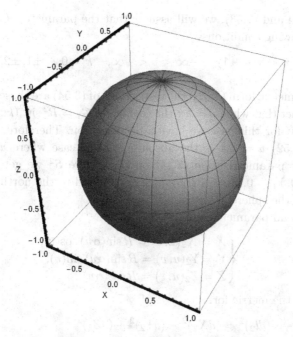

Figure 3.11. Sphere S^2: $X^2 + Y^2 + Z^2 = R^2$.

Consider the two parametrizations of the sphere S^2, induced by the metric (3.50):

(1) The first parametrization

$$\begin{cases} X = X_1(u, v) = R\sin(u)\cos(v), \\ Y = Y_1(u, v) = R\sin(u)\sin(v), \\ Z = Z_1(u, v) = R\cos(u) \end{cases} \qquad (3.51)$$

induces the metric form:

$$(dl_1)^2 = (dX_1)^2 + (dY_1)^2 + (dZ_1)^2 = (ds_1)^2 = R^2[(du)^2 + \sin^2(u)(dv)^2].$$

This metric form:

$$(ds_1)^2 = R^2[(du)^2 + \sin^2(u)(dv)^2] \qquad (3.52)$$

is called the standard spherical metric form. This form has the following coefficients:

$$E_1 = R^2, \quad F_1 = 0, \quad G_1 = R^2\sin^2(u). \qquad (3.53)$$

In (3.52) and (3.53), we will assume that the parameters (u, v) satisfy the following conditions:

$$\pi k < u < \pi(k + 1), \quad -\infty < v < +\infty, \quad k = 0, \quad \pm 1, \pm 2, \pm 3, \ldots$$
$$(3.54)$$

For the metric form (3.52), the restrictions of (3.54) are chosen by virtue of the fact that when $u = k\pi$ the coefficient $G_1 = R^2 \sin^2(k\pi) = 0$, that is to say, for this case we have a degeneration. Therefore, the metric form (3.52) $u = k\pi$ is the absolute. In the case where $u = k\pi$ for the first parametrization (3.44) on the sphere S^2 we get the points $(X_1 = 0, Y_1 = 0, Z_1 = \pm R)$, which correspond to the north and south poles of the sphere.

(2) The second parametrization

$$\begin{cases} X = X_2(u, v) = R \sin(\alpha u) \cos(v), \\ Y = Y_2(u, v) = R \sin(\alpha u) \sin(v), \\ Z = Z_2(u, v) = R \cos(\alpha u) \end{cases} \quad (3.55)$$

induces the metric form:

$$(dl_2)^2 = (dX_2)^2 + (dY_2)^2 + (dZ_2)^2$$
$$= (ds_2)^2 = R^2[\alpha^2(du)^2 + \sin^2(\alpha u)(dv)^2].$$

This metric form:

$$(ds_2)^2 = R^2[\alpha^2(du)^2 + \sin^2(\alpha u)(dv)^2] \quad (3.56)$$

has the following coefficients:

$$E_2 = \alpha^2 R^2, \quad F_2 = 0, \quad G_2 = R^2 \sin^2(\alpha u). \quad (3.57)$$

The metric form (3.56) is called the perturbed spherical metric form.

In (3.56) we will assume that the parameters (u, v) satisfy the following conditions:

$$\frac{k}{\alpha}\pi < u < \frac{k + 1}{\alpha}\pi, \quad -\infty < v < +\infty,$$
$$\alpha > 0, \quad \alpha \neq 1, \quad k = 0, \pm 1, \pm 2, \ldots. \quad (3.58)$$

For the metric form (3.56), the restrictions (3.58) are chosen by virtue of the fact that for the case $u = \frac{k}{\alpha}\pi$ the coefficient $G_2 = R^2 \sin^2(\alpha \times \frac{k}{\alpha}\pi) = 0$. Therefore, for the perturbed metric form (3.56) the case $u = \frac{k}{\alpha}\pi$ is an absolute. On the sphere S^2 for the case $u = \frac{k}{\alpha}\pi$ at the second parametrization (3.55) we again get the points $(X_2 = 0, Y_2 = 0, Z_2 = \pm R)$, which correspond to the north and south poles of the sphere.

Consider the following system of equations:

$$\begin{cases} X_1 = R\sin(u)\cos(v), & Y_1 = R\sin(u)\sin(v), & Z_1 = R\cos(u), \\ X_2 = R\sin(\alpha u)\cos(v), & Y_2 = R\sin(\alpha u)\sin(v), & Z_2 = R\cos(\alpha u). \end{cases}$$
$$(3.59)$$

Exclude the parameters (u, v) from the equations (3.59), then on the sphere S^2 for every $\alpha > 0$ we get the one-to-one point mapping (cascade) f_α: $S^2 \mapsto S^2$ of the form:

$$X_2 = \frac{\sin(\alpha u)}{\sin(u)} X_1, \quad Y_2 = \frac{\sin(\alpha u)}{\sin(u)} Y_1, \quad Z_2 = \frac{\cos(\alpha u)}{\cos(u)} Z_1, \quad (3.60)$$

where

$$u = \text{Arccos}\left(\frac{Z_1}{R}\right) = \pm \arccos\left(\frac{Z_1}{R}\right) + 2\pi k,$$

$$|Z_1| \le R, \quad k = 0, \pm 1, \pm 2, \dots. \quad (3.61)$$

Note that the points $(X^* = 0, Y^* = 0, Z^* = \pm R)$ are the fixed points of the point mapping (3.60) f_α: $S^2 \mapsto S^2$. Also note that the geodesic line L on the sphere S^2: $X^2 + Y^2 + Z^2 = R^2$ is obtained at the intersection of sphere S^2 with the planes $ax + by + cz = 0$, $a^2 + b^2 + c^2 > 0$.

Here we demonstrate that the point mapping f_α: $S^2 \mapsto S^2$ of the form (3.60) on the sphere S^2 is **not isometric**, that is, it does not preserve lengths.

Suppose that the opposite is true, that it is isometric. Then according to the necessary and sufficient conditions for the preservation of lengths (see Table 3.1), the following relations (3.23) must be satisfied:

$$E_2 = E_1, \quad F_2 = F_1, \quad G_2 = G_1.$$

But then in our situation, when $\frac{k}{\alpha}\pi < u < \frac{k+1}{\alpha}\pi, -\infty < v < +\infty, \alpha > 0$, $\alpha \ne 1$, $k = 0, \pm 1, \pm 2, \dots$ for the metric forms:

$$(ds_1)^2 = R^2[(du)^2 + \sin^2(u)(dv)^2],$$
$$(ds_2)^2 = R^2[\alpha^2(du)^2 + \sin^2(\alpha u)(dv)^2]$$

with the coefficients:

$$E_1 = R^2, \quad F_1 = 0, \quad G_1 = R^2\sin^2(u),$$
$$E_2 = \alpha^2 R^2, \quad F_2 = 0, G_2 = R^2\sin^2(\alpha u),$$

the relations (3.23) must be satisfied. From the first relation in (3.23), $E_2 = E_1$, we obtain in this case: $E_2 = \alpha^2 R^2 = E_1 = R^2 \Rightarrow \alpha = 1$ which is impossible.

Next we show that the point mapping f_α: $S^2 \mapsto S^2$ of the form (3.60) on the sphere S^2 is **not conformal.**

Suppose the opposite is true, that it is conformal. Then, according to the necessary and sufficient conditions for the preservation of angles (see Table 3.1), the following relations (3.24) would have to be satisfied:

$$E_2 = \lambda_0 E_1, \quad F_2 = \lambda_0 F_1, \quad G_2 = \lambda_0 G_1, \quad \lambda_0(u, v) > 0.$$

But then by virtue of (3.53) and (3.57), we get the following identities:

$$\begin{cases} E_2 = \alpha^2 R^2 = \lambda_0 E_1 = \lambda_0 R^2 \Rightarrow \lambda_0 = \alpha^2, \\ F_2 = 0 = \lambda_0 F_1 = \lambda_0 \times 0 = 0 \Rightarrow 0 = 0, \\ G_2 = R^2 \sin^2(\alpha u) \equiv \lambda_0 G_1 \equiv \lambda_0 R^2 \sin^2(u). \end{cases} \quad (3.62)$$

Since $\lambda_0 = \alpha^2$, $\frac{k}{\alpha}\pi < u < \frac{k+1}{\alpha}\pi$, $\alpha > 0$, $\alpha \neq 1$, $k = 0, \pm 1, \pm 2, \ldots$ then the last relation in (3.62) leads to the identity:

$$|\sin(\alpha u)| \equiv \alpha|\sin(u)| > 0. \quad (3.63)$$

Since $G_1 = R^2 \sin^2(u)$ is the coefficient of the metric form (3.52), then $u \neq \pi n$, $n = 0, \pm 1, \pm 2, \ldots$. Therefore, the admissible values for the identity (3.63) are determined by the following relations $\alpha u \neq \pi k$, $u \neq \pi n$, $\alpha > 0$, $\alpha \neq 1$, where k, n are any integers.

In particular, the identity (3.63) must also be satisfied for the case $u = 1$, $\alpha = 2$. Then from (3.63) we get:

$$|\sin(2 \times 1)| \equiv 2|\sin(1)| \Rightarrow 0.909297 = 2 \times 0.84141471 = 1.68294,$$

which is impossible.

We show that for the metric form (3.56) under the conditions $\alpha u \neq \pi k, -\infty < v < +\infty$, $\alpha > 0$, $k = 0, \pm 1, \pm 2, \ldots$, the Gaussian curvature equals $K = \frac{1}{R^2} > 0$.

The coefficients of the metric form (3.56) have the form (3.57) $E_2 = \alpha^2 R^2$, $F_2 = 0$, $G_2 = R^2 \sin^2(\alpha u)$. From this it follows that:

$$A = \sqrt{E_2(u, v)} = R\alpha > 0,$$

$$B = \sqrt{G_2(u, v)} = R|\sin(\alpha u)| = R\sigma \sin(\alpha u) > 0,$$

where $\sigma = 1$, if $\sin(\alpha u) > 0$, and $\sigma = -1$, if $\sin(\alpha u) < 0$.

For this given case, the Gaussian curvature is determined by the formula:

$$K = K(u, v) = -\frac{1}{AB}\left[\frac{\partial}{\partial u}\left(\frac{\frac{\partial B}{\partial u}}{A} \right) + \frac{\partial}{\partial v}\left(\frac{\frac{\partial A}{\partial v}}{B} \right) \right].$$

Since $B = B(u)$ and $A = \text{const}$, then K takes the following form:

$$K = -\frac{1}{A^2 B} \times \frac{d^2 B}{(du)^2}.$$

We then get:

$$\frac{dB}{du} = R\alpha\sigma\cos(\alpha u),$$

$$\frac{d^2 B}{(du)^2} = \frac{d}{du}\left(\frac{dB}{du}\right) = \frac{d}{du}(R\alpha\sigma\cos(\alpha u)) = -R\alpha^2\sigma\sin(\alpha u).$$

Hence we have:

$$K = -\frac{1}{A^2 B} \times \frac{d^2 B}{(du)^2} = -\left(\frac{1}{(R\alpha)^2(R\sigma\sin(\alpha u))}\right)$$

$$\times (-R\alpha^2\sigma\sin(\alpha u)) = \frac{1}{R^2} > 0.$$

3.6.3. *Normalized distance between the standard spherical metric form (3.52) and the spherical metric forms (3.56)*

For any $\alpha u \neq \pi k$, $u \neq \pi n$, $\alpha > 0$, where k, n are any integers, the Gaussian curvature of the spherical metric forms (3.56) is preserved and is equal to $K = \frac{1}{R^2} > 0$.

The coefficients of the metric form (3.56) have the form (3.57) $E_2 = \alpha^2 R^2$, $F_2 = 0$, $G_2 = R^2\sin^2(\alpha u)$. From this it follows that:

$$A = \sqrt{E_2(u,v)} = R\alpha > 0,$$

$$B = \sqrt{G_2(u,v)} = R|\sin(\alpha u)| = R\sigma\sin(\alpha u) > 0,$$

where $\sigma = 1$ if $\sin(\alpha u) > 0$, and $\sigma = -1$ if $\sin(\alpha u) < 0$.

For the case $\alpha = 1$, the metric form (3.56) coincides with the standard spherical metric form (3.52). Therefore, we can also introduce the concept of normalized distance between the metric forms (3.56) and (3.52), that is $\bar{\rho}_{12} = |\alpha - 1|$, which does not depend on the particular values of the Gaussian curvature $K = \frac{1}{R^2} > 0$.

The normalized distance $\bar{\rho}_{12}$ has the same properties as the normalized distance for the pseudo-spherical metric forms. Only we replace the word *pseudo-sphere* with the word *sphere* and the Gaussian curvature $K = -\frac{1}{R^2} < 0$ with $K = \frac{1}{R^2} > 0$.

In particular, if we write the distance $\bar{\rho}_{12}$ by using the notion of the hyperbolic metallic proportions $\Phi_\lambda = \frac{\lambda+\sqrt{4+\lambda^2}}{2} > 0$, $\lambda > 0$, we then get:

$$\bar{\rho}_{12} = |\ln(\Phi_\lambda) - 1|.$$

It follows from Table 3.2, that for the case of integer values $\lambda = 1, 2, 3, 4, \ldots$ the metric form (3.56), based on the silver proportion, is closest to the standard spherical metric form (3.52). In this case $\bar{\rho}_{12} = 0.118$ (see Table 3.2).

For the case $\bar{\rho}_{12} = 0$ the spherical metric form (3.56) coincides with the standard spherical metric form (3.52).

3.6.4. Spherical λ-Fibonacci sine and cosine

In Section 3.5.7, a solution to Hilbert's Fourth Problem was set forth for the pseudo-sphere by using the λ-Fibonacci hyperbolic sine,

$$sF_\lambda(u) = \frac{\Phi_\lambda^u - \Phi_\lambda^{-u}}{\sqrt{4 + \lambda^2}},$$

where $\Phi_\lambda = \frac{\lambda+\sqrt{4+\lambda^2}}{2}$.

Therefore, before we derive the solution to Hilbert's Fourth Problem for the sphere, we introduce the following important concepts.

Spherical metallic proportions.   The spherical metallic proportion is identified as the complex number:

$$\tau_\lambda = \Phi_\lambda^i = \cos(\ln \Phi_\lambda) + i\sin(\ln \Phi_\lambda), \quad \lambda \neq 0, \quad i = \sqrt{-1}. \tag{3.64}$$

Note that because $\text{Re}(\tau_\lambda) = \cos(\ln \Phi_\lambda)$, $\text{Im}(\tau_\lambda) = \sin(\ln \Phi_\lambda)$, then the spherical metallic proportion can also be interpreted as a pair of real numbers, which have real and imaginary components of the complex number (3.64). We write this complex number as follows:

$$\tau_\lambda = [\cos(\ln \Phi_\lambda), \quad \sin(\ln \Phi_\lambda)], \quad \lambda \neq 0. \tag{3.65}$$

The graphs of the functions $\text{Re}(\tau_\lambda) = \cos(\ln \Phi_\lambda)$ and $\text{Im}(\tau_\lambda) = \sin(\ln \Phi_\lambda)$ are presented in Figs. 3.12 and 3.13.

We call the spherical metallic proportions of (3.64) the golden, silver, bronze, and copper proportions for the cases where $\lambda = 1, 2, 3, 4$ in (3.65), respectively.

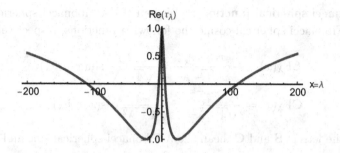

Figure 3.12.   Graph of the function $\mathrm{Re}(\tau_\lambda) = \cos(\ln \Phi_\lambda)$.

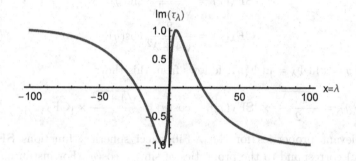

Figure 3.13.   Graph of the function $\mathrm{Im}(\tau_\lambda) = \sin(\ln \Phi_\lambda)$.

We introduce the designation $\tau_1 = \tau$. Then $\Phi_1 = \Phi$ is the golden proportion. By using Φ_λ and (3.65), at $\lambda = 1, 2, 3, 4$ we get the following numerical values for the above spherical metallic proportions (3.65):

golden spherical proportion:
$$\tau = [\cos(\ln \Phi),\ \sin(\ln \Phi)]$$
$$\approx (0.8864,\ 0.4628)$$

silver spherical proportion:
$$\tau_2 = [\cos(\ln \Phi_2),\ \sin(\ln \Phi_2)]$$
$$\approx (0.6969,\ 0.07171)$$

bronze spherical proportion:
$$\tau_3 = [\cos(\ln \Phi_3),\ \sin(\ln \Phi_3)]$$
$$\approx (0.5081,\ 0.8612)$$

copper spherical proportion:
$$\tau_4 = [\cos(\ln \Phi_4),\ \sin(\ln \Phi_4)]$$
$$\approx (0.3361,\ 0.9418).$$

λ-Fibonacci spherical functions. We call the λ-Fibonacci spherical sine and λ-Fibonacci spherical cosine the following functions, respectively:

$$\mathbf{SF}_\lambda(x) = -i\frac{\tau_\lambda^x - \tau_\lambda^{-x}}{\sqrt{4+\lambda^2}} = \frac{2}{\sqrt{4+\lambda^2}}\sin(x\ln\Phi_\lambda) \qquad (3.66)$$

$$\mathbf{CF}_\lambda(x) = \frac{\tau_\lambda^x + \tau_\lambda^{-x}}{\sqrt{4+\lambda^2}} = \frac{2}{\sqrt{4+\lambda^2}}\cos(x\ln\Phi_\lambda) \qquad (3.67)$$

where the letters **S** and **C** mean the λ-Fibonacci spherical sine and cosine, respectively.

We rewrite the formulas (3.66) and (3.67) as follows:

$$\mathbf{SF}_\lambda(x) = \frac{2}{\sqrt{4+\lambda^2}}\sin(y), \qquad (3.68)$$

$$\mathbf{CF}_\lambda(x) = \frac{2}{\sqrt{4+\lambda^2}}\cos(y), \qquad (3.69)$$

where $y = x\ln\Phi_\lambda = \ln\Phi_\lambda^x$. It follows from this that:

$$\sin(y) = \frac{\sqrt{4+\lambda^2}}{2}\times[\mathbf{SF}_\lambda(x)], \quad \cos(y) = \frac{\sqrt{4+\lambda^2}}{2}\times[\mathbf{CF}_\lambda(x)]. \quad (3.70)$$

Relevant properties for the λ-Fibonacci spherical functions $\mathbf{SF}_\lambda(x)$, $\mathbf{CF}_\lambda(x)$ correspond to the properties of $\sin(y)$, $\cos(y)$. For instance, when we consider a well-known property of the classic trigonometric functions $\sin(y)$, $\cos(y)$ (Pythagorean theorem) in comparison with the similar property for the λ-Fibonacci spherical functions, we get:

$$\sin^2(y) + \cos^2(y) = 1 \Leftrightarrow [\mathbf{SF}_\lambda(x)]^2 + [\mathbf{CF}_\lambda(x)]^2 = \frac{4}{4+\lambda^2}.$$

We now compare the λ-Fibonacci spherical functions (3.67) and (3.68) (λ-Fibonacci spherical sine $\mathbf{SF}_\lambda(x)$ and cosine $\mathbf{CF}_\lambda(x)$) with the corresponding λ-Fibonacci hyperbolic functions (λ-Fibonacci hyperbolic sine $\mathrm{sF}_\lambda(x)$ and cosine $\mathrm{cF}_\lambda(x)$).

As Stakhov has established in [17, 18], the λ-Fibonacci hyperbolic sine $\mathrm{sF}_\lambda(x)$ and the λ-Fibonacci hyperbolic cosine $\mathrm{cF}_\lambda(x)$ are defined by the following formulas:

$$\mathrm{sF}_\lambda(x) = \frac{\Phi_\lambda^x - \Phi_\lambda^{-x}}{2} = \frac{2}{\sqrt{4+\lambda^2}}\mathrm{sh}(x\ln\Phi_\lambda) \qquad (3.71)$$

$$\mathrm{cF}_\lambda(x) = \frac{\Phi_\lambda^x + \Phi_\lambda^{-x}}{2} = \frac{2}{\sqrt{4+\lambda^2}}\mathrm{ch}(x\ln\Phi_\lambda). \qquad (3.72)$$

From this, the following relations result between the λ-Fibonacci spherical functions and the λ-Fibonacci hyperbolic functions:

$$\mathrm{SF}_\lambda(x) = \frac{\sin(x\ln\Phi_\lambda)}{\mathrm{sh}(x\ln\Phi_\lambda)}\mathrm{sF}_\lambda(x), \quad \mathrm{CF}_\lambda(x) = \frac{\cos(x\ln\Phi_\lambda)}{\mathrm{ch}(x\ln\Phi_\lambda)}\mathrm{cF}_\lambda(x),$$

$$\text{(3.73)}$$

$$\mathrm{sF}_\lambda(x) = \frac{\mathrm{sh}(x\ln\Phi_\lambda)}{\sin(x\ln\Phi_\lambda)}\mathrm{SF}_\lambda(x), \quad \mathrm{cF}_\lambda(x) = \frac{\mathrm{ch}(x\ln\Phi_\lambda)}{\cos(x\ln\Phi_\lambda)}\mathrm{CF}_\lambda(x).$$

$$\text{(3.74)}$$

The formulas (3.73) and (3.74) allow us to find all relationships for the spherical Mathematics of Harmony by using the λ-Fibonacci hyperbolic functions. Below we demonstrate examples of obtaining the various formulas for the spherical Mathematics of Harmony based on the λ-Fibonacci hyperbolic functions.

We next perform the following replacement $y = x\ln\Phi_\lambda = \ln\Phi_\lambda^x$ in the relations (3.71) and (3.72) which result in:

$$\mathrm{sF}_\lambda(x) = \frac{2}{\sqrt{4+\lambda^2}}\mathrm{sh}(y), \quad \mathrm{cF}_\lambda(x) = \frac{2}{\sqrt{4+\lambda^2}}\mathrm{ch}(y),$$

$$\mathrm{sh}(y) = \frac{\sqrt{4+\lambda^2}}{2}\mathrm{sF}_\lambda(x), \quad \mathrm{ch}(y) = \frac{\sqrt{4+\lambda^2}}{2}\mathrm{cF}_\lambda(x).$$

From the properties of the classic hyperbolic functions $\mathrm{sh}(y)$, $\mathrm{ch}(y)$, the corresponding relations follow for the λ-Fibonacci hyperbolic functions $\mathrm{sF}_\lambda(x)$, $\mathrm{cF}_\lambda(x)$. For example, we consider a well-known property of classic hyperbolic functions in comparison with a similar property for the λ-Fibonacci hyperbolic functions:

$$\mathrm{ch}^2(y) - \mathrm{sh}^2(y) = 1 \Leftrightarrow [\mathrm{cF}_\lambda(x)]^2 - [\mathrm{sF}_\lambda(x)]^2 = \frac{4}{4+\lambda^2}.$$

At the replacement $\alpha = \ln\Phi_\lambda > 0$ ($\lambda > 0$), the parametrized spherical metric form (3.56) can be rewritten as follows:

$$(ds_2)^2 = R^2\left[\ln^2(\Phi_\lambda)(du)^2 + \frac{4+\lambda^2}{4}(\mathrm{SF}_\lambda(u))^2(dv)^2\right], \qquad \text{(3.75)}$$

where $\mathrm{SF}_\lambda(u)$ is the λ-Fibonacci spherical sine, $\frac{k}{\ln(\Phi_\lambda)}\pi < u < \frac{k+1}{\ln(\Phi_\lambda)}\pi$, $-\infty < v < +\infty$, $k \in Z$ (Z is a set of integers).

We call (3.75) the spherical metric λ-form.

3.7.   Spherical Solution to Hilbert's Fourth Problem

By substituting $\alpha = \ln(\Phi_\lambda)$ into (3.60), we conclude that for the case where $\bar{\rho}_{12} > 0$ and $(\lambda > 0)$ on the sphere S^2 there is a nonlinear one-to-one mapping of the form (3.60):

$$f_\lambda^* = f_{\alpha=\ln(\Phi_\lambda)} \colon S^2 \mapsto S^2.$$

If we change the normalized distance $\bar{\rho}_{12} > 0$ (i.e. where the parameter $\lambda > 0$) the metric form (3.75) induces an infinite set of new geometries on the sphere $S^2 \colon X^2 + Y^2 + Z^2 - R^2 = 0$.

These geometries preserve the Gaussian curvature $K = \frac{1}{R^2}$, but do not preserve the rest of the metric values, inherent for the case $\bar{\rho}_{12} = 0$ ($\lambda = 2\mathrm{sh}(1)$) to the standard spherical metric form (3.52) $(ds_1)^2 = R^2[(du)^2 + \sin^2(u)(dv)^2]$.

In this sense, when we change $\bar{\rho}_{12} = |\ln(\Phi_\lambda) - 1| > 0$ ($\lambda > 0$, $\lambda \neq 2\mathrm{sh}(1)$), the metric forms (3.75) are the spherical solution to Hilbert's Fourth Problem.

3.8.   Comparative Table for Hyperbolic and Spherical Solutions of Hilbert's Fourth Problem

Previously, we have published the works [21–24], where we proposed an original solution to Hilbert's Fourth Problem using hyperbolic geometry. This solution includes an endless set of **recursive hyperbolic geometries**, induced by the hyperbolic metric λ-forms of one and the same constant negative Gaussian curvature.

In this chapter we also suggest an original solution to Hilbert's Fourth Problem for the case of spherical geometry. This solution includes an endless set of **recursive spherical geometries**, induced by the spherical metric λ-forms of one and the same constant positive Gaussian curvature.

Both of these solutions are based on the use of the strict and simple formulas of the Mathematics of Harmony [13], which has not only theoretical interest, but also a broad application in the natural sciences.

These two solutions of Hilbert's Fourth Problem are completely correlated with Hilbert's recommendations in finding the solutions to his mathematical problems, which are vaguely formulated. According to Hilbert's recommendations, these solutions, on the one hand, should be strict and simple (the "game with thinking"), and on the other hand, should have, if possible, a practical significance (the "game with experience").

As the solution of Hilbert's Fourth Problem using the spherical metric λ-forms is naturally reduced to the consideration of the induced metric properties of a two-dimensional sphere in three-dimensional space with the Euclidean metric, it will be called the **spherical solution**.

Similarly, because the solution of Hilbert's Fourth Problem using Lobachevsky's hyperbolic metric λ-forms is naturally reduced to the consideration of the induced metric properties of the pseudo-sphere (the upper half of two-sheeted hyperboloid) in three-dimensional space with Minkowski's alternating metrics, it will be called the **pseudo-spherical solution**.

In Table 3.3 we present a comparison of the spherical and pseudo-spherical solutions of Hilbert's Fourth Problem based on the Mathematics of Harmony [13].

3.9.   Searching for New Recursive Hyperbolic and Spherical Worlds of Nature: A New Challenge for the Theoretical Natural Sciences

Thus, the main results of the research described in this chapter are the original solutions to Hilbert's Fourth Problem for the cases of both hyperbolic and spherical geometries and the proofs of the existence of an infinite number of recursive hyperbolic and spherical geometries.

It is important to emphasize that these original solutions to Hilbert's Fourth Problem are based on two fundamental ideas which were introduced when considering hyperbolic and spherical geometries. The first of them is the idea of Harmony, which was developed by prominent Greek philosophers and thinkers, later embodied in Euclid's *Elements*. The second idea is the idea of recursion, which is a distinctive feature of Nature and underlies the recursive hyperbolic and spherical functions.

The new geometric theory of phyllotaxis, created by Oleg Bodnar [33], is a striking example of the practical importance of recursive hyperbolic geometry for studies of Nature. Bodnar demonstrated that the world of phyllotaxis is a specific recursive hyperbolic world, in which its hyperbolicity manifests through Fibonacci spirals on the surface of phyllotaxis objects.

Remember that Bodnar's geometry [33] is based on recursive Fibonacci hyperbolic functions, which were introduced in [15] and are known as the

Table 3.3.   (Steps 1–5). Comparison of the spherical and pseudo-spherical solutions of Hilbert's Fourth Problem.

Step 1	
Spherical solution	Pseudo-spherical solution
Basic surfaces	
Sphere $X^2 + Y^2 + Z^2 = R^2$ in the space (X, Y, Z) with the **Euclidean metrics** $(dl)^2 = (dX)^2 + (dY)^2 + (dZ)^2$.	**Pseudo-sphere**, the upper half of the two-sheeted hyperboloid $Z^2 - X^2 - Y^2 = R^2 > 0, Z \geq R > 0$ in the space (X, Y, Z) with Minkovski's metrics $(dl)^2 = (dZ)^2 - (dX)^2 - (dY)^2$.
Here dl is an element of arc length in the space (X, Y, Z), $R > 0$ is the radius of the sphere, **Gaussian curvature of the** the sphere is equal to $K = \frac{1}{R^2} > 0$.	Here dl is an element of arc length in the space (X, Y, Z), $R > 0$ is the radius of the pseudo-sphere, **Gaussian curvature** of the pseudo-sphere is equal to $K = -\frac{1}{R^2} < 0$.
Parametric form of the basic surfaces	
$X = R\sin(u)\cos(v)$ $Y = R\sin(u)\sin(v)$ $Z = R\cos(u)$	$X = R\,\text{sh}(u)\cos(v)$ $Y = R\,\text{sh}(u)\sin(v)$ $Z = R\,\text{ch}(u)$
Metric basic forms	
Standard spherical metric form of the Gaussian curvature $K = \frac{1}{R^2} > 0$ $(ds_1)^2 = R^2[(du)^2 + \sin^2(u)(dv)^2]$, $\pi k < u < \pi(k+1),\ -\infty < v < +\infty\ (k \in Z)$	Lobachevsky's metric form of the Gaussian curvature $K = -\frac{1}{R^2} < 0$ $(ds_1)^2 = R^2[(du)^2 + \text{sh}^2(u)(dv)^2]$, $0 < u < +\infty,\ -\infty < v < +\infty$

(Continued)

Table 3.3. (*Continued*)

Step 2	
Spherical solution	**Pseudo-spherical solution**
Metallic proportions	

Spherical proportions

$$\tau_\lambda = \Phi_\lambda^i = \cos(\ln\Phi_\lambda) + i\sin(\ln\Phi_\lambda),$$
$$\lambda \neq 0,\; i = \sqrt{-1}$$

Hyperbolic proportions

$$\Phi_\lambda = \frac{\lambda + \sqrt{4+\lambda^2}}{2} = e^{\ln\Phi_\lambda},\; \lambda \neq 0$$

Partial cases of the proportions τ_λ and Φ_λ

$\lambda = 1$: the golden spherical proportion $\tau_1 = \Phi_1^i$
$\lambda = 2$: the silver spherical proportion $\tau_2 = \Phi_2^i$
$\lambda = 3$: the bronze spherical proportion $\tau_3 = \Phi_3^i$
$\lambda = 4$: the copper spherical proportion $\tau_4 = \Phi_4^i$
$\lambda = \lambda^* = 2\,\mathrm{sh}(1) \Rightarrow \tau_{\lambda^*} = e^i$

$\lambda = 1$: the golden hyperbolic proportion Φ_1
$\lambda = 2$: the silver hyperbolic proportion Φ_2
$\lambda = 3$: the bronze hyperbolic proportion Φ_3
$\lambda = 4$: the copper hyperbolic proportion Φ_4
$\lambda = \lambda^* = 2\,\mathrm{sh}(1) \Rightarrow \Phi_{\lambda^*} = e$

λ-sines and λ-cosines

Spherical λ-sine:

$$SF_\lambda(x) = -i\frac{\tau_\lambda^x - \tau_\lambda^{-x}}{\sqrt{4+\lambda^2}} = \frac{2}{\sqrt{4+\lambda^2}}\sin(x\ln\Phi_\lambda)$$

Hyperbolic λ-sine:

$$sF_\lambda(x) = \frac{\Phi_\lambda^x - \Phi_\lambda^{-x}}{\sqrt{4+\lambda^2}} = \frac{2}{\sqrt{4+\lambda^2}}\,\mathrm{sh}(x\ln\Phi_\lambda)$$

Spherical λ-cosine:

$$CF_\lambda(x) = \frac{\tau_\lambda^x + \tau_\lambda^{-x}}{\sqrt{4+\lambda^2}} = \frac{2}{\sqrt{4+\lambda^2}}\cos(x\ln\Phi_\lambda)$$

Hyperbolic λ-cosine:

$$cF_\lambda(x) = \frac{\Phi_\lambda^x + \Phi_\lambda^{-x}}{\sqrt{4+\lambda^2}} = \frac{2}{\sqrt{4+\lambda^2}}\,\mathrm{ch}(x\ln\Phi_\lambda)$$

Basic identity:

$$[CF_\lambda(x)]^2 + [SF_\lambda(x)]^2 = \frac{4}{4+\lambda^2}$$

Basic identity:

$$[cF_\lambda(x)]^2 - [sF_\lambda(x)]^2 = \frac{4}{4+\lambda^2}$$

(*Continued*)

Table 3.3. (Continued)

Step 2	
Spherical solution	**Pseudo-spherical solution**
Parametric forms of the basic surfaces based on λ-sines and λ-cosines	

Spherical solution:

$$X = R\frac{\sqrt{4+\lambda^2}}{2}SF_\lambda(u)\cos(v)$$

$$Y = R\frac{\sqrt{4+\lambda^2}}{2}SF_\lambda(u)\sin(v)$$

$$Z = R\frac{\sqrt{4+\lambda^2}}{2}CF_\lambda(u)$$

Pseudo-spherical solution:

$$X = R\frac{\sqrt{4+\lambda^2}}{2}sF_\lambda(u)\cos(v)$$

$$Y = R\frac{\sqrt{4+\lambda^2}}{2}sF_\lambda(u)\sin(v)$$

$$Z = R\frac{\sqrt{4+\lambda^2}}{2}cF_\lambda(u)$$

Spherical metric forms of the Gaussian curvature $K = \frac{1}{R^2} > 0$

$$(ds_2)^2 = R^2\left[\ln^2\Phi_\lambda(du)^2 + \frac{4+\lambda^2}{4}(SF_\lambda(u))^2(dv)^2\right]$$

$$\frac{\pi k}{\ln\Phi_\lambda} < u < \frac{\pi(k+1)}{\ln\Phi_\lambda}, \ -\infty < v < +\infty \ (k \in Z), \ \lambda > 0$$

Lobachevsky's hyperbolic metric forms of the Gaussian curvature $K = -\frac{1}{R^2} < 0$

$$(ds_2)^2 = R^2\left[\ln^2\Phi_\lambda(du)^2 + \frac{4+\lambda^2}{4}(sF_\lambda(u))^2(dv)^2\right]$$

$$0 < u < +\infty, \ -\infty < v < +\infty, \ \lambda > 0$$

(Continued)

Table 3.3. (*Continued*)

Step 3

Spherical solution	Pseudo-spherical solution
Connection between spherical and hyperbolic λ-functions	

$$SF_\lambda(x) = \frac{\sin(x\ln\Phi_\lambda)}{\operatorname{sh}(x\ln\Phi_\lambda)}\, sF_\lambda(x);$$

$$sF_\lambda(x) = \frac{\operatorname{sh}(x\ln\Phi_\lambda)}{\sin(x\ln\Phi_\lambda)}\, SF_\lambda(x)$$

$$CF_\lambda(x) = \frac{\cos(x\ln\Phi_\lambda)}{\operatorname{ch}(x\ln\Phi_\lambda)}\, cF_\lambda(x);$$

$$cF_\lambda(x) = \frac{\operatorname{ch}(x\ln\Phi_\lambda)}{\cos(x\ln\Phi_\lambda)}\, CF_\lambda(x)$$

Distances between metric forms	

Distance between the standard spherical metric form and spherical metric λ-form:

$$\rho_{12} = R|(\ln\Phi_\lambda) - 1|, \lambda > 0$$

Normalized distance:

$$\bar\rho_{12} = \frac{\rho_{12}}{R} = |(\ln\Phi_\lambda) - 1|, \lambda > 0$$

At $\bar\rho_{12} = 0$

$$\Rightarrow \lambda^* = 2\operatorname{sh}(1) \Rightarrow \Phi_{\lambda^*} = e \Rightarrow \ln\Phi_{\lambda^*} = 1,$$

the spherical metric form coincides with the standard spherical metric form; at $\bar\rho_{12} > 0$, these forms do not coincide.

Distance between Lobachevski's hyperbolic metric form and Lobachevski's metric λ-form:

$$\rho_{12} = R|(\ln\Phi_\lambda) - 1|, \lambda > 0$$

Normalized distance:

$$\bar\rho_{12} = \frac{\rho_{12}}{R} = |(\ln\Phi_\lambda) - 1|, \lambda > 0$$

At $\bar\rho_{12} = 0$

$$\Rightarrow \lambda^* = 2\operatorname{sh}(1) \Rightarrow \Phi_{\lambda^*} = e \Rightarrow \ln\Phi_{\lambda^*} = 1,$$

Lobachevski's metric form coincides with Lobachevski's hyperbolic metric λ-form; at $\bar\rho_{12} > 0$, these forms do not coincide.

The minimal distance	

Comparing the *metallic proportions* with the *normalized distance*, we are led to the conclusion that the *silver proportion* $1+\sqrt{2} \approx 2.414$ ($\lambda = 2$) has the **minimal distance** $\bar\rho_{12} = 0.118$ amongst all metallic proportions Φ_λ for the positive integer values of the parameter $\lambda = 1, 2, 3, \ldots$. This means that among the *spherical metric* λ-forms, generated by the *metallic proportions* Φ_λ for the integer values of $\lambda = 1, 2, 3, \ldots$, the *silver spherical metric* λ-form, generated by the *silver proportion* $1+\sqrt{2} \approx 2.414$ ($\lambda = 2$), is the closest spherical metric λ-form to the standard metric spherical form, in the sense of the normalized distance $\bar\rho_{12}$.

(*Continued*)

Table 3.3.   (*Continued*)

Step 4
Comparison of the metric properties of the mappings f_λ of the sphere S^2: $X^2 + Y^2 + Z^2 = R^2$ and the pseudo-sphere M^2: $Z^2 - Y^2 - X^2 = R^2$, $Z \geq R > 0$ under different parametrizations of the sphere and pseudo-sphere by using factors of induced metric forms

The mappings $f_\lambda\colon S^2 \to S^2$ when comparing the standard spherical parametrization (Step 1) with the spherical λ-parametrizations (Step 2) have the following form:

$$f_\lambda = \left\{ \begin{array}{l} X_2 = \dfrac{\sqrt{4+\lambda^2}}{2} \times \dfrac{\mathrm{SF}_\lambda(u)}{\sin(u)} X_1 \\[2mm] Y_2 = \dfrac{\sqrt{4+\lambda^2}}{2} \times \dfrac{\mathrm{SF}_\lambda(u)}{\sin(u)} Y_1 \\[2mm] Z_2 = \dfrac{\sqrt{4+\lambda^2}}{2} \times \dfrac{\mathrm{CF}_\lambda(u)}{\cos(u)} Z_1 \end{array} \right.$$

where

$$u = \mathrm{Arccos}\left(\frac{Z_1}{R}\right) = \pm \arccos\left(\frac{Z_1}{R}\right) + 2\pi k,$$

$$\lambda > 0; (X_1, Y_1, Z_1) \in S^2; (X_2, Y_2, Z_2) \in S^2.$$

For the case $\lambda^* = 2\mathrm{sh}(1)$ we have:

$$\bar\rho_{12} = 0 \Rightarrow \Phi_{\lambda^*} = e \approx 2.718 \Rightarrow \ln \Phi_{\lambda^*} = 1,$$

the mapping $f_{\lambda^*}\colon S^2 \to S^2$ is **identical** (the record $f_{\lambda^*} =$ id), i.e. for any point $(X_1, Y_1, Z_1) \in S^2$ we get $(X_2 = X_1, Y_2 = Y_1, Z_2 = Z_1) \in S^2$.

The mappings $f_\lambda\colon M^2 \to M^2$ when comparing the standard spherical parametrization (Step 1) with the spherical λ-parametrizations (Step 2) have the following form:

$$f_\lambda = \left\{ \begin{array}{l} X_2 = \dfrac{\sqrt{4+\lambda^2}}{2} \times \dfrac{\mathrm{sF}_\lambda(u)}{\mathrm{sh}(u)} X_1 \\[2mm] Y_2 = \dfrac{\sqrt{4+\lambda^2}}{2} \times \dfrac{\mathrm{sF}_\lambda(u)}{\mathrm{sh}(u)} Y_1 \\[2mm] Z_2 = \dfrac{\sqrt{4+\lambda^2}}{2} \times \dfrac{\mathrm{cF}_\lambda(u)}{\mathrm{ch}(u)} Z_1 \end{array} \right.$$

where

$$u = \mathrm{Arch}\left(\frac{Z_1}{R}\right) = \ln\left(\frac{Z_1 + \sqrt{(Z_1)^2 - R^2}}{R}\right),$$

$$Z_1 \geq R > \lambda > 0; \quad (X_1, Y_1, Z_1) \in M^2; (X_2, Y_2, Z_2) \in M^2.$$

For the case $\lambda^* = 2\mathrm{sh}(1)$ we have:

$$\bar\rho_{12} = 0 \Rightarrow \Phi_{\lambda^*} = e \approx 2.718 \Rightarrow \ln \Phi_{\lambda^*} = 1,$$

the mapping $f_{\lambda^*}\colon M^2 \to M^2$ is **identical** (the record $f_{\lambda^*} =$ id), i.e. for any point $(X_1, Y_1, Z_1) \in M^2$ we get $(X_2 = X_1, Y_2 = Y_1, Z_2 = Z_1) \in M^2$.

(*Continued*)

Table 3.3.   (*Continued*)

	Step 4
For the case $\bar\rho_{12} > 0$ ($\lambda^* \neq 2\mathrm{sh}(1)$) the mapping $f_{\lambda^*}: S^2 \to S^2$ is **not identical**.	For the case $\bar\rho_{12} > 0$ ($\lambda^* \neq 2\mathrm{sh}(1)$) the mapping $f_{\lambda^*}: S^2 \to S^2$ is **not identical**.
Here $\bar\rho_{12} = |(\ln\Phi_\lambda) - 1|$ is a normalized distance (Step 3) between the standard spherical metric form (Step 1) and metric spherical λ-form (Step 2).	Here $\bar\rho_{12} = |(\ln\Phi_\lambda) - 1|$ is a normalized distance (Step 3) between Lobachevsky's metric form (Step 1) and Lobachevsky's metric λ-form (Step 2).
For the case $\bar\rho_{12} = 0$ the metric spherical λ-form coincides with the standard spherical form.	For the case $\bar\rho_{12} = 0$ Lobachevsky's metric λ-form coincides with the standard Lobachevsky's form.
For the case $\bar\rho_{12} > 0$ the metric spherical λ-form does not coincide with the standard spherical form.	For the case $\bar\rho_{12} > 0$ Lobachevsky's metric λ-form does not coincide with the standard Lobachevsky's form.
The mappings $f_\lambda: S^2 \to S^2$ have the following properties:	The mappings $f_\lambda: M^2 \to M^2$ have the following properties:
(1) For the case $\bar\rho_{12} = 0$ the **Gaussian curvature** of the sphere $S^2: X^2 + Y^2 + Z^2 = R^2$ **is preserved** and is equal to $K = \frac{1}{R^2} > 0$.	(1) For the case $\bar\rho_{12} = 0$ the **Gaussian curvature** of the pseudosphere $M^2; Z^2 - Y^2 - X^2 = R^2$ **is preserved** and is equal to $K = -\frac{1}{R^2} < 0$.
(2) When comparing the case $\bar\rho_{12} > 0$ with $\bar\rho_{12} = 0$, all other metric properties are **not preserved**, i.e. the mappings $f_\lambda: S^2 \to S^2$ are **non-isometric** (does not preserve arc lengths), **non-conformal** (does not preserve angles between arcs) and **non-equiareal** (does not preserve areas of figures).	(2) When comparing the case $\bar\rho_{12} > 0$ with $\bar\rho_{12} = 0$, all other metric properties are **not preserved**, i.e. the mappings $f_\lambda: M^2 \to M^2$ are **non-isometric** (does not preserve arc lengths), **non-conformal** (does not preserve angles between arcs) and **non-equiareal** (does not preserve areas of figures).

(*Continued*)

Table 3.3.   (Continued)

Step 5	
The main results of the study	
The spherical solution of Hilbert's Fourth Problem	**The hyperbolic solution of Hilbert's Fourth Problem**
(1) There are infinite geometries, induced on the sphere S^2: $X^2 + Y^2 + Z^2 = R^2$ by the spherical metric λ-forms of the same constant positive curvature.	(1) There are infinite geometries, induced on the pseudo-sphere M^2: $Z^2 - X^2 - Y^2 = R^2$ by the hyperbolic metric λ-forms of the same constant negative curvature.
(2) Among the *spherical metric* λ-forms generated by the *metallic proportions* Φ_λ for the integer values of $\lambda = 1, 2, 3, ...$, the *silver spherical metric* λ-form, generated by the *silver proportion* $1 + \sqrt{2} \approx 2.414$ ($\lambda = 2$), is the closest spherical metric λ-form to the standard metric spherical form, in the sense of the normalized distance $\bar{\rho}_{12}$.	(2) Among the *hyperbolic metric* λ-forms generated by the *metallic proportions* Φ_λ for the integer values of $\lambda = 1, 2, 3, ...$, the *silver hyperbolic metric* λ-form, generated by the *silver proportion* $1 + \sqrt{2} \approx 2.414$ ($\lambda = 2$), is the closest spherical metric λ-form to the standard metric spherical form, in the sense of the normalized distance $\bar{\rho}_{12}$.
(3) The silver spherical geometry is of great interest for theoretical physics and can apply directly to the physical world.	(3) The silver hyperbolic geometry is of great interest for theoretical physics and can apply directly to the physical world.

"golden" hyperbolic functions:

$$\begin{cases} sF(x) = \dfrac{\Phi^x - \Phi^{-x}}{\sqrt{5}} \\[2mm] cF(x) = \dfrac{\Phi^x + \Phi^{-x}}{\sqrt{5}} \end{cases}. \tag{3.76}$$

These are connected with Fibonacci numbers F_n $(n = 0, \pm1, \pm2, \pm3, \ldots)$ by the simple relation:

$$F_n = \begin{cases} sF(n) & \text{for } n = 2k \\ cF(n) & \text{for } n = 2k+1 \end{cases}. \tag{3.77}$$

It should be noted that relation (3.77) captures the distinctive features of the "golden" hyperbolic function (3.76) in comparison to the classic hyperbolic functions

$$sh(x) = \frac{e^x - e^{-x}}{2}; \quad ch(x) = \frac{e^x + e^{-x}}{2}. \tag{3.78}$$

According to (3.77), the "golden" hyperbolic functions (3.76), like Fibonacci numbers, have recursive properties, which give us the right to ascribe the "golden" hyperbolic functions (3.76) to a new class of hyperbolic functions, called recursive hyperbolic functions.

The so-called λ-Fibonacci hyperbolic functions, introduced in [18, 19]:

$$\begin{cases} sF_\lambda(x) = \dfrac{\Phi_\lambda^x - \Phi_\lambda^{-x}}{\sqrt{4 + \lambda^2}} \\[2mm] cF_\lambda(x) = \dfrac{\Phi_\lambda^x + \Phi_\lambda^{-x}}{\sqrt{4 + \lambda^2}} \end{cases} \tag{3.79}$$

where $\Phi_\lambda = \frac{\lambda + \sqrt{4+\lambda^2}}{2}$ is the metallic proportion [20], have similar recursive properties. It turned out that when $\lambda = 1, 2, 3, \ldots$ the hyperbolic functions (3.79), are fundamentally connected with the Fibonacci λ-numbers, given by the following recurrence relation:

$$F_\lambda(n + 2) = \lambda F_\lambda(n + 1) + F_\lambda(n)(n = 0, \pm1, \pm2, \pm3; \lambda = 1, 2, 3, \ldots) \tag{3.80}$$

with the seeds:

$$F_\lambda(0) = 0, \quad F_\lambda(1) = 1. \tag{3.81}$$

The recurrence relation (3.80) with the (3.81) seeds "generates" an infinite number of new numerical sequences, because every integer

$\lambda = 1, 2, 3, \ldots$ "generates" its own numerical sequence. Note that the classic Fibonacci numbers and Pell numbers are special limiting cases of the Fibonacci λ-numbers for the cases $\lambda = 1$ and $\lambda = 2$, respectively. Thus, the λ-Fibonacci hyperbolic functions are an essentially new (theoretically infinite) fundamental class of hyperbolic functions, having recursive properties and are called **recursive hyperbolic functions**. It should be noted that the classic hyperbolic functions, which underlie Lobachevsky's geometry, do not possess this recursive property.

However, the "golden" hyperbolic functions (3.76), which underlie the hyperbolic phyllotaxis world [33], are a special limiting case of the λ-Fibonacci hyperbolic functions ($\lambda = 1$). In this respect, there is every reason to suppose that other types of recursive hyperbolic functions — *λ-Fibonacci hyperbolic functions* — can be the basis for proposing new recursive hyperbolic worlds that may actually exist in Nature.

Modern science could not find these special recursive hyperbolic worlds, because the λ-Fibonacci hyperbolic functions [17, 18] were unknown until the 21st century. Based on the success of Bodnar's geometry [33], we can challenge theoretical physics, chemistry, crystallography, botany, biology, and other branches of the theoretical natural sciences to find new recursive hyperbolic worlds of Nature, based on the λ-Fibonacci hyperbolic functions (3.79). In this case, perhaps, the first candidate for a new recursive hyperbolic world of Nature may be the "silver" hyperbolic functions:

$$\begin{cases} sF_2(x) = \dfrac{\Phi_2^x - \Phi_2^{-x}}{\sqrt{8}} = \dfrac{1}{2\sqrt{2}}[(1 + \sqrt{2})^x - (1 + \sqrt{2})^{-x}] \\[4mm] cF_2(x) = \dfrac{\Phi_2^x + \Phi_2^{-x}}{\sqrt{8}} = \dfrac{1}{2\sqrt{2}}[(1 + \sqrt{2})^x + (1 + \sqrt{2})^{-x}] \end{cases} \qquad (3.82)$$

which are uniquely connected with Pell numbers and are based on the silver proportion $\Phi_2 = 1 + \sqrt{2} \approx 2.414$, connected, of course, with the fundamental mathematical constant $\sqrt{2}$.

In this regard, we should pay special attention to the fact that the new recursive hyperbolic geometry, based on the "silver" hyperbolic functions (3.82), is the closest to Lobachevsky's geometry, based on the classic hyperbolic functions (3.77) with the base of $e \approx 2.71$. Its distance to Lobachevsky's geometry is equal to $\bar{\rho}_{12} \approx 0.1677$ which has the minimum distance for Lobachevsky's metric λ-forms, where $\lambda = 1, 2, 3, \ldots$.

We may assume that the "silver" hyperbolic functions (3.82) and their generation by recursive "silver" hyperbolic geometry will soon be found in Nature and this geometry can take its rightful place in modern science next to Bodnar's geometry [33], based on the "golden" hyperbolic functions (3.76).

Similar arguments can be made about the recursive spherical worlds of Nature. In this chapter we first introduced the new mathematical notions:

(1) Spherical metallic proportions:

$$\tau_\lambda = \Phi_\lambda^i = \cos(\ln \Phi_\lambda) + i \sin(\ln \Phi_\lambda), \lambda \neq 0, \quad i = \sqrt{-1}. \qquad (3.83)$$

(2) Spherical Fibonacci λ-functions:

$$
\begin{cases}
\mathbf{SF}_\lambda(x) = -i \dfrac{\tau_\lambda^x - \tau_\lambda^{-x}}{\sqrt{4 + \lambda^2}} = \dfrac{2}{\sqrt{4 + \lambda^2}} \sin(x \ln \Phi_\lambda) \\[4mm]
\mathbf{CF}_\lambda(x) = \dfrac{\tau_\lambda^x + \tau_\lambda^{-x}}{\sqrt{4 + \lambda^2}} = \dfrac{2}{\sqrt{4 + \lambda^2}} \cos(x \ln \Phi_\lambda)
\end{cases}
\qquad (3.84)
$$

where τ_λ is the spherical metallic proportion, given above by (3.83).

This chapter establishes the deep mathematical connections of the λ-Fibonacci spherical functions (3.84) with the classic hyperbolic functions (3.78) and the λ-Fibonacci hyperbolic functions (3.79).

In general, the mathematical materials contained in this chapter and presented in the form of remarkable mathematical formulas (3.76)–(3.84), open a new stage in the development of non-Euclidean geometry that has already led to the solution of Hilbert's Fourth Problem. In the future, developments in this direction can lead to the creation of a new recursive non-Euclidean geometry, which can become the foundation for a broad process of finding **new recursive hyperbolic and spherical worlds of Nature**. This could well lead to a rethinking of the entire theoretical natural sciences, including the areas of theoretical physics, chemistry, biology, botany, genetics, etc., and will serve to more deeply connect mathematics with the natural sciences.

3.10. Hilbert's Fourth Problem as a Possible Candidate for the Millennium Problem in Geometry

In recent years, the so-called Millennium Problems have inspired mathematicians and physicists. The outstanding mathematician David Hilbert started this in 1900 when he presented twenty-three great mathematical problems at the International Congress of Mathematicians in Paris.

Explaining the purpose for his formulation of these mathematical problems [9], Hilbert writes:

"For the close of a great epoch not only invites us to look back into the past but also directs our thoughts to the unknown future."

Thus, Hilbert invites us to not only look to the past, but also to direct our efforts into the unknown future. As outlined in [36]:

"Hilbert's address of 1900 to the International Congress of Mathematicians in Paris is perhaps the most influential speech ever given to mathematicians, given by a mathematician, or given about mathematics. In it, Hilbert outlined 23 major mathematical problems to be studied in the coming century. . . . Hilbert's address was more than a collection of problems. It outlined his philosophy of mathematics and proposed problems important to his philosophy."

Modern mathematicians decided to continue this great tradition of Hilbert's. In May 2000, emulating Hilbert, the Clay Mathematics Institute of Cambridge announced (in Paris, for full effect) seven "Millennium Prize Problems", each with a bounty of $1 million [37].

Modern physicists too decided to join the Millennium Madness by formulating their own *Physics Problems for the Next Millennium* [38].

An analysis of Clay Mathematics Institute's list of the Millennium Prize Problems [37], leaves some dissatisfaction. For example, only one unsolved mathematical problem from Hilbert's list of 23 mathematical problems has been included in the list; this is the Riemann hypothesis, which is well known as Hilbert's Eighth Problem [25].

Of course, we do not have any intention to question the inclusion of the Riemann hypothesis, which has fundamental interest. However, it is surprising that some of Hilbert's other unsolved mathematical problems, which have the same fundamental interest, are excluded from the Clay Mathematics Institute's list.

We simply would like to show that in Hilbert's list there are important mathematical problems, which deserve to be included as Millennium Problems. As an example, we chose Hilbert's Fourth Problem, which concerns hyperbolic geometry and have interdisciplinary significance for many branches of mathematics and theoretical natural sciences.

Hilbert's Fourth Problem is considered by the modern mathematical community as unsolved. Wikipedia [25] has reflected the opinion of the modern mathematical community on the solution to this problem as follows: "Too vague to be stated resolved or not." This means that the modern mathematical community has placed all responsibility for solution (or rather, for the lack of solution) to this problem on Hilbert himself, who formulated this problem too vaguely.

Our studies into this problem are set out in [21–24] and the present book. The purpose of these publications is to present the original solution to Hilbert's Fourth Problem in a popular form, accessible to all mathematicians, teachers and students of mathematics, as well as representatives of theoretical natural sciences, interested in new classes of non-Euclidean geometries.

Taking into consideration the importance of Hilbert's Fourth Problem not only for geometry, but also for the entire theoretical natural sciences, we wish to assert that Hilbert's Fourth Problem was omitted from the Clay Mathematics Institute's list of Millennium Problems, by mistake (the modern mathematicians could not understand and evaluate this problem) and therefore **Hilbert's Fourth Problem deserves to be recognized as the Millennium Problem in Geometry.**

References

[1] Dubrovin, B. A., Novikov, S. P. and Fomenko, A. T. *Modern Geometry. Methods and Applications.* Moscow: Nauka (1979) (Russian).

[2] Arnold, V. I., Il'yashenko, Yu. S., Anosov, D. V., Bronstein, I. U., Aranson, S. Kh. and Grines, V. Z. "Dynamical systems — I: Ordinary differential equations and smooth dynamical systems", in *Encyclopedia of Mathematical Sciences.* Berlin: Springer-Verlag (1988).

[3] Anosov, D. V., Aranson, S. Kh., Arnold, V. I., Bronshtein, I. U., Grines, V. Z. and Il'yashenko, Yu. S. *Dynamical Systems — I. Ordinary Differential Equations. Smooth Dynamical Systems.* Moscow: Publisher VINITI (1985) (Russian).

[4] Aranson, S. Kh., Belitsky, G. R. and Zhuzhoma, E. V. *Introduction to the Qualitative Theory of Dynamical Systems on Surfaces.* Providence, RI: American Mathematical Society (1996).

[5] Anosov, D. V., Aranson, S. Kh., Arnold, V. I., Bronshtein, I. U., Grines, V. Z. and Il'yashenko, Yu. S. *Ordinary Differential Equations and Smooth Dynamical Systems.* Berlin: Springer (1997).

[6] Aranson S. Kh., Medvedev, V. and Zhuzhoma, E. "Collapse and continuity of geodesic frameworks of surface foliations", in *Methods of Qualitative Theory of Differential Equations and Related Topics.* Providence, RI: American Mathematical Society (2000).

[7] Aranson, S. Kh. "Qualitative properties of foliations on closed surfaces", *Journal of Dynamical and Control Systems.* (2000) Vol. 6, No. 1: 127–157.

[8] Aranson, S. Kh. and Zhuzhoma, E. V. "Nonlocal properties of analytic flows on closed orientable surfaces", in *Proceedings of the Steklov Institute of Mathematics* (2004) Vol. 244: 2–17.

[9] Hilbert, D. "Mathematical problems", lecture delivered before the International Congress of Mathematicians, Paris 1900 (translated by Maby Winton Newson). http://aleph0.clarku.edu/~djoyce/hilbert/problems.html#prob4 (accessed November 4, 2015).

[10] Klein, F. *Lectures on the Icosahedron.* New York: Courier Dover Publications (1956).

[11] Stakhov, A. P. "The mathematics of harmony: clarifying the origins and development of mathematics", *Congressus Numerantium* (2008) Vol. CXCIII: 5–48.

[12] Soroko, E. M. *Structural Harmony of Systems.* Minsk: Nauka i Technika (1984) (Russian).

[13] Stakhov, A. P., assisted by S. Olsen. *The Mathematics of Harmony. From Euclid to Contemporary Mathematics and Computer Science.* Singapore: World Scientific (2009).

[14] Stakhov, A. P. and Tkachenko, I. S. "Hyperbolic Fibonacci trigonometry", *Reports of the Ukrainian Academy of Sciences* (1993) Vol. 208, No. 7: 9–14 (Russian).

[15] Stakhov, A. P. and Rozin, B. N. "On a new class of hyperbolic functions", *Chaos, Solitons & Fractals* (2004) Vol. 23: 379–389.

[16] Stakhov, A. P. and Rozin, B. N. "The 'golden' hyperbolic models of universe", *Chaos, Solitons & Fractals* (2007) Vol. 34, Issue 2: 159–171.

[17] Stakhov, A. P. and Rozin, B. N. "The golden section, Fibonacci series and new hyperbolic models of nature", *Visual Mathematics* (2006) Vol. 8, No. 3. http://www.mi.sanu.ac.rs/vismath/stakhov/index.html (accessed November 4, 2015).

[18] Stakhov, A. P. "Gazale's formulas, a new class of hyperbolic Fibonacci and Lucas Functions and the improved method of the 'golden' cryptography", Academy of Trinitarism, Moscow: Electronic number 77-6567, publication 14098 (2006) (Russian). http://www.trinitas.ru/rus/doc/0232/004a/02321063.htm (accessed November 4, 2015).

[19] Stakhov, A. P. "On the general theory of hyperbolic functions based on the hyperbolic Fibonacci and Lucas functions and on Hilbert's Fourth Problem", *Visual Mathematics* (2013), Vol. 15, No. 1. http://www.mi.sanu.ac.rs/vismath/2013stakhov/hyp.pdf (accessed November 4, 2015).

[20] de Spinadel, V. W. *From the Golden Mean to Chaos.* Nueva Libreria (first edition 1998); Nobuko (second edition 2004).

[21] Stakhov, A. P. and Aranson, S. Kh. "Hyperbolic Fibonacci and Lucas functions, 'golden' Fibonacci goniometry, Bodnar's geometry, and Hilbert's fourth problem. Part I. Hyperbolic Fibonacci and Lucas functions and 'golden' Fibonacci goniometry", *Applied Mathematics* (2011) Vol. 2, No. 1: 74–84.

[22] Stakhov, A. P. and Aranson, S. Kh. "Hyperbolic Fibonacci and Lucas functions, 'golden' Fibonacci goniometry, Bodnar's geometry, and Hilbert's fourth problem. Part II", *Applied Mathematics* (2011) Vol. 2, No. 2: 181–188.

[23] Stakhov, A. P. and Aranson, S. Kh. "Hyperbolic Fibonacci and Lucas functions, 'golden' Fibonacci goniometry, Bodnar's geometry, and Hilbert's fourth problem. Part III", *Applied Mathematics* (2011) Vol. 2, No. 3: 283–293.

[24] Stakhov, A. P. "Hilbert's fourth problem: Searching for harmonic hyperbolic worlds of nature", *Journal of Applied Mathematics and Physics* (2013) Vol. 1, No. 3: 60–66.

[25] "Hilbert's problems", *Wikipedia, The Free Encyclopedia*, http://en.wikipedia.org/wiki/Hilbert's_problems (accessed November 4, 2015).

[26] "Hilbert's fourth problem", *Wikipedia. The Free Encyclopedia.* http://en.wikipedia.org/wiki/Hilbert's_fourth_problem (accessed November 4, 2015).

[27] Stakhov, A. P. "Non-Euclidean geometries. From the 'game of postulates' to the 'game of functions'", Academy of Trinitarism, Moscow: Electronic number 77-6567, publication 18048 (2013) (Russian). http://www.trinitas.ru/rus/doc/0016/001d/00162125.htm (accessed November 4, 2015).

[28] Pogorelov, A. V. *Hilbert's Fourth Problem.* Moscow: Nauka (1974) (Russian).

[29] Aranson, S. Kh. "Once again on Hilbert's fourth problem", Academy of Trinitarism, Moscow: Electronic number 77-6567, publication 15677 (2009) (Russian). http://www.trinitas.ru/rus/doc/0232/009a/02321180.htm (accessed November 4, 2015).

[30] Aleksandrov, P. S. (General Editor). *Hilbert's Problems.* Moscow: Nauka (1969) (Russian).

[31] Busemann, H. "On Hilbert's fourth problem", *Uspechi Mathematicheskich Nauk* (1966) Vol. 21, No. 1(27): 155–164 (Russian).

[32] Yandell, B. H. *The honors class: Hilbert's problems and their solvers.* Massachusetts: AK Peters/CRC Press (2001).

[33] Bodnar, O. Y. *The Golden Section and Non-Euclidean Geometry in Nature and Art.* Lvov: Publishing House "Svit" (1994) (Russian).

[34] Bronstein, I. N. and Semendyaev, K. A. *Handbook of Mathematics.* Moscow: Nauka (1986) (Russian).

[35] "Spherical geometry", *Wikipedia, The Free Encyclopedia*, http://en.wikipedia.org/wiki/Spherical_geometry (accessed November 4, 2015).

[36] Joyce, D. E. "The mathematical problems of David Hilbert. http://aleph0.clarku.edu/~djoyce/hilbert/index.html (accessed November 4, 2015).

[37] "Millennium prize problems", *Wikipedia, The Free Encyclopedia*, https://en.wikipedia.org/wiki/Millennium_Prize_Problems (accessed November 4, 2015).

[38] "'Millennium Madness' — Physics problems for the next millennium". http://www.theory.caltech.edu/~preskill/millennium.html (accessed November 4, 2015).

Chapter 4

Introduction to the "Golden" Qualitative Theory of Dynamical Systems Based on the Mathematics of Harmony

4.1. Beauty and Aesthetics of Mathematics

4.1.1. *Dirac's principle of mathematical beauty*

At the moment, no one doubts that science is beautiful, because it reflects and refracts in our minds the beauty, harmony and unity of the universe. But physicists have gone even further. English physicist and Nobel Laureate Paul Dirac proposed the thesis: "Beauty is the criterion of the truth of physical theory." Dirac not only recognized the beauty of mathematical formulations of a theory, but also understood the heuristic, regulatory role of beauty as a methodological principle of construction for scientific knowledge.

4.1.2. *Aesthetics of the Mathematics of Harmony*

Outstanding English mathematician, philosopher and Nobel Prize winner Bertrand Russell (1872–1970) stated in [1]:

> *"Mathematics, rightly viewed, possesses not only truth, but supreme beauty — a beauty cold and austere, like that of sculpture, without appeal to any part of our weaker nature, without the gorgeous trappings of painting or music, yet sublimely pure, and capable of a stern perfection such as only the greatest art can show. The true spirit of delight, the exaltation, the sense of being more than Man, which is the touchstone of the highest excellence, is to be found in mathematics as surely as poetry."*

This quote is taken as an epigraph from Wenninger's book *Polyhedron Models* [2].

Note that this quote is in full accord with the Mathematics of Harmony as described in [3]. Nearly all of the basic mathematical formulas, obtained within the framework of the Mathematics of Harmony, including the theory of Fibonacci numbers [4–7], are expressing aesthetic qualities evoking feelings of beauty and pleasure. Here are some examples.

Aesthetics of the golden section.   Early in Book II of the *Elements*, Euclid introduces the problem of the division of a line segment into extreme and mean ratio (DEMR), well known as the golden section in modern science. Traditionally the golden section has been admired for its unique mathematical and geometric properties.

Aesthetics of Fibonacci and Lucas numbers.   Table 4.2 includes several well-known mathematical properties of Fibonacci and Lucas numbers. The aesthetic nature of these expressions is obvious.

The λ-Fibonacci and λ-Lucas numbers and metallic proportions.   Mathematical identities for the λ-Fibonacci and λ-Lucas numbers and metallic proportions can have a great aesthetic effect (Table 4.3).

Thus it follows from an analysis of the mathematical formulas in Tables 4.1–4.3 that the Mathematics of Harmony [3], developed since ancient times through generations of outstanding scientists and thinkers, displays exceptional mathematical and aesthetic features.

It is important to emphasize that Dirac's Principle of Mathematical Beauty in connection with the Mathematics of Harmony is exemplified in

Table 4.1.   Aesthetics of the golden section (GS).

1	The equation of the GS	$x^2 - x - 1 = 0$
2	The golden proportion	$\Phi = \dfrac{1 + \sqrt{5}}{2}$
3	The basic identity	$\Phi^n = \Phi^{n-1} + \Phi^{n-2} = \Phi \times \Phi^{n-1};$ $n = 0, \pm 1, \pm 2, \pm 3, \ldots$
4	The simplest continued fraction	$\Phi = 1 + \cfrac{1}{1 + \cfrac{1}{1 + \cfrac{1}{1 + \cdots}}}$
5	The simplest nested radical	$\Phi = \sqrt{1 + \sqrt{1 + \sqrt{1 + \sqrt{1 + \cdots}}}}$

Table 4.2. Aesthetics of Fibonacci and Lucas numbers.

1	Recurrent relations for Fibonacci and Lucas numbers	$F_n = F_{n-1} + F_{n-2}$; $F_1 = F_2 = 1$ $L_n = L_{n-1} + L_{n-2}$; $F_1 = 1$, $L_2 = 3$
2	The extended Fibonacci and Lucas numbers	(see table below)
3	Cassini's formula	$F_n^2 - F_{n-1}F_{n+1} = (-1)^{n+1}$; $n = 0, \pm 1, \pm 2, \pm 3, \ldots$
4	Kepler's formula	$\displaystyle \lim_{n \to \infty} \frac{F_n}{F_{n-1}} = \Phi = \frac{1+\sqrt{5}}{2}$
5	Binet's formulas	$F_n = \begin{cases} \dfrac{\Phi^n + \Phi^{-n}}{\sqrt{5}} & \text{for } n = 2k+1 \\[2mm] \dfrac{\Phi^n - \Phi^{-n}}{\sqrt{5}} & \text{for } n = 2k \end{cases}$ $L_n = \begin{cases} \Phi^n + \Phi^{-n} & \text{for } n = 2k \\[2mm] \Phi^n - \Phi^{-n} & \text{for } n = 2k+1 \end{cases}$
6	Hyperbolic Fibonacci and Lucas functions	$sFs = \dfrac{\Phi^x - \Phi^{-x}}{\sqrt{5}}$; $\quad cFs = \dfrac{\Phi^x + \Phi^{-x}}{\sqrt{5}}$ $sLs = \Phi^x - \Phi^{-x}$; $\quad cLs = \Phi^x + \Phi^{-x}$
7	Fibonacci matrices (Hoggat)	$Q = \begin{pmatrix} 1 & 1 \\ 1 & 0 \end{pmatrix}$; $\quad Q^n = \begin{pmatrix} F_{n+1} & F_n \\ F_n & F_{n-1} \end{pmatrix}$; $\det(Q^n) = (-1)^n$
8	The "golden" matrices (Stakhov)	$Q_0(x) = \begin{pmatrix} cFs(2x+1) & sFs(2x) \\ sFs(2x) & cFs(2x-1) \end{pmatrix}$; $\det Q_0(x) = +1$ $Q_1(x) = \begin{pmatrix} sFs(2x+2) & cFs(2x+1) \\ cFs(2x+1) & sFs(2x) \end{pmatrix}$; $\det Q_0(x) = -1$

Extended Fibonacci and Lucas numbers (item 2):

n	0	1	2	3	4	5	6
F_n	0	1	1	2	3	5	8
F_{-n}	0	1	-1	2	-3	5	-8
L_n	2	1	3	4	7	11	18
L_{-n}	2	-1	3	-4	7	-11	18

Table 4.3.   Aesthetics of the λ-Fibonacci and λ-Lucas numbers ($\lambda > 0$) and metallic proportions.

1	Recurrent relations for the λ-Fibonacci and λ-Lucas numbers	$F_\lambda(n+2) = \lambda F_\lambda(n+1) + F_\lambda(n);\ F_\lambda(0) = 0,\ F_\lambda(1) = 1$ $L_\lambda(n+2) = \lambda L_\lambda(n+1) + L_\lambda(n);\ L_\lambda(0) = 2,\ F_\lambda(1) = \lambda$
2	Generalized Cassini formula	$F_\lambda^2(n) - F_\lambda(n-1)F_\lambda(n+1) = (-1)^{n+1};$ $n = 0, \pm1, \pm2, \pm3, \ldots$
3	Metallic proportions	$x^2 - \lambda x - 1 = 0 \to \Phi_\lambda = \dfrac{\lambda + \sqrt{4 + \lambda^2}}{2}$
4	Basic identity	$\Phi_\lambda^n = \lambda\Phi_\lambda^{n-1} + \Phi_\lambda^{n-2} = \Phi_\lambda \times \Phi_\lambda^{n-1}$
5	Metallic proportion represented as a continued fraction	$\Phi_\lambda = \lambda + \dfrac{1}{\lambda + \dfrac{1}{\lambda + \frac{1}{\lambda + \cdots}}}$
6	Metallic proportion represented as a nested radical	$\Phi_\lambda = \sqrt{1 + \lambda\sqrt{1 + \lambda\sqrt{1 + \lambda\sqrt{1 + \cdots}}}}$
7	Gazale's formulas	$F_\lambda(n) = \dfrac{\Phi_\lambda^n - (-1)^n\Phi_\lambda^{-n}}{\sqrt{4 + \lambda^2}};\ L_\lambda(n) = \Phi_\lambda^n + (-1)^n\Phi_\lambda^{-n}$
8	λ-Fibonacci and λ-Lucas hyperbolic functions (Stakhov)	$sF_\lambda(x) = \dfrac{\Phi_\lambda^x - \Phi_\lambda^{-x}}{\sqrt{4 + \lambda^2}};\quad cF_\lambda(x) = \dfrac{\Phi_\lambda^x + \Phi_\lambda^{-x}}{\sqrt{4 + \lambda^2}}$ $sL_\lambda(x) = \Phi_\lambda^x - \Phi_\lambda^{-x};\quad cL_\lambda(x) = \Phi_\lambda^x + \Phi_\lambda^{-x}$

several modern scientific discoveries rooted in ancient mathematical science (e.g. the golden ratio and Platonic solids). These include the discoveries of quasi-crystals and fullerenes, with the respective discoverers each being awarded the Nobel Prize.

A purpose of Chapter 4.   We have discussed our scientific cooperation in the introductory pages of this book. Once again we want to emphasize our deep belief in the objective beauty of Nature and share the sentiments of Pythagoras, Plato, Euclid, Pacioli, Kepler, and Einstein regarding a Cosmic Harmony. It has become the primary motivation behind our scientific cooperation.

Shevelev, the well-respected Russian architect and researcher of Harmony, wrote the following in his 1995 book, *Shaping: Number, Shape, Art,*

Life [8]:

> "*We want to abstract the general principle of the universal connection of each thing with everything, which could be expanded to any phenomena of the Being, that is, we want to understand the essential nature of 'Harmony', that is, we want to abstract the idea of communication, which determines the existence of the phenomenal world, that is, of the universal communication, without which all integrity could be disintegrated into its component parts, and Nature would cease to exist.*"

We share the widespread view that "beauty is an expression or manifestation of harmony." This introduction of the idea of "Universal Harmony" in critical areas of modern mathematics became the primary focus of our successful collaboration.

Stakhov came to mathematics from the technical sciences. In this field he developed several applied mathematical theories (algorithmic measurement theory [9], numeral systems with irrational bases [10], the "golden" number theory [11], etc.), all of which have been of significant value for computer science and number theory. And the primary scientific result has been the establishment of the Mathematics of Harmony as a new interdisciplinary development in modern science [1, 12, 13]. This interdisciplinary approach to the Mathematics of Harmony has led to a number of new developments in hyperbolic functions and their applications [14–22]. These were the mathematical results that attracted the attention of co-author Samuil Aranson.

It should be noted that Aranson is a professional mathematician specializing in an area related to Stakhov's field of research in hyperbolic functions [14–22]. His contribution to the development of the qualitative theory of differential equations and dynamical systems is exemplified by an impressive number of scientific articles and books, published in the USSR, USA and Germany [23–51] being recognized throughout the scientific world.

The combined scientific efforts of Aranson and Stakhov led to several significant mathematical discoveries. This includes the original solution to Hilbert's Fourth Problem. Publication of a number of joint articles [52–55] and their book [56] followed shortly thereafter.

In this chapter, we further extend the application of the Mathematics of Harmony into the qualitative theory of dynamical systems. One of the main problems in the classic qualitative theory of dynamical systems is the study of their local and global topological structures. In describing the global topological structure of dynamical systems and foliations on surfaces,

Poincaré's rotation numbers and homotopic classes of rotation are known to be the primary topological invariants.

In order to outline the initial steps toward creation of the "golden" qualitative theory of dynamical systems, we first select two surfaces of the small genuses, the closed orientable surface of the genus "one", i.e. the two-dimensional torus T^2, and the closed orientable surface of the genus "zero", i.e. the two-dimensional sphere S^2.

We have already studied the golden, silver, bronze, copper, and other metallic irrational foliations on these surfaces. Then, by using the decomposition of Poincaré's irrational rotation numbers into continued fractions, we approximate these foliations by using suitable rational foliations. We solve similar problems for dynamical systems with continuous time (flows), whose integral curves are just such foliations.

In this chapter, we also consider other tasks of the "golden" qualitative theory of dynamical systems, in order to update the eigenvalues of Anosov's automorphisms (hyperbolic automorphisms) on the two-dimensional torus T^2 by employing the metallic and golden proportions.

4.2. Preliminaries

4.2.1. Definitions and assumptions

The concept of a dynamical system. A dynamical system is the mathematical model of an object, process or phenomenon which is characterized by an initial state and law, under which the system moves from the initial state to another state.

Many phenomena of science and technology are dynamical by their very nature. For some time, stationary states, periodic orbits and heartbeats, caused by modulations, were considered to be the only possible observable states. However, discoveries during the second half of the 20th century dramatically changed our traditional views on the nature of dynamic processes. The discovery of a new kind of oscillation, called dynamical chaos, was an extremely significant achievement.

A deeper understanding of dynamic processes led to a better understanding of the world's nonlinear nature. Out of this, nonlinear dynamics emerged as a scientific discipline aimed at studying the general rules (laws) of nonlinear dynamic processes.

A mathematical model of nonlinear dynamics is typically represented by a system of equations with analytically given nonlinearities and a finite number of parameters. The system may be described by ordinary differential equations, differential equations with partial derivatives, delay equations, integro-differential equations, etc.

Foliations. Let M^2 be a closed surface. We can then introduce the following definitions.

Definition 4.1. A foliation Q without singularities on M^2 is a partition of M^2 into disjoint curves (layers), which are locally homeomorphic to the family of parallel lines:

$$\xi = \xi(t) = t, \quad \eta = \eta(t) = C, \tag{4.1}$$

where (ξ, η) are the local coordinates, $t \in (-\infty, +\infty)$ is a parameter, C is a constant.

Definition 4.2. A foliation Q with a finite set of saddle singularities Σ on M^2 is a partition of M^2 into the curves (layers), which satisfies the following properties:

(1) On the set $M^2 \backslash \Sigma$, this partition is a foliation without singularities.
(2) For the local coordinates (ξ, η) in the neighborhood w $(\xi = 0, \eta = 0) \in \Sigma$ this partition on curves is described by the equations:

$$(\xi + i\eta)^v = (t + iC)^2, \tag{4.2}$$

if v is a positive odd number, and

$$(\xi + i\eta)^{\frac{v}{2}} = (t + iC), \tag{4.3}$$

if v is a positive even number $(v \neq 2)$.

Here $t \in (-\infty, +\infty)$ is a parameter, C is a constant, $i = \sqrt{-1}$ is the imaginary unit. If we rewrite (4.2) and (4.3) in the form: $\xi + i\eta = (t + iC)^{\frac{2}{v}}, v = 1, 3, 4, 5, \ldots$ then we have:

$$t + iC = \sqrt{t^2 + C^2} e^{i\varphi} = \sqrt{t^2 + C^2}[\cos(\varphi) + i\sin(\varphi)].$$

Hence we have:

$$\cos(\varphi) = \frac{t}{\sqrt{t^2 + C^2}}, \quad \sin(\varphi) = \frac{C}{\sqrt{t^2 + C^2}}.$$

From the condition $\sin(\varphi) = \frac{C}{\sqrt{t^2 + C^2}}$ we get:

$$\varphi = (-1)^k \arcsin\left(\frac{C}{\sqrt{t^2 + C^2}}\right) + \pi k, \quad k \in Z, \tag{4.4}$$

where $-\frac{\pi}{2} \le \arcsin(\frac{C}{\sqrt{t^2+C^2}}) \le \frac{\pi}{2}$, Z is a set of integers.

As

$$\xi + i\eta = (t+iC)^{\frac{2}{v}} = (t^2+C^2)^{\frac{1}{v}}\left[\cos\left(\frac{2}{v}\varphi\right) + i\sin\left(\frac{2}{v}\varphi\right)\right],$$

then

$$\xi = (t^2+C^2)^{\frac{1}{v}}\cos\left(\frac{2}{v}\varphi\right), \quad \eta = (t^2+C^2)^{\frac{1}{v}}\sin\left(\frac{2}{v}\varphi\right).$$

Formula (4.4) is then substituted into the above expression. Then the layers of the foliation Q with saddle singularities on M^2 in the local coordinates (ξ, η) in the neighborhood of the saddle singularity ($\xi = 0, \eta = 0$), which has v separatrices ($v \ne 2$), are given in the following form:

$$
\begin{cases}
\xi = (t^2+C^2)^{\frac{1}{v}}\cos\left(\frac{2}{v}\varphi\right) \\
\quad = (t^2+C^2)^{\frac{1}{v}}\cos\left[\frac{2}{v}(-1)^k\arcsin\left(\frac{C}{\sqrt{t^2+C^2}}\right) + \frac{2\pi k}{v}\right], \\
\eta = (t^2+C^2)^{\frac{1}{v}}\sin\left(\frac{2}{v}\varphi\right) \\
\quad = (t^2+C^2)^{\frac{1}{v}}\sin\left[\frac{2}{v}(-1)^k\arcsin\left(\frac{C}{\sqrt{t^2+C^2}}\right) + \frac{2\pi k}{v}\right],
\end{cases}
\tag{4.5}
$$

where $k = 0, 1, 2, \ldots, v-1$.

Separatrix L of a saddle for the foliation Q with a finite set of saddle singularities Σ is called a layer of this foliation, where in the local coordinates (ξ, η) the constant C equals 0.

For the given value of v ($v \ne 2$), the equations (4.5) have v solutions $L_1, L_2, \ldots, L_v$, which are separatrices. In the local coordinates (ξ, η) these separatrices are half-lines, starting at ($\xi = 0, \eta = 0$).

We substitute $C = 0$ in (4.5). Then we get the parametric equation of the separatrices:

$$\xi = t^{\frac{2}{v}}\cos\left(\frac{2\pi k}{v}\right), \quad \eta = t^{\frac{2}{v}}\sin\left(\frac{2\pi k}{v}\right), \tag{4.6}$$

where $k = 0, 1, 2, \ldots, v-1$.

Particular cases of saddle singularities.   Consider the following particular cases of saddle singularities: (1) needle; (2) tripod; (3) ordinary saddle; (4) false saddle.

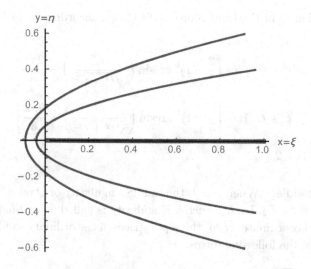

Figure 4.1(a).   A saddle of foliation with one separatrix (needle) ($v = 1$).

Needle. When $v = 1$, the saddle singularity of Q on M^2 has one separatrix L_1. Such singularity is called a needle. In the local coordinates (ξ, η), the separatrix L_1 of the needle ($\xi = 0$, $\eta = 0$) has the form L_1: $\xi = t^2$, $\eta = 0$, and the foliation Q has the form shown in Fig. 4.1(a):

$$\xi = (t^2 + C^2) \cos \left[2 \arcsin \left(\frac{C}{\sqrt{t^2 + C^2}} \right) \right] = t^2 - C^2,$$

$$\eta = (t^2 + C^2) \sin \left[2 \arcsin \left(\frac{C}{\sqrt{t^2 + C^2}} \right) \right] = 2tC. \tag{4.7}$$

Tripod. When $v = 3$, the saddle singularity of Q on M^2 has three separatrices L_1, L_2, L_3. This singularity is called a tripod. In the local coordinates (ξ, η) the separatrices of the tripod ($\xi = 0$, $\eta = 0$) have the following forms:

$$\begin{cases} L_1 : \xi = t^{\frac{2}{3}}, \quad \eta = 0; \quad (k = 0) \\[2mm] L_2 : \xi = t^{\frac{2}{3}} \cos \left(\frac{2\pi}{3} \right), \quad \eta = \sin \left(\frac{2\pi}{3} \right), \quad (k = 1) \\[2mm] L_3 : \xi = t^{\frac{2}{3}} \cos \left(\frac{4\pi}{3} \right), \quad \eta = \sin \left(\frac{4\pi}{3} \right), \quad (k = 2). \end{cases} \tag{4.8}$$

The foliation Q in the local coordinates (ξ, η), according to (4.5), has the form:

$$\begin{cases} \xi = (t^2 + C^2) \cos\left[\dfrac{2}{3}(-1)^k \arcsin\left(\dfrac{C}{\sqrt{t^2 + C^2}}\right) + \dfrac{2\pi k}{3}\right], \\[4mm] \eta = (t^2 + C^2) \sin\left[\dfrac{2}{3}(-1)^k \arcsin\left(\dfrac{C}{\sqrt{t^2 + C^2}}\right) + \dfrac{2\pi k}{3}\right], \end{cases} \qquad (4.9)$$

where $k = 0, 1, 2$.

Ordinary saddle. When $v = 4$, the saddle singularity of Q on M^2 has four separatrices L_1, L_2, L_3, L_4. Such a singularity is called an ordinary saddle. In the local coordinates (ξ, η) the separatrices of the ordinary saddle ($\xi = 0$, $\eta = 0$) have the following forms:

$$L_1 : \xi = \sqrt{|t|}, \quad \eta = 0 \quad (k = 0); \quad L_2 : \xi = 0, \quad \eta = \sqrt{|t|} \quad (k = 1);$$

$$L_3 : \xi = -\sqrt{|t|}, \quad \eta = 0 \quad (k = 2); \quad L_4 : \xi = 0, \quad \eta = -\sqrt{|t|} \quad (k = 3).$$

The foliation Q in the local coordinates (ξ, η), according to (4.5), has the following form (see Fig. 4.1(b)):

$$\begin{cases} \xi = \sqrt[4]{t^2 + C^2} \cos\left[\dfrac{1}{2}(-1)^k \arcsin\left(\dfrac{C}{\sqrt{t^2 + C^2}}\right) + \dfrac{\pi k}{2}\right], \\[4mm] \eta = \sqrt[4]{t^2 + C^2} \sin\left[\dfrac{1}{2}(-1)^k \arcsin\left(\dfrac{C}{\sqrt{t^2 + C^2}}\right) + \dfrac{\pi k}{2}\right], \end{cases} \qquad (4.10)$$

where $k = 0, 1, 2, 3$.

False saddle. When $v = 2$, the saddle singularity of Q on M^2 has two separatrices L_1, L_2. Such a singularity is called a false saddle. In the local coordinates (ξ, η) the separatrices of the false saddle ($\xi = 0$, $\eta = 0$) have the following forms:

$$L_1 : \xi = |t|, \quad \eta = 0 \quad (k = 0); \quad L_2 : \xi = -|t|, \quad \eta = 0 \quad (k = 1).$$

The foliation Q in the local coordinates (ξ, η), according to (4.5), has the form:

$$\xi = t, \quad \eta = C.$$

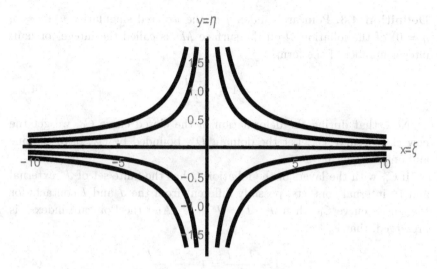

Figure 4.1(b).   A saddle of foliation with four separatrices ($v = 4$).

Poincaré's index [57].   The notion of the Poincaré index j plays an important role in the theory of dynamical systems for the isolated singularities (not necessarily saddle) of the foliation Q on the surface M^2.

Let $w(\xi = 0,\ \eta = 0)$ be an isolated singularity of the foliation Q on the surface M^2 in the local coordinates $(\xi,\ \eta)$. Consider the closed curve C_0 (without self-intersections) on M^2 for the following conditions:

(1) C_0 belongs to the local area Δ on M^2, which has the one and only singularity w ($\xi = 0,\ \eta = 0$), with the case that the Euclidean metric has been introduced into the local area Δ.

(2) C_0 is homeomorphic to a circle.

(3) The singularity w ($\xi = 0,\ \eta = 0$) lies in the area D, and is achievable inside the border ∂D, where the boundary is the closed curve $C_0 = \partial D$. Hence we have the inclusions:

$$w(\xi = 0, \eta = 0) \in D \subset (D \cup \partial D) \subset \Delta.$$

(4) The layers of the foliation Q on the surface C_0^* with the closed curve C_0 have a finite (possibly empty) set of contacts.

We denote through J a number of external contacts and through I a number of internal contacts of the layers of the foliation Q with the closed curve C_0.

Definition 4.3. Poincaré's index j of the isolated singularity w ($\xi = 0$, $\eta = 0$) of the foliation Q on the surface M^2 is called the integer or half-integer number of the form:

$$j = 1 - \frac{J - I}{2}. \tag{4.11}$$

Note that during the deformation of the closed curve C_0, we get the new closed curve C_0^*. Let the domain D^*, bounded by C_0^*, have the one and only singularity w.

If C_0^* with the layers of the foliation Q has the finite set of J^* external and I^* internal contacts, possibly different from the J and I contacts for the closed curve C_0, then $J - I = J^* - I^*$. And the Poincaré index j is preserved, that is,

$$j = 1 - \frac{J - I}{2} = 1 - \frac{J^* - I^*}{2}. \tag{4.12}$$

In particular, the index of the saddle with v separatrices is equal to $j = 1 - \frac{v}{2}$.

For example, for the foliation Q with the neighborhood of the needle $w(\xi = 0, \eta = 0)$ of the form (4.7) with the separatrix L_1: $\xi = t^2$, $\eta = 0$, the circle C_0: $\xi^2 + \eta^2 = C^4$ has one external contact at the point ($\xi = -C^2$, $\eta = 0$), that is, $J = 1$, and does not have any internal contacts, that is, $I = 0$.

Here $C \neq 0$ is any fixed real number. In this case, for any needle, Poincaré's index is $j = 1 - \frac{J-I}{2} = 1 - \frac{1-0}{2} = 0.5$.

Euler characteristic.   Let us introduce the notion of the Euler characteristic. In mathematics, and more specifically in algebraic topology and polyhedral combinatorics, the Euler characteristic is a topological invariant, a number which describes the topological space's form or structure regardless of the way it is bent. It is commonly denoted by χ.

It is known that the sum of Poincaré's indices of the singularities of the foliation Q on the closed surface M^2 of the genus of p is equal to the Euler characteristic $\chi(M^2)$ of the surface M^2:

$$\sum j = \chi(M^2).$$

Recall that for the closed orientable surfaces, the Euler characteristic is equal to $\chi(M^2) = 2 - 2p$ ($p \geq 0$), but for the closed non-orientable surfaces the Euler characteristic is equal to $\chi(M^2) = 2 - p$ ($p \geq 1$). Hence, for the closed orientable surface of M^2 its genus is equal to $p = 1 - \frac{\chi(M^2)}{2} \geq 0$, but for the closed non-orientable surface of M^2 its genus is equal to $p = 2 - \chi(M^2) \geq 1$.

Therefore, if M^2 is a closed orientable surface of the genus $p \geq 0$, then the Euler characteristic $\chi(M^2) = 2, 0, -2, -4, -6, \ldots$, respectively for the sphere S^2 (the genus $p = 0$, $\chi(S^2) = 2$), for the torus T^2 (the genus $p = 1$, $\chi(T^2) = 0$) and $\chi(M^2) < 0$ for the closed orientable surfaces M^2 of the genuses $p = 2, 3, 4, \ldots$.

If $\chi(M^2) = -2$, then the closed orientable surface M^2 has the genus $p = 2$ and is called a *pretzel*.

If M^2 is the closed non-orientable surface of the genus $p \geq 1$, then the Euler characteristic $\chi(M^2) = 1, 0, -1, -2, -3, \ldots$, respectively, for the projective plane P^2 (the genus $p = 1$, $\chi(P^2) = 1$), for the Klein bottle K^2 (the genus $p = 2$, $\chi(K^2) = 0$) and $\chi(M^2) < 0$ for the closed non-orientable surfaces M^2 of the genuses $p = 3, 4, 5, \ldots$.

It is known that the sum of indices of the singularities of the foliation of Q on the closed surface M^2 is equal to the Euler characteristic $\chi(M^2)$ of this surface. Therefore, if on the closed surface M^2 the foliation Q has no singularities, then $\chi(M^2) = 0$. In this case, the surface M^2 is the torus T^2, the closed orientable surface of the genus $p = 1$, or the Klein bottle K^2, the closed non-orientable surface of the genus $p = 2$.

Suppose that for the closed surfaces M^2 only the saddles with non-zero or half-integer Poincaré's indices may be singularities of foliations (see, for example, [31, 34]). In particular, for the saddles with the numbers of the separatrices $v = 1, 3, 4, 5, 6$, Poincaré's indices are equal, that is $j = 0.5, -0.5, -1, -1.5, -2$ respectively.

Let M^2 be the closed surface (orientable or not) of the genus of p. For the case of an orientable surface, it is a sphere with $p \geq 0$ pasting handles, but for the case of a non-orientable surface, it is a sphere with $p \geq 1$ pasting Mobius bands. In particular, the sphere of S^2 has the genus of $p = 0$, the torus of T^2 has the genus of $p = 1$, and the projective plane of P^2 has the genus of $p = 1$.

Assumption 4.1. With regard to the set of singularities of $\Sigma \in M^2$, then in the future (unless otherwise stated) we will consider two situations: $\Sigma = \phi$, that is, the foliation Q on M^2 has no singularities; or $\Sigma \neq \phi$, that is, the foliation Q on M^2 has singularities. Here, the symbol ϕ means the empty set.

Then, we assume that the set of singularities Σ is finite, and each such singularity is a saddle with v separatrices, where $v \geq 1$ and $v \neq 2$ (see Fig. 4.1(a) and (b)).

4.2.2. Terminology

As mentioned above, the saddle with one separatrix is a needle, with three separatrices is a tripod and with four separatrices is an ordinary saddle.

If $v = 2$, then such singularity of foliation has two separatrices and is a traversable crumb, and its Poincaré index is equal to $j = 0$. Although this regularity is important, we will not consider it when looking at dynamical systems on surfaces (see, for example, [35]).

4.3. Metallic Irrational Foliations without Singularities on the Two-dimensional Torus T^2

4.3.1. Definitions and assumptions

The two-dimensional torus T^2 (i.e. the closed orientable surface of the genus $p = 1$) in the three-dimensional Euclidean space R^3 (x, y, z) can be set either in the implicit form:

$$F(x, y, z) = x^2 + \left(\sqrt{y^2 + z^2} - R\right)^2 - r^2 = 0 \qquad (4.13)$$

or in the parametric form:

$$\begin{cases} x = x(\varphi, \psi) = r \cdot \sin(2\pi\varphi) \\ y = y(\varphi, \psi) = [R + r\cos(2\pi\varphi)] \cdot \sin(2\pi\psi) \\ z = z(\varphi, \psi) = [R + r\cos(2\pi\varphi)] \cdot \cos(2\pi\psi) \end{cases} \qquad (4.14)$$

where $0 < r < R$, $-\infty < \varphi < +\infty$, $-\infty < \psi < +\infty$ (see Fig. 4.2).

Representation of T^2 in the parametric form (4.14) can be considered to be a covering $f \colon \Pi^2 \mapsto T^2 = \Pi^2/G$ of the universal covering plane Π^2 (φ, ψ) on the torus T^2, where G is the covering group, whose elements $g \in G$ are integer shifts of the form:

$$g \colon \varphi' = \varphi + m, \quad \psi' = \psi + n \quad (m, n = 0, \pm 1, \pm 2, \pm 3, \ldots).$$

This interpretation follows from the fact that the right-hand parts of (4.14) are periodic functions of the variables φ, ψ with the period $T = 1$.

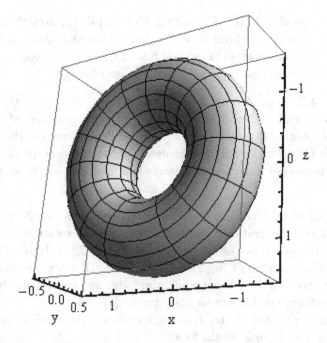

Figure 4.2.   Surface of the torus T^2.

Definition 4.4. Linear irrational foliation Q without singularities on the torus T^2 is a partition of the torus T^2 on the disjoint curves (layers) of the form:

$$\begin{cases} x = x(\varphi) = r \cdot \sin(2\pi\varphi) \\ y = y(\varphi) = [R + r\cos(2\pi\varphi)] \cdot \sin(2\pi\psi(\varphi)) \\ z = z(\varphi) = [R + r\cos(2\pi\varphi)] \cdot \cos(2\pi\psi(\varphi)) \end{cases} \qquad (4.15)$$

where

$$\psi = \psi(\varphi) = \omega\varphi + C, \quad C = \text{const}, \quad -\infty < \varphi < +\infty, \qquad (4.16)$$

and ω is an irrational number.

The partition of the universal covering plane $\Pi^2(\varphi, \psi)$ on the parallel lines (4.16) is called a linear foliation $\overline{Q}$, covering the linear irrational foliation Q on the torus T^2 by virtue of the covering $f \colon \Pi^2 \mapsto T^2 = \Pi^2/G$.

Poincaré's rotation number.

Remark 4.1. Onto the universal covering plane $\Pi^2(\varphi, \psi)$ with an irrational number ω and the constant $C = C_0$, fix the straight line $\overline{L}_0 \colon \psi = \omega\varphi + C_0$.

At the covering $f\colon \Pi^2 \mapsto T^2$ of the kind (4.14), the straight line $\overline{L}_0$ is projected into a certain layer $L_0 = f(\overline{L}_0)$ of the foliation Q on the torus T^2. But then the countable set of straight lines $\overline{L}_{m,n}\colon \psi = \omega\varphi + C_0 + \omega \cdot m + n$, when $m,\, n = 0, \pm1, \pm2, \ldots$ is projected also into the layer L_0.

Remark 4.2. Because $\lim_{\varphi \to \pm\infty} \frac{\psi(\varphi)}{\varphi} = \lim_{\varphi \to \pm\infty} \frac{\omega\varphi + C}{\varphi} = \omega$, then in this situation the number ω is Poincaré's irrational rotation number of the linear irrational foliation Q on the torus T^2, independently from the choice of the constants C in (4.16). For further detail about the concept of Poincaré rotation numbers for dynamical systems and foliations on the torus T^2 see [48–50].

Next on torus T^2 assume the linear irrational foliation Q without singularities and the irrational number ω is Poincaré's rotation number. By virtue of the properties of the covering $f\colon \Pi^2 \mapsto T^2 = \Pi^2/G$, the linear foliation $\overline{Q}$ of the kind (4.16), given on the plane Π^2 (φ, ψ), is "embedded" on Π^2 into the flow $\overline{F^t}$ without equilibrium; this flow covers some transitive flow F^t without equilibrium on the torus T^2 [57–60].

The flow $\overline{F^t}$, that is, the dynamic system without equilibrium on Π^2, can be set, for example, in the form:

$$\begin{cases} \dfrac{d\varphi}{dt} = P(\varphi, \psi) = 1, \\[2mm] \dfrac{d\psi}{dt} = Q(\varphi, \psi) = \omega. \end{cases} \tag{4.17}$$

We define the following initial conditions for the trajectories $\varphi = \varphi(t)$, $\psi = \psi(t)$ of this system:

$$\varphi = \varphi(0) = 0, \quad \psi = \psi(0) = C = \text{const.}$$

Then the solution of this system is the following:

$$\varphi = \varphi(t) = t, \quad \psi = \psi(t) = \omega \cdot t + C.$$

For the arbitrary trajectory $\varphi = \varphi(t)$, $\psi = \psi(t)$ of the flow $\overline{F^t}$, the following limit exists:

$$\lim_{t \to \pm\infty} \frac{\psi(t)}{\varphi(t)} = \lim_{t \to \pm\infty} \frac{\omega \cdot t + C}{t} = \omega, \tag{4.18}$$

which does not depend on the choice of the flow path $\overline{F^t}$, where ω coincides with Poincaré's irrational rotation number of the linear irrational foliation Q on the torus T^2.

Therefore, the number ω is also called Poincaré's irrational rotation number of the transitive flow F^t without equilibrium on the torus T^2, where the flow $\overline{F^t}$ of the kind (4.17) is a covering flow on the universal covering plane Π^2.

Note that the flow $\overline{F^t}$ can be set in a form other than (4.17). For example, in the form:

$$\begin{cases} \dfrac{d\varphi}{dt} = P(\varphi, \psi) = \lambda(\varphi, \psi) \\[2mm] \dfrac{d\psi}{dt} = Q(\varphi, \psi) = \lambda(\varphi, \psi) \cdot \omega \end{cases} \qquad (4.19)$$

where $\lambda(\varphi, \psi) \neq 0$ is any smooth periodic function of the variables φ, ψ with the period $T = 1$.

But then the transitive flow F^t, without equilibrium states, on the torus T^2, for which the covering flow $\overline{F^t}$ on the plan Π^2 is the flow of the kind (4.19), will *differ* from the transitive flow, without equilibrium states, on the torus T^2, for which the covering flow on the plane Π^2 is the flow of the kind (4.17).

But the flow $\overline{F^t}$ of the kind (4.19) for its trajectories $\varphi = \varphi(t)$, $\psi = \psi(t)$ has the same limit $\lim_{t \to \pm\infty} \frac{\psi(t)}{\varphi(t)} = \omega$, as for the trajectories of the flow (4.17).

We also note that as for (4.17) and (4.19), the linear foliation $\overline{Q}$ of the kind (4.16), defined on the universal covering plane Π^2, does not change. This follows from the fact that the layers of the foliation $\overline{Q}$ of the kind (4.16) are *integral curves* for the dynamical systems (4.17) and (4.19), because they satisfy the differential equation:

$$\frac{d\psi}{d\varphi} = \frac{Q(\varphi, \psi)}{P(\varphi, \psi)} = \omega. \qquad (4.20)$$

The general solution of the differential equation (4.20) has the form:

$$\psi = \psi(\varphi) = \omega\varphi + C,$$

that is, it coincides with the views of the layers of the foliation $\overline{Q}$ of the kind (4.16).

4.3.2.   Terminology

The terms "flow" and "cascade" were introduced by D. V. Anosov in [61]. The dynamical system with continuous time is called a flow. The dynamical system with discrete time is called a cascade.

Any dynamical system is a mathematical model of some object, process or phenomenon and is characterized by its initial state and law under which the system switches from the initial state to another state.

For flows, the state of the system is defined for each point of time t on the real axis. The flow, which has a dense semi-trajectory everywhere, is called a transitive flow.

Integral curves of flows are foliations. However, the foliations with singularities and without singularities as mathematical objects can be considered independently from the flows.

For cascades, the behavior of the system is described by a sequence of states.

Definition 4.5. Foliations Q_1, Q_2 (possibly with singularities), given on the closed surface M^2, are topologically equivalent, if there exists a homeomorphism $f\colon M^2 \to M^2$, which transforms the layers of the foliation Q_1 into the layers of the foliation Q_2.

A necessary and sufficient condition in order for the linear irrational foliations Q_1, Q_2 without singularities, given on the torus T^2 and having Poincaré's irrational rotation numbers ω_1, ω_2, to be topologically equivalent, is the fact that there exists at least one integer unimodular matrix $A = \begin{pmatrix} a_{11} & a_{12} \\ a_{21} & a_{22} \end{pmatrix}$ so that Poincaré's irrational rotation numbers ω_1 and ω_2 satisfy the relation (see [34, 37, 42]):

$$\omega_2 = \frac{a_{21} + a_{22}\omega_1}{a_{11} + a_{12}\omega_1}. \tag{4.21}$$

The unimodularity of matrix A means that its determinant $\det(A) = \pm 1$.

4.3.3. Examples

Example 4.1. We show that two linear irrational foliations Q_1 and Q_2 without singularities, defined on the torus T^2 and having Poincaré's irrational rotation numbers $\omega_1 = \sqrt{2}$, $\omega_2 = \frac{\sqrt{2}}{2}$, are topologically equivalent.

We can use the integer unimodular matrix $A = \begin{pmatrix} a_{11} & a_{12} \\ a_{21} & a_{22} \end{pmatrix} = \begin{pmatrix} 2 & 1 \\ 1 & 1 \end{pmatrix}$ to verify the condition (4.21). Then the right-hand side of (4.21) takes the form:

$$\frac{a_{21} + a_{22}\omega_1}{a_{11} + a_{12}\omega_1} = \frac{1 + \sqrt{2}}{2 + \sqrt{2}}$$

$$= \frac{(1+\sqrt{2})(2-\sqrt{2})}{(2+\sqrt{2})(2-\sqrt{2})}$$

$$= \frac{2-\sqrt{2}+2\sqrt{2}-2}{4-2}$$

$$= \frac{\sqrt{2}}{2} = \omega_2.$$

Consequently, the relation (4.21) is satisfied. And then, according to the above, the linear irrational foliations Q_1, Q_2, defined on the torus T^2, are topologically equivalent.

Definition 4.6. [62–64]. The real numbers Φ_λ, defined by the formula

$$\Phi_\lambda = \frac{\lambda + \sqrt{4+\lambda^2}}{2}, \tag{4.22}$$

where $\lambda \neq 0$ are real numbers, are called metallic proportions.

For the cases $\lambda = 1, 2, 3, 4$ the metallic proportions (4.22) are called golden, silver, bronze, and copper proportions, respectively.

Note that for any integer values $\lambda \neq 0$ the metallic proportion Φ_λ is an irrational number.

Definition 4.7. We say that the linear irrational foliation Q without singularities on the torus T^2 is called a m-metallic irrational foliation on the torus T^2, if Poincaré's rotation number of the foliation Q satisfies the condition $\omega = \Phi_\lambda$, where $\lambda = m$ is a non-zero integer. For the cases $m = 1, 2, 3, 4$ the m-metallic irrational foliations Q without singularities on the torus T^2 are called golden, silver, bronze, and copper m-metallic irrational foliations, respectively.

Further, let Q be an m-metallic irrational foliation on the torus T^2.

Example 4.2. Now we show that the golden irrational foliation Q_1, which has Poincaré's irrational rotation number $\omega_1 = \frac{1+\sqrt{5}}{2}$, is not topologically equivalent to the silver irrational foliation Q_2, which has Poincaré's irrational rotation number $\omega_2 = 1 + \sqrt{2}$.

Assume the contrary, that they are equivalent. Then we must find at least one integral unimodular matrix $A = \begin{pmatrix} a_{11} & a_{12} \\ a_{21} & a_{22} \end{pmatrix}$, $\Delta = \det(A) = \pm 1$, which connects Poincaré's rotation irrational numbers $\omega_1 = \frac{1+\sqrt{5}}{2}$ and $\omega_2 = 1 + \sqrt{2}$ by the relation (4.21).

For the given values $\omega_1 = \frac{1+\sqrt{5}}{2}$ and $\omega_2 = 1 + \sqrt{2}$, we get from (4.21) the following equality:

$$n = k_1\sqrt{2} + k_2\sqrt{5} + k_3\sqrt{10}, \qquad (4.23)$$

where n, k_1, k_2, k_3 are the following integers:

$$n = 2a_{11} + a_{12} - 2a_{21} - 2a_{22}, \quad k_1 = -(2a_{11} + a_{12}),$$
$$k_2 = a_{22} - a_{12}, \quad k_3 = -a_{12}. \qquad (4.24)$$

But then from (4.21) and $\Delta = \det(A) = \pm 1$ we get the following relations:

$$a_{11} = \frac{k_1 - k_3}{2}, \; a_{12} = -k_3, \; a_{21} = \frac{2\Delta - (k_1 - k_3)(k_2 - k_3)}{2k_3}, \; a_{22} = k_2 - k_3. \qquad (4.25)$$

The right-hand side $k_1\sqrt{2} + k_2\sqrt{5} + k_3\sqrt{10}$ of equality (4.23) is either equal to 0, when $k_1 = k_2 = k_3 = 0$, or is an irrational number, when $k_1^2 + k_2^2 + k_3^2 > 0$.

If $k_1 = k_2 = k_3 = 0$, then we get from (4.23) that $a_{11} = 0$, $a_{12} = 0$, $a_{21} = \infty$, $a_{22} = 0$, which is impossible.

If $k_1^2 + k_2^2 + k_3^2 > 0$, then the right-hand side of (4.23) is an irrational number, that is, the left-hand side n of the equality (4.23) is an irrational number, which is impossible, because $n = 2a_{11} + a_{12} - 2a_{21} - 2a_{22}$ is an integer.

Definition 4.8. The linear rational foliation Q^* without singularities on the torus T^2 is called a partition of the torus T^2 into the disjoint closed curves (closed layers), having the form (4.14) for the additional condition that φ, ψ satisfy the relation:

$$\Omega(\varphi, \psi) = p\varphi - q\psi = C = \text{const}, \qquad (4.26)$$

where $p, q(p^2 + q^2 > 0)$ are co-prime integers.

A partition of the universal covering plane $\Pi^2(\varphi, \psi)$ on parallel lines of the form (4.26) is called a linear foliation $\overline{Q}^*$, covering the linear rational foliation Q^* on the torus T^2 by virtue of the covering (4.14) $f \colon \Pi^2 \mapsto T^2$.

Remark 4.3. Poincaré's rotation number ω^* of the linear rational foliation Q^* without singularities on the torus T^2, is calculated by the formula:

$$\omega^* = \lim_{\varphi \to \pm\infty} \frac{\psi(\varphi)}{\varphi}. \qquad (4.27)$$

According to Definition 4.8, p, q are co-prime numbers, $p^2 + q^2 > 0$. Therefore, taking into consideration the formula (4.26), we get for ω^* the following possible values:

(1) $\omega^* = \frac{p}{q}$ is the rational irreducible fraction $\frac{p}{q}$, if $p \neq nq$, $q \neq 0$, and n is an integer.

(2) $\omega^* = n$, if $p = nq$, $q \neq 0$, and n is an integer.

(3) $\omega^* = \infty$, if $q = 0$.

Note that the layers (4.26) of the linear foliation $\overline{Q}^*$ are the integral curves of the flow on the plane Π^2 (φ, ψ), given by the system:

$$\begin{cases} \dfrac{d\varphi}{dt} = P(\varphi, \psi) = q \\ \dfrac{d\psi}{dt} = Q(\varphi, \psi) = p \end{cases} \qquad (4.28)$$

The general solution of the system (4.28) has the form:

$$\varphi = \varphi(t) = q \cdot t + C_1, \quad \psi = \psi(t) = p \cdot t + C_2, \qquad (4.29)$$

where C_1, C_2 are constants.

Any trajectory $\varphi = \varphi(t)$, $\psi = \psi(t)$ of the system (4.28) has the following limit:

$$\lim_{t \to \pm\infty} \frac{\psi(t)}{\varphi(t)} = \lim_{t \to \pm\infty} \frac{p \cdot t + C_2}{q \cdot t + C_1} = \omega^*. \qquad (4.30)$$

Note that the possible values of ω^* in (4.30) are given in Remark 4.3 and they do not depend on the choice of the flow path (4.29).

Note also that ω^* coincides with Poincaré's rotation number of the rational linear foliation Q^* without singularities, given on the torus T^2. Therefore, the number ω^* is also called Poincaré's rotation number of the flow without equilibrium on the torus T^2; moreover, all trajectories of this flow are closed curves.

Definition 4.9. We say that the linear rational foliation Q^* without singularities on the torus T^2 is called a suitable linear rational foliation for the linear rational foliation Q without singularities on the torus T^2, if Poincaré's rotation number ω^* for the foliation Q^* is a suitable fraction for Poincaré's irrational rotation number ω of the foliation Q at the expansion of ω into a continued fraction.

We can see in [65] and [66] the expansion of irrational numbers ω into continued fractions and finding suitable fractions for ω^*.

Example 4.3. Let us find the suitable linear rational foliation Q^* without singularities on the torus T^2 for the golden irrational foliation Q on the torus T^2. For such a foliation, Poincaré's irrational rotation number is the golden ratio $\omega = \Phi = \frac{1+\sqrt{5}}{2}$.

The expansion of $\omega = \Phi$ into a continued fraction has the following form:

$$\Phi = 1 + \cfrac{1}{1 + \cfrac{1}{1 + \cfrac{1}{1 + \cfrac{1}{1 + \cdots}}}}, \tag{4.31}$$

that is, the golden ratio is a periodic continued fraction with a period of 1.

Note that there is also another method of presenting a continued fraction. For example, the golden ratio (4.31) can be represented as $\omega = \Phi = [1; (1)]$.

The suitable fractions for the golden ratio are rational numbers of the form $\omega^* = \frac{p}{q}$, where $p > 0$, $q > 0$ are Fibonacci numbers.

Fibonacci numbers are based on the following recursion (first series):

$$F_n = F_{n-1} + F_{n-2}; \quad F_0 = 0, \quad F_1 = 1 \quad (n = 0, 1, 2, 3, \ldots). \tag{4.32}$$

Hence we get a second series of Fibonacci numbers:

$$F_{-n} = (-1)^n F_n \quad (n = 0, 1, 2, 3, \ldots). \tag{4.33}$$

The first and second series of Fibonacci numbers are presented in Table 4.4 where they are also called "extended" Fibonacci numbers (a name already used in Chapter 1).

The following remarkable relations link Fibonacci numbers F_n with the golden ratio Φ:

$$\begin{cases} \Phi = \lim\limits_{n \to +\infty} \dfrac{F_n}{F_{n-1}}, \\ 1 = \dfrac{F_2}{F_1} < \dfrac{F_4}{F_3} < \dfrac{F_6}{F_5} < \cdots < \Phi < \dfrac{F_{2k+1}}{F_{2k}} < \cdots < \dfrac{F_7}{F_6} < \dfrac{F_5}{F_4} < \dfrac{F_3}{F_2} < \dfrac{F_1}{F_0} \\ = +\infty. \end{cases} \tag{4.34}$$

Table 4.4.   The "extended" Fibonacci numbers.

n	0	1	2	3	4	5	6	7	8	9	10
F_n	0	1	1	2	3	5	8	13	21	34	55
F_{-n}	0	1	-1	2	-3	5	-8	13	-21	34	-55

According to [65], the golden ratio Φ among all irrationalities are approximated by suitable fractions least efficiently. This follows from the fact that, if any irrational number α in the form of the continued fraction $\alpha = [a_0; a_1, a_2, \ldots, a_k, \ldots]$ is present, then, by virtue of [65] we have the following inequalities:

$$\frac{1}{q_k^2(a_{k+1} + 2)} < \left| \alpha - \frac{p_k}{q_k} \right| \leq \frac{1}{q_k^2 a_{k+1}} \quad (k = 1, 2, 3, \ldots). \qquad (4.35)$$

For the golden ratio Φ, when all the values $a_k = 1$, then the inequalities (4.35) can be rewritten as follows:

$$\frac{1}{3q_k^2} < \left| \Phi - \frac{p_k}{q_k} \right| \leq \frac{1}{q_k^2},$$

where the suitable fractions have the following form:

$$\omega_k^* = \frac{p_k}{q_k} = \frac{F_{k+1}}{F_k} \quad (k = 1, 2, 3, \ldots).$$

For the case $k = 0$, we have:

$$\omega_0^* = \frac{p_0}{q_0} = \frac{F_1}{F_0} = \frac{1}{0} = \infty.$$

Regarding this, Aleksandr Khinchin writes [65]:

"The following conclusion follows from this: the fraction $\frac{p_k}{q_k}$ approximates closer to the number α, the greater the element a_{k+1}. Since the suitable fractions in all cases are the best approximations, we come to the conclusion that the best approximation by suitable rational fractions should allow such irrationalities, among whose elements large integers may occur.

In particular, the lowest order of approximation can allow irrationalities with bounded elements. Thus, it becomes clear why we have repeatedly chosen for this purpose the golden ratio $\Phi = \frac{1+\sqrt{5}}{2} = [1; 1, 1, \ldots]$, when we want to get an irrational number, where the approximation is not allowed above a certain given order. Among all irrationalities, the golden ratio obviously is made up of the smallest possible elements, meaning that the golden ratio is approximated by suitable rational fractions in the slowest possible manner."

Recall the algorithm of the expansion of any real number $\alpha \neq 0$ into the continued fraction $\alpha = [a_0; a_1, a_2, \ldots, a_k, \ldots]$ and obtaining the suitable rational fractions for the continued fraction [65, 66]. We will designate by the symbol $[\alpha]$ the integer part of the number α, meaning by this the largest integer, which is less than or equal to α, that is, $\alpha - 1 < [\alpha] \leq \alpha$. The integer $[\alpha]$ is called a floor.[a]

For this case the integer a_0 (possible negative) of the continued fraction $\alpha = [a_0; a_1, a_2, \ldots, a_k, \ldots]$, is the floor, and the numbers $a_1, a_2, \ldots, a_k, \ldots$ are natural numbers. For the case when the integer $[\alpha]$ is the floor, the fractional part $\{\alpha\}$ of the number α has the form $\{\alpha\} = \alpha - [\alpha] \geq 0$.

The algorithm for obtaining the continued fraction $[a_0; a_1, a_2, \ldots, a_k, \ldots]$ for the real number $\alpha \neq 0$ is

$$
\begin{cases}
a_0 = [\alpha] \Rightarrow \alpha_0 = \alpha - a_0, \\[2mm]
a_1 = \left[\dfrac{1}{\alpha_0}\right] \Rightarrow \alpha_1 = \dfrac{1}{\alpha_0} - a_1, \\[2mm]
\qquad \vdots \\[2mm]
a_k = \left[\dfrac{1}{\alpha_{k-1}}\right] \Rightarrow \alpha_k = \dfrac{1}{\alpha_{k-1}} - a_k, \\[2mm]
\qquad \vdots
\end{cases}
\tag{4.36}
$$

The algorithm for obtaining the suitable fractions $\frac{P_k}{Q_k}$ for the given continued fraction $[a_0; a_1, a_2, \ldots, a_k, \ldots]$ is

$$
\begin{cases}
P_{-1} = 1, & P_0 = a_0, & P_1 = a_1 P_0 + P_{-1}, \ldots, P_k = a_k P_{k-1} + P_{k-2}, \ldots \\
Q_{-1} = 0, & Q_0 = 1, & Q_1 = a_1 Q_0 + Q_{-1}, \ldots, Q_k = a_k Q_{k-1} + Q_{k-2}, \ldots
\end{cases}
\tag{4.37}
$$

4.3.4.   Conclusions to Section 4.3

(1) The golden linear irrational foliation Q without singularities on the torus T^2 is approximated least efficiently on the torus T^2 by the suitable linear rational foliations Q^*, in which all the layers are the closed curves: here the numerators and denominators of the suitable fractions $\omega^* = \frac{p}{q}$, which are Poincaré's rotation numbers, coinciding with Fibonacci numbers.

(2) The same procedure of estimation for approximation by suitable rational foliations can be applied to silver, bronze, copper and other metallic

---

[a]Not to be confused with the definition of the integer $[\alpha]$, called a ceiling, for the condition $\alpha \leq [\alpha] < \alpha + 1$. Sometimes the concept of a ceiling is used in calculators. For the ceiling the fractional part is $\{\alpha\} = \alpha - [\alpha] \leq 0$.

foliations without singularities on the torus T^2; also the same procedure can be applied to the *flows* on the torus T^2, where these foliations are inserted into these flows and are integral curves of these flows.

4.4.   The Metallic Irrational Foliations with Four Needle Type Singularities on the Two-dimensional Sphere S^2

Mark the four points on the torus T^2 of the form (4.14) as follows:

$$\begin{cases} s_1(x=0, y=0, z=R+r) \Leftrightarrow \varphi_1 = 0, & \psi_1 = 0; \\[2mm] s_2(x=0, y=0, z=R-r) \Leftrightarrow \varphi_2 = \dfrac{1}{2}, & \psi_2 = 0; \\[2mm] s_3(x=0, y=0, z=-R+r) \Leftrightarrow \varphi_3 = \dfrac{1}{2}, & \psi_3 = \dfrac{1}{2}; \\[2mm] s_4(x=0, y=0, z=-R-r) \Leftrightarrow \varphi_4 = 0, & \psi_4 = \dfrac{1}{2}. \end{cases} \qquad (4.38)$$

Fix any irrational number ω and consider on the torus T^2 the linear irrational foliation Q with Poincaré's rotation number ω.

Then, on the torus T^2 the foliation Q has four layers $L_1^0, L_2^0, L_3^0, L_4^0$ such that the marked points $s_1 \in L_1^0$, $s_2 \in L_2^0$, $s_3 \in L_3^0$, $s_4 \in L_4^0$.

As Q is the linear foliation on the torus T^2 with the Poincaré irrational rotation number ω, the layers of the covering foliation $\overline{Q}$, by virtue of the projection (4.14), have on the universal covering surface Π^2 (φ, ψ) the form (4.16) $\psi = \omega\varphi + C$, where $C = $ const.

By virtue of (4.38), the pre-image of the points $s_1, s_2, s_3, s_4 \in T^2$ on the surface Π^2 (φ, ψ) are the following:

$$(\varphi_1 = 0, \ \psi_1 = 0), \quad \left(\varphi_2 = \frac{1}{2}, \ \psi_2 = 0 \right),$$

$$\left(\varphi_3 = \frac{1}{2}, \ \psi_3 = 0 \right), \quad \left(\varphi_4 = 0, \ \psi_4 = \frac{1}{2} \right).$$

Therefore, the layers of the covering foliation $\overline{Q}$, passing through those points, are of the following form:

$$\overline{L_1^0}: \ \psi = \omega\varphi, \quad \overline{L_2^0}: \ \psi = \omega\varphi - \frac{\omega}{2},$$

$$\overline{L_3^0}: \ \psi = \omega\varphi - \frac{\omega-1}{2}, \quad \overline{L_4^0}: \ \psi = \omega\varphi + \frac{1}{2}. \qquad (4.39)$$

Then under the action of continuous (i.e. not one-to-one) mapping:

$$f\colon X = X(x,y,z) = x^2 - y^2,$$
$$Y = Y(x,y,z) = 2xy, \quad Z = Z(x,y,z) = z, \tag{4.40}$$

the torus (4.14) T^2 is converted into the spherical surface $S^2 = f(T^2)$, and is homeomorphic to the standard sphere: $X^2 + Y^2 + Z^2 = 1$.

Later this custom spherical surface S^2 will, for the sake of brevity, be called a *field*. In the parametric form of the equation, the resulting sphere S^2 has the form:

$$\begin{cases} X = X(\varphi,\psi) = [x(\varphi,\psi)]^2 - [y(\varphi,\psi)]^2 \\ Y = Y(\varphi,\psi) = 2x(\varphi,\psi) \cdot y(\varphi,\psi) \\ Z = Z(\varphi,\psi) = z(\varphi,\psi) \end{cases} \tag{4.41}$$

where $x = x(\varphi,\psi)$, $y = y(\varphi,\psi)$, $z = z(\varphi,\psi)$ satisfy the conditions of (4.14).

The sphere S^2 of the form (4.41) is shown in Fig. 4.3.

At the mapping $f\colon T^2 \mapsto S^2$ of the form (4.41), each of the marked points $\{s_1, s_2, s_3, s_4\} \in T^2$ of the form (4.38) pass bijectively to the corresponding

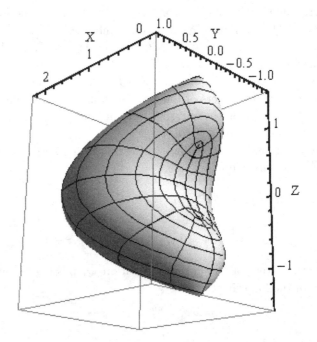

Figure 4.3.   The sphere S^2 of the form (4.41).

points $\{q_1, q_2, q_3, q_4\} \in S^2$ of the form:

$$\begin{cases} q_1(X=0,\ Y=0, Z=R+r), & q_2(X=0,\ Y=0,\ Z=R-r), \\ q_3(X=0,\ Y=0, Z=-R+r), & q_4(X=0,\ Y=0,\ Z=-R-r). \end{cases}$$

$$(4.42)$$

We call the points at (4.42) the marked points of the sphere S^2.

To now find the inverse mapping $f^{-1}\colon S^2 \mapsto T^2$ we represent the original mapping (4.40) $f\colon T^2 \mapsto S^2$ in complex form:

$$f\colon X = x^2 - y^2, \quad Y = 2xy,$$
$$Z = z \Leftrightarrow X + iY = (x+iy)^2, Z = z, \quad i = \sqrt{-1}.$$

Hence we obtain the inverse mapping $f^{-1}\colon S^2 \mapsto T^2$ of the form:

$$f^{-1} : (x+iy) = \sqrt{X+iY} = \sqrt[4]{X^2+Y^2}\left[\cos\left(\frac{\theta}{2}\right) + i\sin(\frac{\theta}{2})\right], z = Z.$$

Then, we have:

$$x = \sqrt[4]{X^2+Y^2}\cos\left(\frac{\theta}{2}\right), \quad y = \sqrt[4]{X^2+Y^2}\sin\left(\frac{\theta}{2}\right), \quad z = Z,$$

where

$$\cos(\theta) = \frac{X}{\sqrt{X^2+Y^2}}, \quad \sin(\theta) = \frac{Y}{\sqrt{X^2+Y^2}}.$$

As

$$\sin^2\left(\frac{\theta}{2}\right) = \frac{1-\cos(\theta)}{2} = \frac{\sqrt{X^2+Y^2}-X}{2\cdot\sqrt{X^2+Y^2}},$$

$$\cos^2\left(\frac{\theta}{2}\right) = \frac{1+\cos(\theta)}{2} = \frac{\sqrt{X^2+Y^2}+X}{2\cdot\sqrt{X^2+Y^2}},$$

we then get:

$$x = \sigma_1\frac{1}{\sqrt{2}}\sqrt{\sqrt{X^2+Y^2}+X}, \quad y = \sigma_2\frac{1}{\sqrt{2}}\sqrt{\sqrt{X^2+Y^2}-X}, \quad z+Z,$$

$$\sigma_1 = \text{sign}(x), \quad \sigma_2 = \text{sign}(y).$$

We denote $\sigma = \text{sign}(Y)$. Then we get from the condition $Y = 2xy$ the following:

$$\sigma = \sigma_1\sigma_2.$$

Recall that for any real number μ the sign(μ) is defined by the formula:

$$\text{sign}(\mu) = 1, \quad \text{if } \mu > 0$$
$$\text{sign}(\mu) = -1, \quad \text{if } \mu < 0$$
$$\text{sign}(\mu) = 0, \quad \text{if } \mu = 0.$$

As for the other points of the torus T^2, through the action of the inverse (not one-to-one) mapping $f^{-1} \colon S^2 \mapsto T^2$, each point $q \in S^2$, different from the marked points $\{q_1, q_2, q_3, q_4\} \in S^2$ on the torus T^2, corresponds to *two and only two points* s, s' ($s \neq s'$) and these points do not belong to the marked points $\{s_1, s_2, s_3, s_4\}$ of the form (4.38).

In this case the torus T^2 is the two-sheeted ramified covering of the sphere S^2 with the branching of the genus 2 in the marked points $\{s_1, s_2, s_3, s_4\} \in T^2$, which is completely consistent with Adolf Hurwitz's formula [67].

We apply the mapping $f \colon T^2 \mapsto S^2$ of the form (4.40) on the irrational foliation without singularities Q of the form (4.15), given on the torus T^2 with four marked points $\{s_1, s_2, s_3, s_4\} \in T^2$ of the form (4.38) and with the Poincaré irrational rotation number ω.

Then the foliation Q is projected onto the sphere S^2 into the irrational foliation $P = f(Q)$ with the four singularities $\{q_1, q_2, q_3, q_4\} \in S^2$ of the *needle* kind with the same Poincaré irrational rotation number ω, where the points $\{q_1, q_2, q_3, q_4\}$ have the form (4.42) and are the marked points on the sphere S^2.

Remark 4.4. The irrational foliation $P = f(Q)$ has the following form on the sphere S^2:

$$\begin{cases} X = X(\varphi) = r^2 \sin^2(2\pi\varphi) - [R + r\cos(2\pi\varphi)]^2 \sin^2[2\pi\psi(\varphi)], \\ Y = Y(\varphi) = 2r\,[R + r\cos(2\pi\varphi)] \sin(2\pi\varphi)\cos[2\pi\psi(\varphi)], \qquad (4.43) \\ Z = Z(\varphi) = [R + r\cos(2\pi\varphi)]\cos[2\pi\psi(\varphi)], \end{cases}$$

where $0 < r < R, \psi(\varphi) = \omega\varphi + C, C = \text{const}, -\infty < \varphi < +\infty$, and ω is an irrational number.

The separatrix $f(L_i^0)$ ($i = 1, 2, 3, 4$) of the *needle* $q_i \in \{q_1, q_2, q_3, q_4\} \in S^2$ of the irrational foliation $P = f(Q)$ on the sphere S^2 has the form (4.43), where $\psi(\varphi) = \omega\varphi + C_i^0$ ($i = 1, 2, 3, 4$), ω is an irrational number, and $C_1^0 = 0$, $C_2^0 = -\frac{\omega}{2}, C_3^0 = -\frac{\omega-1}{2}, C_4^0 = \frac{1}{2}$.

Similarly, the metallic irrational foliations on the torus T^2 (Definition 4.7) of the foliation $P = f(Q)$ on the sphere S^2 with four regularities of the needle kind are called the metallic irrational foliations on the sphere S^2.

For the cases $m = 1, 2, 3, 4$, the metallic irrational foliations $P = f(Q)$ on the sphere S^2 with four singularities of the needle kind are called, respectively, the golden, silver, bronze, and copper foliations on the sphere S^2.

Note that each point of the torus T^2, given in the form (4.14), remains *stationary*, when changing parameters:

$$(\varphi, \psi) \mapsto (\varphi + m, \psi + n),$$

where m, n are any integers.

Similarly, each point on the sphere S^2, defined in the form (4.41), remains *stationary*, when changing parameters

$$(\varphi, \psi) \mapsto \left[(-1)^k \varphi + m, (-1)^k \psi + n\right],$$

where k, m, and n are any integers.

For the irrational foliation Q on the torus T^2, for which $\psi = \omega \cdot \varphi + C$, the Poincaré rotation number equals $\omega = \lim_{\varphi \to \pm\infty} \frac{\psi}{\varphi} = \lim_{\varphi \to \pm\infty} \frac{\omega\varphi + C}{\varphi}$; then at the projection $f: T^2 \mapsto S^2$ of the form (4.40) we get the Poincaré rotation number ω on the sphere S^2 that remains unchanged, because

$$\lim_{\varphi \to \pm\infty} \frac{(-1)^k \psi + n}{(-1)^k \varphi + m} = \lim_{\varphi \to \pm\infty} \frac{(-1)^k(\omega\varphi + C) + n}{(-1)^k \varphi + m} = \omega. \qquad (4.44)$$

For the foliation $P = f(Q)$ on the sphere S^2, the Poincaré rotation number ω is irrational and any layer of the foliation P (including separatrix of each needle q_1, q_2, q_3, and q_4) are everywhere dense on the sphere S^2.

This property of the foliation $P = f(Q)$ on the sphere S^2 follows from the fact that the layers of the irrational linear foliation Q on the torus T^2, having the same Poincaré irrational rotation number ω, are everywhere dense on the torus T^2.

In this connection, the torus T^2, relative to the projection $f: T^2 \mapsto S^2$, is the two-sheeted branched covering of the sphere S^2 with the branching of order 2 in the marked points $\{s_1, s_2, s_3, s_4\} \in T^2$, which are projected into the needles $\{q_1, q_2, q_3, q_4\} \in S^3$.

Note that, due to the density of layers P everywhere on the sphere S^2, two different needles cannot belong to the same layer. For more information about the geometric and topologic properties of the foliations with four needles on the sphere S^2, which everywhere have dense layers on the sphere S^2, and the necessary and sufficient conditions for the topological equivalence of such foliations, see [26, 28, 30, 34, 35].

Note the following important property of the *flows* on the sphere S^2 (genus $p = 0$). This property consists of the fact that on the sphere S^2 (genus $p = 0$), on the projective plane P^2 (genus $p = 1$) and the Klein bottle K^2 (genus $p = 2$), flows *cannot exist* which have at least one non-closed Poisson-stable semi-trajectory.

This means that on the sphere S^2 the non-closed Poisson curve without self-intersections cannot be embedded into the continuous or smooth flow f^t as its trajectory or semi-trajectory. Consequently, on sphere S^2, the irrational foliation P with the four singularities, which are the needles q_1, q_2, q_3, and q_4, cannot be embedded into the flow f^t.

This fact follows directly from the estimations of the maximum possible number N of the independent (i.e. not bounded for each other) non-closed Poisson-stable semi-trajectories of the flows f^t, defined on the closed orientable surfaces M^2 of the genus $p \geq 0$ and the closed non-orientable surfaces of the genus $p \geq 1$.

These estimations for the closed orientable surfaces M^2 of any genus $p \geq 0$ were obtained by the Russian mathematician Artemy Mayer [68] and have the form $N = p$. Thus, on the sphere S^2 (the genus $p = 0$), flows do not exist with open Poisson-stable semi-trajectories.

For the closed orientable surfaces M^2 of any genus of $p = 1, 2, 3, 4, 5, \ldots$, the estimation for the maximum possible number N of the independent non-closed Poisson-stable semi-trajectories of the flows f^t was obtained in 1969 by Samuil Aranson [35] and independently in the same year by Nielsen Markley [69] for the cases $p = 1$ or $p = 2$. The estimation for the cases $p = 3, 4, 5, \ldots$, was obtained independently in 1970 by Nielsen Markley [70]. This estimation has the following form:

$$N = \left[\frac{p-1}{2}\right], \tag{4.45}$$

where the bracket symbol $[\ ]$ means the integer part, understood in the sense of the floor

$$p - 1 < N = \left[\frac{p-1}{2}\right] \leq p$$

but not the ceiling

$$p \leq N < p + 1.$$

Hence, in particular, it follows that for $p = 1$ (the projective plane P^2) and $p = 2$ (Klein bottle), the number $N = 0$. This means that the flows f^t with open Poisson-stable semi-trajectories cannot exist on such surfaces.

Currently, this important estimation $N = \left[\frac{p-1}{2}\right]$ for the maximum possible number N of independent non-closed Poisson-stable semi-trajectories of the flows f^t on the closed orientable surfaces M^2 of the genus of $p \geq 1$ is called the Aranson–Markley theorem.

4.5. Anosov's Automorphisms (Hyperbolic Automorphisms) on the Two-dimensional Torus T^2 and the Metallic Proportions

4.5.1. Preliminaries

In this section we consider an important part of modern theoretical natural sciences, having not only scientific, but also great practical significance. We are talking about the dynamical systems associated with U-flows and U-cascades, defined on manifolds of any dimension [61].

Currently, instead of the term U-cascades, the more commonly used terms are Anosov's diffeomorphisms (proposed by the American mathematician Stephen Smale) as well as hyperbolic diffeomorphisms. Such diffeomorphisms were the subject of research of many mathematicians including Dmitry Anosov, Yakov Sinai, Vladimir Arnold, Andre Avez, Zbigniew Nitezki, Issac Kornfeld, Sergey Fomin, and Sergei Pilyugin (see [61, 71–79]).

Here we restrict ourselves to the simplest variant of such diffeomorphisms, namely, Anosov's automorphisms on the torus T^2, for which we point out their unexpected connection with the metallic proportions and the golden ratio. To our knowledge, we were the first to establish this relationship.

For visual perception, we will present the two-dimensional torus T^2 as a geometric surface in three-dimensional Euclidean space $R^3(x, y, z)$, analytically in the form (4.13) (implicit form), or in the form (4.14) (parametric form).

Representation of the torus T^2 in the parametric form (4.14) can be considered to be the covering $f: \Pi^2 \mapsto T^2 = \Pi^2/G$ of the universal covering plane $\Pi^2(\varphi, \psi)$ on the torus T^2, where G is a group of the covering, whose elements $g \in G$ are the integer shifts of the form:

$$g: \varphi' = \varphi + m, \psi' = \psi + n \ (m, n = 0, \pm 1, \pm 2, \ldots).$$

Definition 4.10. Anosov's automorphism on the two-dimensional torus T^2 is just such a diffeomorphism[b] $h\colon T^2 \mapsto T^2$, which has the covering transformation $\bar{h}\colon \Pi^2 \mapsto \Pi^2$ on the universal covering plane $\Pi^2(\varphi, \psi)$ of the form:

$$\bar{h} = \begin{pmatrix} \varphi' \\ \psi' \end{pmatrix} = \begin{pmatrix} a_{11} & a_{12} \\ a_{21} & a_{22} \end{pmatrix} \begin{pmatrix} \varphi \\ \psi \end{pmatrix}, \qquad (4.46)$$

where $A = \begin{pmatrix} a_{11} & a_{12} \\ a_{21} & a_{22} \end{pmatrix}$ is an integer unimodular matrix, for which the eigenvalues λ_1, λ_2 satisfy the conditions:

$$|\lambda_1| < 1, \quad |\lambda_2| > 1. \qquad (4.47)$$

Recall that the roots of the equation $\det(A - \lambda E) = 0$ are called the eigenvalues λ_1, λ_2 of the matrix, that is, λ_1, λ_2 are the roots of the quadratic equation:

$$\lambda^2 - p\lambda + \Delta = 0, \qquad (4.48)$$

where $p = \mathrm{Tr}(A) = a_{11} + a_{22}$ is a trace of the matrix $A, \Delta = \det(A)$.

As for the torus T^2, the matrix A is integer and unimodular, that is, $a_{ij} \in Z$ (Z is the set of integers), $\Delta = \det(A) = \pm 1$, where in (4.48) the coefficient p, Δ are integers, where $\Delta = \pm 1$.

The roots λ_1, λ_2 of the equation (4.48) have the following form:

$$\lambda_1 = \frac{1}{2}(p - \sigma\sqrt{p^2 - 4\Delta}), \quad \lambda_2 = \frac{1}{2}(p + \sigma\sqrt{p^2 - 4\Delta}), \quad \sigma = \mathrm{sign}(p) = \pm 1. \qquad (4.49)$$

Let the diffeomorphism $h\colon T^2 \mapsto T^2$ be Anosov's automorphism. Then, according to Definition 4.10, depending on the choice $\Delta = \pm 1, \sigma = \pm 1$, the additional condition (4.47) is observed.

4.5.2. *Properties of Anosov's automorphisms on the torus T^2*

The following Proposition 4.1 can be taken, for example, from the works [61, 76, 79].

---

[b]Diffeomorphism on the torus T^2 is the following one-to-one and smooth mapping $h\colon T^2 \mapsto T^2$; and its inverse mapping $h^{-1}\colon T^2 \mapsto T^2$ is also a smooth one-to-one mapping.

Proposition 4.1. [61, 76, 79]. *Let the diffeomorphism* $h: T^2 \mapsto T^2$ *be Anosov's automorphism. Then the covering transformation* $\bar{h}: \Pi^2 \mapsto \Pi^2$ *of the kind* (4.46), *satisfies the following properties:*

(1) $p = a_{11} + a_{22} \neq 0, a_{12} \neq 0, a_{21} \neq 0, p^2 - 4\Delta > 0.$

(2) *The eigenvalues* λ_1, λ_2 *of Anosov's automorphisms on the torus* T^2 *are irrational numbers.*

(3) *Anosov's automorphism* $h: T^2 \mapsto T^2$ *is structurally stable in the* C^1-*topology.*

(4) *Anosov's automorphism* $h: T^2 \mapsto T^2$ *on the torus* T^2 *has two invariant transverse transitive foliations:* compressive *foliation* F^s, *corresponding to the eigenvalue* $\lambda_1, |\lambda_1| < 1$, *and the* expanding *foliation* F^u, *corresponding to the eigenvalue* $\lambda_2, |\lambda_2| > 1$.

(5) *Pre-image on* $\Pi^2(\varphi, \psi)$ *of the foliations* F^s, F^u *are invariant transversal foliations of the form:*

$$\overline{F}^s : \psi = k_1\varphi + C_1, \quad \overline{F}^u : \psi = k_2\varphi + C_2, \quad k_1 = \frac{\lambda_1 - a_{11}}{a_{12}},$$

$$k_2 = \frac{\lambda_2 - a_{11}}{a_{12}}, \quad C_1, C_2 = \text{const.}$$

(6) *Periodic points of Anosov's automorphism* $h: T^2 \mapsto T^2$ *are everywhere dense on the torus* T^2.

(7) *All points on the torus* T^2, *whose pre-images on the universal covering plane* $\Pi^2(\varphi, \psi)$ *have rational coordinates, are periodic points on the torus* T^2 *of Anosov's automorphism and they alone are periodic points.*

(8) *The coefficients* k_1, k_2 *are Poincaré irrational rotation numbers for the transversal linear irrational foliations* F^s, F^u *on the torus* T^2.

Example 4.4(a) and (b). Consider Anosov's diffeomorphism $h: T^2 \mapsto T^2$, where the covering mapping $\bar{h}: \Pi^2 \mapsto \Pi^2$ on the universal covering plane $\Pi^2(\varphi, \psi)$ has the form:

$$\bar{h} : \begin{pmatrix} \varphi' \\ \psi' \end{pmatrix} = \begin{pmatrix} 2 & 1 \\ 1 & 1 \end{pmatrix} \begin{pmatrix} \varphi \\ \psi \end{pmatrix}.$$

In this case, according to (4.49), for the matrix $A = \begin{pmatrix} a_{11} & a_{12} \\ a_{21} & a_{22} \end{pmatrix} = \begin{pmatrix} 2 & 1 \\ 1 & 1 \end{pmatrix}$, with the determinant $\Delta = \det(A) = 1$ we get the following eigenvalues:

$$\lambda_1 = \frac{3 - \sqrt{5}}{2} = \Phi^{-2} \approx 0.382, \quad \lambda_2 = \frac{3 + \sqrt{5}}{2} = \Phi^2 \approx 2.618,$$

where $\Phi = \frac{1+\sqrt{5}}{2} \approx 1.618$ is the golden ratio.

Hence the coefficients k_1, k_2 of the transversal foliations $\overline{F}^s, \overline{F}^u$ on the universal covering plane $\Pi^2(\varphi, \psi)$ take the following values:

$$k_1 = \frac{\lambda_1 - a_{11}}{a_{12}} = \frac{\Phi^{-2} - 2}{1} = -\Phi \approx -1.618,$$

$$k_2 = \frac{\lambda_2 - a_{11}}{a_{12}} = \frac{\Phi^2 - 2}{1} = \Phi^{-1} \approx 0.618.$$

In this case, transversal foliations $\overline{F}^s$, $\overline{F}^u$ on the plane $\Pi^2(\varphi, \psi)$ have the forms:

(1) Compressive foliation $\overline{F}^s$: $\psi = k_1\varphi + C_1 = -1.618\varphi + C_1$ (Fig. 4.4(a));
(2) Expanding foliation $\overline{F}^u$: $\psi = k_2\varphi + C_2 = 0.618\varphi + C_2$ (Fig. 4.4(b)).

Proposition 4.2. *Let the diffeomorphism* $h\colon T^2 \mapsto T^2$ *be Anosov's automorphism. Then, for the covering transformation* $\bar{h}\colon \Pi^2 \mapsto \Pi^2$ *of the form*

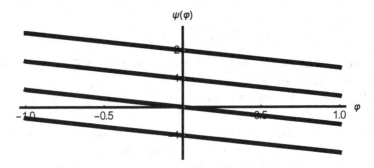

Figure 4.4(a).   Compressive foliation $\overline{F}^s$: $\psi = k_1\varphi + C_1 = -1.618\varphi + C_1(\lambda_1 = 0.382)$.

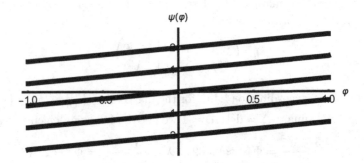

Figure 4.4(b).   Expanding foliation $\overline{F}^u$: $\psi = k_2\varphi + C_2 = 0.618\varphi + C_2(\lambda_2 = 2.618)$.

(4.46) *the following properties are satisfied:*

(1) *The admissible values for $p = a_{11} + a_{22}$ for the case $\Delta = 1$ are the numbers $p = \pm 3, \pm 4, \pm 5, \ldots$ resulting in $|\lambda_1| \leq \Phi^{-2} < 1, |\lambda_2| \geq \Phi^2 > 1$; for the case $\Delta = -1$, the numbers $p = \pm 1, \pm 2, \pm 3, \ldots$ are admissible resulting in $|\lambda_1| \leq \Phi^{-1} < 1, |\lambda_2| \geq \Phi > 1$, where $\Phi = \frac{1+\sqrt{5}}{2} \approx 1.61803$ is the golden ratio.*

(2) λ_1, λ_2 *are the fixed points of the pointwise transformation:*

$$\bar{x} = w(x) = p - \frac{\Delta}{x}, \tag{4.50}$$

here λ_1 ($|\lambda_1| < 1$) is unstable and λ_2 ($|\lambda_2| > 1$) is stable for the transformation (4.50).

(3) $\lambda^n = p\lambda^{n-1} - \Delta\lambda^{n-2}, n \in Z, \lambda = \lambda_1, \lambda_2$, *where $|\lambda_1| < 1, |\lambda_2| > 1$.*

Before giving the proof of Proposition 4.2, we make two remarks:

Remark 4.5. For the case $\Delta = \det(A) = -1$, the quadratic equation (4.48) is the equation for the so-called metallic proportions [62–64]. For this given case, the metallic proportion is called the positive root $\Phi_p = \frac{p+\sqrt{p^2+4}}{2}$ of the equation $x^2 - px - 1 = 0$, where $p = 1, 2, 3, \ldots$.

Recall the common names for the first four metallic proportions:

$$\Phi_1 = \frac{1+\sqrt{5}}{2} \approx 1.618 \text{ (the golden proportion)},$$

$$\Phi_2 = 1 + \sqrt{2} \approx 2.414 \text{ (the silver proportion)},$$

$$\Phi_3 = \frac{3+\sqrt{13}}{2} \approx 3.303 \text{ (the bronze proportion)},$$

$$\Phi_4 = 2 + \sqrt{5} \approx 4.236 \text{ (the copper proportion)}.$$

When $p > 4$ the metallic proportions do not have individual names.

The numerical values of the eigenvalues $\lambda_1, \lambda_2(|\lambda_1| < 1, |\lambda_2| > 1)$ of Anosov's automorphism, depending on the integer values of the coefficient p of the equation (4.48) and the choice of the sign $\sigma = \pm 1$ in (4.37), are presented in Table 4.5 ($\Delta = 1$) and Table 4.6 ($\Delta = -1$).

Remark 4.6. It follows from Propositions 4.1 and 4.2, that for any positive integer n the following alternatives are fulfilled:

(1) If $\Delta = 1$, then $|(\lambda_1)^n| \leq \Phi^{-2n} < 1, |(\lambda_2)^n| \geq \Phi^{2n} > 1$.

(2) If $\Delta = -1$, then $|(\lambda_1)^n| \leq \Phi^{-n} < 1, |(\lambda_2)^n| \geq \Phi^n > 1$, where $\Phi = \frac{1+\sqrt{5}}{2} \approx 1.61803$ is the golden ratio.

Table 4.5. $\Delta = 1$.

$p \geq 0$ $\sigma = 1$	$\lambda_1 \in (0,1)$ $|\lambda_1| < 1$	$\lambda_2 \in (1,+\infty)$ $|\lambda_2| > 1$	$p \leq 0$ $\sigma = -1$	$\lambda_1 \in (-1,0)$ $|\lambda_1| < 1$	$\lambda_2 \in (-\infty,-1)$ $|\lambda_2| > 1$
0	no	no	$-\infty$	$0-0$	$-\infty$
1	no	no	-8	-0.127017	-7.87298
2	no	no	-7	-0.14898	-6.85410
3	**0.381966**	**2.61803**	-6	0.171573	-5.82843
4	0.267949	3.73205	-5	0.208712	-4.79129
5	0.208712	4.79129	-4	-0.267949	-3.73205
6	0.171573	5.82843	-3	**-0.381966**	**-2.61803**
7	0.145898	6.85410	-2	no	no
8	0.127017	7.87298	-1	no	no
$+\infty$	$0+0$	$+\infty$	0	no	no

Table 4.6. $\Delta = -1$.

$p \geq 0$ $\sigma = 1$	$\lambda_1 \in (0,1)$ $|\lambda_1| < 1$	$\lambda_2 \in (1,+\infty)$ $|\lambda_2| > 1$	$p \leq 0$ $\sigma = -1$	$\lambda_1 \in (-1,0)$ $|\lambda_1| < 1$	$\lambda_2 \in (-\infty,-1)$ $|\lambda_2| > 1$
0	no	no	$-\infty$	$0+0$	$-\infty$
1	**-0.618034**	**1.61803**	-8	0.123106	-8.12311
2	-0.414214	2.41421	-7	0.14005	-7.14005
3	-0.302776	3.30278	-6	0.162278	-6.16228
4	-0.236068	4.23607	-5	0.192582	-5.19258
5	-0.192582	5.19258	-4	0.236068	-4.23607
6	-0.162278	6.16228	-3	0.302776	-3.30278
7	-0.140055	7.14005	-2	0.414214	-2.41421
8	-0.123106	8.12311	-1	0.618034	-1.61803
$+\infty$	$0-0$	$+\infty$	0	no	no

Proof of Property 1 of Proposition 4.2. At this point we need to prove that the admissible values of $p = a_{11} + a_{22} \neq 0$ for the case $\Delta = 1$ are the integers $p = \pm 3, \pm 4, \pm 5, \ldots$ and then we have $|\lambda_1| \leq \Phi^{-2} < 1, |\lambda_2| \geq \Phi^2 > 1$, but for the case $\Delta = -1$ the admissible values of p are equal to $p = \pm 1, \pm 2, \pm 3, \ldots$ and then we have $|\lambda_1| \leq \Phi^{-1} < 1, |\lambda_2| \geq \Phi > 1$, where $\Phi = \frac{1+\sqrt{5}}{2} \approx 1.61803$ is the golden ratio.

First we observe that from the recursion $\Phi^n = \Phi^{n-1} + \Phi^{n-2}$ ($n = 0,$ $\pm 1, \pm 2, \pm 3, \ldots$) we get the following values for the golden ratio:

$$
\begin{cases}
\Phi^{-2} = \dfrac{-3 + \sqrt{5}}{2} \approx 0.381 \\[2mm]
\Phi^{-1} = \dfrac{-1 + \sqrt{5}}{2} \approx 0.618 \\[2mm]
\Phi = \dfrac{1 + \sqrt{5}}{2} \approx 1.618 \\[2mm]
\Phi^2 = \dfrac{3 + \sqrt{5}}{2} \approx 2.618
\end{cases}
\qquad (4.51)
$$

Step 1. We show that the numbers $p = \pm 3, \pm 4, \pm 5, \ldots$ are admissible values for $p = a_{11} + a_{22}$ for the case $\Delta = 1$.

By virtue of Property 1 of Proposition 4.1, $p = a_{11} + a_{22} \neq 0, p^2 - 4\Delta > 0$; here the formulas (4.49) must meet the additional requirements $|\lambda_1| < 1, |\lambda_2| > 1$, because $\sigma = \text{sign}(p) = \pm 1$. Then for the case $\Delta = 1$, we find $p^2 > 4$, whence we have $|p| > 2$. Since $p \in Z$, then we have $|p| > 2 \Rightarrow |p| \geq 3 \Rightarrow p = \pm 3, \pm 4, \pm 5, \ldots$. It follows from this, that for the cases $p = 3, 4, 5, \ldots$, by virtue of the requirements $|\lambda_1| < 1, |\lambda_2| > 1$, we get from formulas (4.49) that the sign $\sigma = \text{sign}(p) = 1$, and hence,

$$
0 < \lambda_1 = \frac{1}{2}\left(p - \sqrt{p^2 - 4}\right) < 1 \Rightarrow |\lambda_1| < 1,
$$
$$
1 < \lambda_2 = \frac{1}{2}\left(p + \sqrt{p^2 - 4}\right) < +\infty \Rightarrow |\lambda_2| > 1.
\qquad (4.52)
$$

If $p = -3, -4, -5, \ldots$, then by virtue of the requirements $|\lambda_1| < 1, |\lambda_2| > 1$, we get from the formulas (4.49), that the sign $\sigma = \text{sign}(p) = -1$, and hence,

$$
-1 < \lambda_1 = \frac{1}{2}\left(-|p| + \sqrt{p^2 - 4}\right) < 0 \Rightarrow |\lambda_1| < 1,
$$
$$
-\infty < \lambda_2 = \frac{1}{2}\left(-|p| - \sqrt{p^2 - 4}\right) < -1 \Rightarrow |\lambda_2| > 1.
\qquad (4.53)
$$

Step 2. We show that the numbers $p = \pm 1, \pm 2, \pm 3, \ldots$ are admissible values for $p = a_{11} + a_{22} \neq 0$ for the case $\Delta = -1$.

Since $a_{11}, a_{22} \in Z, p = a_{11} + a_{22} \neq 0$ and $p^2 - 4\Delta > 0$, then for the case $\Delta = -1$ we get:

$$
p^2 - 4\Delta = p^2 + 4 > 0 \Rightarrow p = 0, \pm 1, \pm 2, \pm 3, \ldots.
$$

However, we should exclude the case where $p = 0$, because we get from formulas (4.49):

$$\lambda_1 = \left(p - \sigma\sqrt{p^2 - 4\Delta}\right) = \frac{1}{2}\left(-\sigma\sqrt{4}\right) = -\sigma \Rightarrow |\lambda_1| = 1$$

$$\lambda_2 = \left(p + \sigma\sqrt{p^2 - 4\Delta}\right) = \frac{1}{2}\left(\sigma\sqrt{4}\right) = \sigma \Rightarrow |\lambda_2| = 1$$

which is impossible, because for Anosov's automorphisms on the torus T^2 we have $|\lambda_1| \neq 1, |\lambda_2| \neq 1$.

Therefore, when $\Delta = -1$, we get the admissible values of p are the following integers: $p = 1, 2, 3, \ldots$, for the eigenvalues $\lambda_1, |\lambda_1| < 1$; but for the case $\lambda_2, |\lambda_2| > 1$ the admissible values of p are the following integers: $p = -1, -2, -3, \ldots$.

As for the case $p = 1, 2, 3, \ldots$ the sign $\sigma = \text{sign}(p) = 1$, then for the case $\Delta = -1$ we get:

$$-1 < \lambda_1 = \frac{1}{2}\left(p - \sqrt{p^2 + 4}\right) < 0 \Rightarrow |\lambda_1| < 1,$$
$$1 < \lambda_2 = \frac{1}{2}\left(p + \sqrt{p^2 + 4}\right) < +\infty \Rightarrow |\lambda_2| > 1. \tag{4.54}$$

For the case $p = -1, -2, -3, \ldots$, the sign $\sigma = \text{sign}(p) = -1$, and because of $\Delta = -1$, we get:

$$0 < \lambda_1 = \frac{1}{2}\left(-|p| + \sqrt{p^2 + 4}\right) < 1 \Rightarrow |\lambda_1| < 1,$$
$$-\infty < \lambda_2 = \frac{1}{2}\left(-|p| - \sqrt{p^2 + 4}\right) < -1 \Rightarrow |\lambda_2| > 1. \tag{4.55}$$

Note also for the limit values of eigenvalues λ_1, λ_2 where $p \to \pm\infty$ we have:

$$\Delta = 1 \Rightarrow \lim_{\substack{\sigma=1 \\ p\to+\infty}} \lambda_1 = 0 + 0, \quad \lim_{\substack{\sigma=1 \\ p\to+\infty}} \lambda_2 = +\infty,$$

$$\lim_{\substack{\sigma=-1 \\ p\to-\infty}} \lambda_1 = 0 - 0, \quad \lim_{\substack{\sigma=-1 \\ p\to-\infty}} \lambda_2 = -\infty,$$

$$\Delta = -1 \Rightarrow \lim_{\substack{\sigma=1 \\ p\to+\infty}} \lambda_1 = 0 - 0, \quad \lim_{\substack{\sigma=1 \\ p\to+\infty}} \lambda_2 = +\infty,$$

$$\lim_{\substack{\sigma=-1 \\ p\to-\infty}} \lambda_1 = 0 + 0, \quad \lim_{\substack{\sigma=-1 \\ p\to-\infty}} \lambda_2 = -\infty.$$

Step 3. We show that for the case $\Delta = 1$ we have $|\lambda_1| \leq \Phi^{-2} < 1, |\lambda_2| \geq \Phi^2 > 1$, but for the case $\Delta = -1$ we have $|\lambda_1| \leq \Phi^{-1} < 1, |\lambda_2| \geq \Phi > 1$, where $\Phi = \frac{1+\sqrt{5}}{2}$ is the golden ratio.

For the admissible values of $p = a_{11} + a_{22}$ (p is the trace of the matrix A), $p^2 - 4\Delta > 0, \sigma = \text{sign}(p) = \pm 1$, the derivatives $\frac{\partial \lambda_1}{\partial p}, \frac{\partial \lambda_2}{\partial p}$ of the functions (4.49) have the form:

$$\frac{\partial \lambda_1}{\partial p} = \frac{1}{2}\left(1 - \frac{\sigma p}{\sqrt{p^2 - 4\Delta}}\right), \qquad \frac{\partial \lambda_2}{\partial p} = \frac{1}{2}\left(1 + \frac{\sigma p}{\sqrt{p^2 - 4\Delta}}\right). \quad (4.56)$$

Recall that, if the derivative is less than zero, then the corresponding function is decreasing, and if greater than zero, it is increasing.

Option 1. (Numerical values, see Table 4.5). Let $\Delta = 1, \sigma = 1, p = 3, 4, 5, \ldots$. Then we get the following expressions:

$$\frac{\partial \lambda_1}{\partial p} = \frac{1}{2}\left(1 - \frac{p}{\sqrt{p^2 - 4\Delta}}\right) \leq -0.17082,$$

$$\frac{\partial \lambda_2}{\partial p} = \frac{1}{2}\left(1 + \frac{p}{\sqrt{p^2 - 4\Delta}}\right) > -0.17082 > 0. \quad (4.57)$$

As $\frac{\partial \lambda_1}{\partial p} < 0$, then the function $\lambda_1(p) = \frac{1}{2}(p - \sqrt{p^2 - 4})$ decreases and the maximum of the function $\lambda_1(p)$ is reached at $p = 3$. We then obtain the following value for the **maximum**:

$$\lambda_1(p) = \frac{1}{2}\left(p - \sqrt{p^2 - 4}\right) = \lambda_1(p = 3) = \frac{3 - \sqrt{5}}{2} = \Phi^{-2} \approx 0.381 < 1. \quad (4.58)$$

As for $p \to \pm\infty$ we have $\lambda_1(p) \to 0 + 0$, then $\lambda_1(p)$ satisfies the following inequalities:

$$0 < \lambda_1(p) \leq \frac{3 - \sqrt{5}}{2} < 1 \Rightarrow \lambda_1(p) \leq \frac{3 - \sqrt{5}}{2} = \Phi^{-2} < 1. \quad (4.59)$$

As $\frac{\partial \lambda_2}{\partial p} > 0$, the function $\lambda_2(p) = \frac{1}{2}\left(p + \sqrt{p^2 - 4}\right)$ increases and its minimum is attained at $p = 3$. From this we get the following value for the **minimum**:

$$\left[\lambda_2(p) = \frac{1}{2}\left(p + \sqrt{p^2 - 4}\right)\right] = \lambda_2(p = 3) = \frac{3 + \sqrt{5}}{2} = \Phi^2 \approx 2.618 > 1. \quad (4.60)$$

As $p \to +\infty$, the function $\lambda_2(p) \to +\infty$, then the function $\lambda_2(p)$ satisfies the following inequalities:

$$1 < \frac{3 + \sqrt{5}}{2} \leq \lambda_2(p) < +\infty \Rightarrow |\lambda_2(p)| \geq \frac{3 + \sqrt{5}}{2} = \Phi^2 > 1. \quad (4.61)$$

Thus, when $\Delta = \pm 1, \sigma = 1, p = \pm 3, \pm 4, \pm 5, \ldots$, we get the following inequalities from (4.58) and (4.60):

$$|\lambda_1(p)| \leq \frac{3 - \sqrt{5}}{2} = \Phi^{-2} < 1; \quad |\lambda_2(p)| \geq \frac{3 + \sqrt{5}}{2} = \Phi^2 > 1. \quad (4.62)$$

Option 2. (Numerical values, see Table 4.6). Using arguments similar to Option 1, for the case $\Delta = -1, \sigma = 1, p = 1, 2, 3, \ldots$, we get:

$$|\lambda_1(p)| \leq \frac{-1 + \sqrt{5}}{2} = \Phi^{-1} < 1, \quad |\lambda_2(p)| \geq \frac{1 + \sqrt{5}}{2} = \Phi > 1. \quad (4.63)$$

For the case $\Delta = -1, \sigma = -1, p = -1, -2, -3, \ldots$, we get:

$$|\lambda_1(p)| \leq \frac{-1 + \sqrt{5}}{2} = \Phi^{-1} < 1, \quad |\lambda_2(p)| \geq \frac{1 + \sqrt{5}}{2} = \Phi > 1. \quad (4.64)$$

Proof of Property 2 of Proposition 4.2. A geometric method for obtaining the eigenvalues λ_1, λ_2 ($|\lambda_1| < 1, |\lambda_2| > 1$) by using point mapping (4.50) is demonstrated in Fig. 4.5 (the case $p = 3, \Delta = 1, \sigma = 1$, where $\lambda_1 \approx 0.381966, \lambda_2 \approx 2.61803$).

Recall in [80], that the point x^* is the fixed point of the conversion $\bar{x} = w(x)$, if $x^* = w(x^*)$. The fixed point x^* of the conversion $\bar{x} = w(x)$ is called a stable point, if on the curve $\Gamma : \bar{x} = w(x)$ there exists a one-dimensional neighborhood $U = U(x^*)$, where all sequences of the points $x_n = w(x_{n-1}), n = 1, 2, 3, \ldots$ with starting points in this neighborhood *converge* to the fixed point x^*.

The fixed point x^* of the conversion $\bar{x} = w(x)$ is called an unstable point, if for an arbitrarily small one-dimensional neighborhood $U_\varepsilon(x^*)(\varepsilon > 0)$ on the curve $\Gamma : \bar{x} = w(x)$ there is at least one point x_0 such that the sequence $x_n = w(x_{n-1}), n = 1, 2, 3, \ldots$ does not converge to the fixed point x^*.

Sufficient conditions for the stability or instability of the fixed point x^* of the conversion $\bar{x} = w(x)$ are presented in the following Koenig's Theorem (see, for example, [80]).

Theorem 4.1 (Koenig's Theorem). *If at the fixed point x^* of the point conversion $\bar{x} = w(x)$ the condition $|\frac{dw}{dx}| < 1$ is met, then the fixed point x^* is a stable point; otherwise, when $|\frac{dw}{dx}| > 1$, the fixed point x^* is an unstable point.*

In particular, for the point conversion (4.50), we have $|\frac{dw}{dx}| = x^{-2}$. Therefore, for the conversion (4.50), in the case of $x^* = \lambda_1$ ($|\lambda_1| < 1$) we get

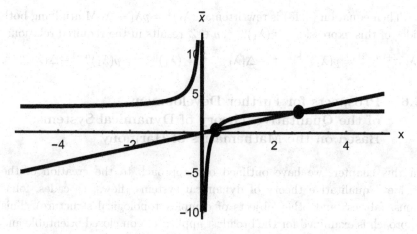

Figure 4.5. The point transformation $\bar{x} = w(x) = p - \frac{\Delta}{x} (p = 3, \Delta = 1)$.

$|\frac{dw}{dx}| = |\lambda_1|^{-2} > 1$, that is, the point λ_1 is an unstable fixed point. Similarly, if $x^* = \lambda_2$ ($|\lambda_2| > 1$) then $|\frac{dw}{dx}| = |\lambda_2|^{-2} < 1$, that is, the point λ_2 is a stable fixed point of the conversion (4.50). For the point mapping (4.50) $\bar{x} = w(x) = p - \frac{\Delta}{x}$ for the values $p = 3$ and $\Delta = 1$ we have two fixed points: $M_1(\lambda_1; \lambda_1)$ and $M_2(\lambda_2; \lambda_2)$, where $\lambda_1 \approx 0.382$ and $\lambda_2 \approx 2.618$. Since at point M_1 the derivative $|\frac{dw}{dx}| = |\lambda_1|^{-2} = 6.853 > 1$, then by Koenig's Theorem the fixed point $M_1(\lambda_1; \lambda_1)$ is unstable. By analogy, at the point M_2 the derivative $|\frac{dw}{dx}| = |\lambda_2|^{-2} = 0.1459 < 1$, and thus, the fixed point $M_2(\lambda_2; \lambda_2)$ is stable. The situation with the geometric derivation of the fixed points $M_1(\lambda_1; \lambda_1)$ and $M_2(\lambda_2; \lambda_2)$ of the point transformation (4.50) $\bar{x} = w(x) = p - \frac{\Delta}{x}$ at the values $p = 3$ and $\Delta = 1$ is presented in Fig. 4.5.

Here the point $M_1(\lambda_1; \lambda_1) = M_1(0.382; 0.382)$ is *unstable* where $\lambda_1 \approx 0.382$; and the point $M_2(\lambda_2; \lambda_2) = M_2(2.618; 2.618)$ is *stable* where $\lambda_2 \approx 2.618$.

Proof of Property 3 of Proposition 4.2. For Property 3 we should establish the relation:

$$\lambda^n = p\lambda^{n-1} - \Delta\lambda^{n-2}, \quad n \in Z, \quad \lambda = \lambda_1, \lambda_2, \quad |\lambda_1| < 1, \quad |\lambda_2| > 1.$$

We demonstrate that this relation follows directly from equation (4.48) $\lambda^2 - p\lambda + \Delta = 0$, whose roots are the eigenvalues λ_1, λ_2. Prove, for example, the relation for the root $\lambda_1(\lambda_1 < 1)$; the proof for the root $\lambda_2(\lambda_2 > 1)$ is analogous.

Then equation (4.48) is rewritten as $(\lambda_1)^2 = p\lambda_1 - \Delta$. Multiplying both sides of this expression by $(\lambda_1)^{n-2}, n \in Z$ results in the required relation:

$$(\lambda_1)^{2+(n-2)} = p(\lambda_1)^{1+(n-2)} - \Delta(\lambda_1)^{n-2} \Rightarrow (\lambda_1)^{n-2} = p(\lambda_1)^{n-1} - \Delta(\lambda_1)^{n-2}.$$

4.6. Prospects for Further Development of the Qualitative Theory of Dynamical Systems, Based on the Mathematics of Harmony

In this chapter, we have outlined our approach to the creation of the "golden" qualitative theory of dynamical systems (flows, cascades, foliations, fabrics^c and other objects of complex topological structure). This approach is examined for the broadest application on closed orientable and non-orientable surfaces M of any kind of unbranched or branched surfaces over any set of points on these surfaces.

Scientific material, as outlined in this chapter, can be a source for further development of the "golden" qualitative theory of dynamical systems in the following directions.

4.6.1. Homotopic class of rotation

In terms of the Mathematics of Harmony, primarily for such objects as defined on M, we must begin with the study of the "golden properties" of the global topological invariant, the homotopic class of rotation, introduced by Samuil Aranson and Vyacheslav Grines in 1973 [24] for the flows f^t, defined on the closed orientable manifolds M of any kind $p \geq 2$. The homotopic class of rotation characterizes the rotation of semi-trajectories of the flows f^t on the generators of the fundamental group G of manifolds M. In this sense, the homotopic class of rotation is a generalization of the Poincaré rotation number.

4.6.2. A notion of the absolute E and the set of E rational and irrational points

We introduce on the closed orientable manifold M of the kind where $p \geq 2$ the analytic structure that turns M into a Riemann surface. Consider the

---

^{c}The theory of fabrics is a promising new direction in the qualitative theory of dynamical systems and foliations on manifolds from both theoretical and applied points of view. For more details about fabrics, see [81–84].

conformal mapping $\pi: \overline{M} \to M$, where $\overline{M}$ is the universal covering for M, which is Lobachevsky's plane for Poincaré's realization on the inside circle $|z| < 1$ of the complex z-plane. The circle $E: |z| = 1$ is called an absolute.

It is known that when the surface M corresponds to the univalently defined discrete group Γ of the non-Euclidean translations, M is conformally equivalent to $\frac{\overline{M}}{\Gamma}$ and Γ is isomorphic to the fundamental group G of the surface M.

Each element of $g \in T$ has two and only two fixed points: the stable point g^+ and the unstable point g^-, and these points lie on the absolute E.

The set of the *rational* points on the absolute E, denoted by I, is the set of all fixed points of the elements of the group Γ. The set I is countable and elsewhere dense on the absolute E. The appendix $J = E \backslash I$ to the set of rational points on the absolute E is called the set of *irrational* points. It has the power of being a continuum.

4.6.3. A study of the "golden properties" of the homotopic class of rotation

To study the "golden properties" of the homotopic class of rotation of the semi-trajectories of flows, as well as other objects, we must associate these "golden properties" with the phenomenon of asymptotic attraction of the pre-images of trajectories of flows, layers, foliations, transversal layers of fabrics on the universal covering $\overline{M}$, implemented in the circle $|z| < 1$, which are points of the absolute E.

We should understand which points of the absolute E correspond for such objects to the decomposition in a continued fraction of metallic proportions, in particular, the golden, silver, bronze, copper proportions [62–64], and so on.

We must find the absolute E among rational points, which are analogs of suitable fractions, and also find the order of approximation of the irrational points of the absolute by rational points, which are in this case, the analog of the suitable fractions.

4.6.4. Regarding small denominators

In more general terms, we need to create such mathematical objects on the basis of the Mathematics of Harmony, defined on the surfaces M, which are an analog of the field of science called small denominators. For the mapping of circles and flows on the torus T^2, a superstructure for such mappings, the small denominators are considered from other points of view in [85].

4.6.5.   Reeb foliations, Pfaff's equations and others

Similar problems, based on the Mathematics of Harmony, must also arise when considering the appropriate objects on manifolds of dimensions greater than two. For example, such objects may be the so-called Reeb foliations of the co-dimension of one on the three-dimensional torus T^3 [86], the foliations of the co-dimension of one on the three-dimensional manifolds [87], Pfaffian differential equations of the n-dimensional torus and three-dimensional manifold $M^3 = M^2 \times S^1$, where M^2 is the closed orientable two-dimensional manifold of the genus $p \geq 2$ and S^1 is a circle [88].

Note that the list of the applications of the Mathematics of Harmony for the creation of the "golden" qualitative theory of differential equations and other objects is extensive.

4.6.6.   Implementation of the closed orientable surface M^2 of any genus $p \geq 0$ in the Euclidean space $R^3(x, y, z)$ [42]

In this section, we propose a technique of construction in the Euclidean space $R^3(x, y, z)$ of the closed orientable surface M^2 of any genus $p \geq 0$ in the form:

$$F(x, y, z) = \prod_{k=0}^{p} f_k(x, y) + z^2 = 0. \qquad (4.65)$$

We define the function $f_k(x, y)$ in the form:

$$f_k(x, y = (x - x_k)^2 + y^2 - r_k^2 \quad (k = 0, 1, 2, \ldots, p). \qquad (4.66)$$

By virtue of (4.66), on the plane R^2 $(-\infty < x < +\infty, -\infty < y < +\infty, z = 0)$ the curves:

$$f_k(x, y) = 0 \quad (k = 0, 1, 2, \ldots, p) \qquad (4.67)$$

are the circles, respectively, with radii $r_k > 0$ and centers at $(x_k; 0)$.

Assume that $x_0 = 0$, that is, for the case $k = 0$ the circle $f_0(x, y) = 0$ has the form:

$$f_0(x, y) = x^2 + y^2 - r_0^2 = 0. \qquad (4.68)$$

We further define the numbers $r_0, r_1, \ldots, r_p, x_1, \ldots x_p$ so that the closed circumferences:

$$D_k : 0 \le x^2 + y^2 \le r_k^2 \quad (k = 0, 1, 2, \ldots, p), \tag{4.69}$$

bounded by the circles (4.68), have the following properties:

(1) $\bigcup_{k=1}^{p} D_k \subset D_0$, that is, the areas $D_1, \ldots, D_k$ entirely belong to the domain D_0.

(2) $D_i \cap D_j = \phi$, where ϕ is an empty set, $i \ne j, i, j = 1, 2, \ldots, p$, that is, any two different areas $D_1, \ldots, D_k$ do not intersect.

For example, when $p = 0, 1, 2$ the circles of form (4.68) are the following:

$$p = 0 \Rightarrow f_0(x, y) = x^2 + y^2 - 16 = 0$$

$$p = 1 \Rightarrow f_0(x, y) = x^2 + y^2 - 16 = 0$$

$$f_1(x, y) = x^2 + y^2 - 4 = 0$$

$$p = 2 \Rightarrow f_0(x, y) = x^2 + y^2 - 16 = 0$$

$$f_1(x, y) = (x + 2)^2 + y^2 - 1 = 0$$

$$f_2(x, y) = (x - 2)^2 + y^2 - 1 = 0.$$

Hence we get these equations for the following surfaces:

(1) **Sphere** (the genus of $p = 0$) $\Rightarrow F(x, y, z) = x^2 + y^2 - 16 + z^2 = 0$.

(2) **Torus** (the genus of $p = 1$) $\Rightarrow F(x, y, z) = (x^2 + y^2 - 16)(x^2 + y^2 - 4) + z^2 = 0$.

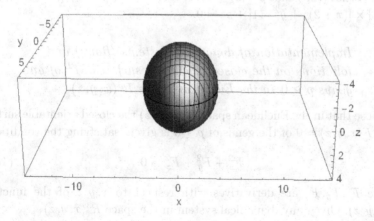

Figure 4.6(a). Sphere.

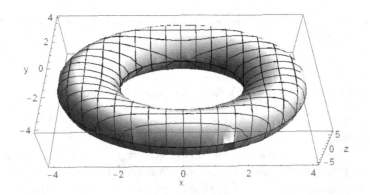

Figure 4.6(b). Torus.

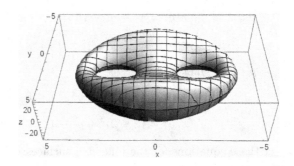

Figure 4.6(c). Pretzel.

(3) **Pretzel** (the genus $p = 2$) $\Rightarrow F(x, y, z) = (x^2 + y^2 - 16)[(x - 2)^2 + y^2 - 1] \times [(x + 2)^2 + y^2 - 1] + z^2 = 0$.

4.6.7. *Implementation of dynamical systems (flows) and foliations on the closed orientable surfaces M^2 of any genus $p \geq 0$ in the Euclidean space $R^3(x, y, z)$*

Suppose that in the Euclidean space $R^3(x, y, z)$ the closed orientable surface M^2: $F(x, y, z) = 0$ of the genus of $p \geq 0$ is given, satisfying the condition:

$$F_x^2 + F_y^2 + F_z^2 > 0, \qquad (4.70)$$

where F_x, F_y, F_z are derivatives with respect to x, y, z of the function $F(x, y, z)$. Given any dynamical system in the space $R^3(x, y, z)$

$$\frac{dx}{dt} = P(x, y, z), \quad \frac{dy}{dt} = Q(x, y, z), \quad \frac{dz}{dt} = R(x, y, z), \qquad (4.71)$$

where $P = P(x, y, z), Q = Q(x, y, z), R = R(x, y, z)$ are continuous or smooth functions of the variables x, y, z.

We further consider in the space $R^3(x, y, z)$ the vector field $\vec{V} = (P, Q, R)$. We project this vector field $\vec{V}$ onto the surface $M^2 : F(x, y, z) = 0$ so that the resulting vector field $\vec{W} = (P^*, Q^*, R^*)$ becomes the vector field tangent to this surface. Then this vector field $\vec{W}$ will have the form:

$$\vec{W} = \vec{V} + k\vec{N}, (\vec{W} \cdot \vec{N}) = 0, \qquad (4.72)$$

where $\vec{N} = (F_x, F_y, F_z)$ is a vector field orthogonal to the surface $M^2 : F(x, y, z) = 0, k$ is a real number (scalar), and $(\vec{W} \cdot \vec{N})$ is the scalar product of the vectors.

To find a scalar k, we scalarly multiply both sides of (4.72) by the vector $\vec{N}$, resulting in:

$$(\vec{W} \cdot \vec{N}) = (\vec{V} \cdot \vec{N}) + k(\vec{N} \cdot \vec{N}) \qquad (4.73)$$

where $(\vec{W} \cdot \vec{N}) = 0, (\vec{N} \cdot \vec{N}) = |\vec{N}|^2, |\vec{N}|$ is a modulus of the vector $\vec{N}$. From (4.73) we obtain the value of $k = -\frac{(\vec{V} \cdot \vec{N})}{|\vec{N}|^2}$. Therefore, the vector field (4.71), tangent to the surface $M^2 : F(x, y, z) = 0$, will have the form:

$$\vec{W} = \vec{V} - \frac{(\vec{V} \cdot \vec{N})}{|\vec{N}|^2}\vec{N}. \qquad (4.74)$$

But then, by virtue of (4.73), the coordinates P^*, Q^*, R^* of the vector field $\vec{W} = (P^*, Q^*, R^*)$ tangent to the surface $M^2 : F(x, y, z) = 0$, will have the form:

$$P^* = P - \lambda F_x, \quad Q^* = Q - \lambda F_y, \quad R^* = R - \lambda F_z, \qquad (4.75)$$

where

$$\lambda = \lambda(x, y, z) = \frac{PF_x + QF_y + RF_z}{(F_x)^2 + (F_y)^2 + (F_z)^2}. \qquad (4.76)$$

Therefore, the dynamical system on the closed orientable surface $M^2 : F(x, y, z) = 0$ of the genus $p \geq 0$ has the form:

$$\frac{dx}{dt} = P^* = P - \lambda F_x, \quad \frac{dy}{dt} = Q^* = Q - \lambda F_y,$$

$$\frac{dz}{dt} = R^* = R - \lambda F_z, \qquad (4.77)$$

where $\lambda = \lambda(x, y, z)$ satisfies the condition (4.76).

The integral curves of the dynamical system (4.77) define foliations on the surface M^2. Layers of these foliations have trajectories without a direction of movement. However, not every foliation on the surface M^2, as indicated above, produces a dynamical system on this surface.

We demonstrate this technique on the example of the sphere of S^2: $F(x, y, z) = x^2 + y^2 - 1 + z^2 = 0$ in three-dimensional Euclidean space $R^3(x, y, z)$. Then we have $F_x = 2x, F_y = 2y, F_z = 2z$.

We show that on the sphere S^2 the condition $F_x^2 + F_y^2 + F_z^2 > 0$ is satisfied. Suppose the contrary to be the case. Then we get:

$$F_x = 2x = 0, \quad F_y = 2y = 0, \quad F_z = 2z = 0 \Rightarrow x = 0, \quad y = 0, \quad z = 0.$$

We then substitute $x = 0$, $y = 0$, $z = 0$ into the equation of the sphere S^2: $F(x, y, z) = x^2 + y^2 - 1 + z^2 = 0$, resulting in the relation $-1 = 0$, which is impossible.

Next we place any dynamical system of the form (4.71) in the space $R^3(x, y, z)$. Then on the sphere S^2 we derive the following:

$$\lambda = \lambda(x, y, z) = \frac{PF_x + QF_y + RF_y}{(F_x)^2 + (F_y)^2 + (F_z)^2} = \frac{xP + yQ + zR}{2}.$$

Then, by virtue of (4.77), any dynamical system on the sphere S^2: $F(x, y, z) = x^2 + y^2 - 1 + z^2 = 0$ has the form:

$$\begin{cases} \dfrac{dx}{dt} = P - \lambda F_x = P - (xP + yQ + zR)x, \\[2mm] \dfrac{dy}{dt} = Q - \lambda F_y = Q - (xP + yQ + zR)y, \\[2mm] \dfrac{dz}{dt} = R - \lambda F_z = R - (xP + yQ + zR)z. \end{cases} \qquad (4.78)$$

It immediately follows from (4.78) that on the sphere S^2 the right-hand parts of the dynamical system (4.78) in at least one point $(x^*, y^*, z^*) \in S^2$ vanish. This means that any dynamical system on S^2 has at least one equilibrium state.

For example, if $P = 0$, $Q = 0$, and $R = -1$, then the dynamical system (4.78) on the sphere S^2 has the form:

$$\frac{dx}{dt} = xz, \quad \frac{dy}{dt} = yz, \quad \frac{dz}{dt} = z^2 - 1, \quad x^2 + y^2 + z^2 = 1. \qquad (4.79)$$

The dynamical system (4.79) on the sphere S^2 has two equilibrium states:

$$M_1(x_1 = 0, y_1 = 0, z_1 = 1) \text{ (North pole)},$$
$$M_2(x_2 = 0, y_2 = 0, z_2 = -1) \text{ (South pole)}.$$

We integrate the system (4.79), and then on the sphere S^2: $x^2 + y^2 + z^2 = 1$ we get the following trajectories:

$$L: x(t) = \frac{C_1}{\sqrt{C_1^2 + C_2^2}} \frac{1}{\operatorname{ch}(t)}, \quad y(t) = \frac{C_2}{\sqrt{C_1^2 + C_2^2}} \frac{1}{\operatorname{ch}(t)}, \quad z(t) = -\operatorname{th}(t),$$

$$(4.80)$$

where C_1, C_2 are any constants satisfying the condition $C_1^2 + C_2^2 > 0$. At $t = 0$, each trajectory L passes through the point:

$$M_0 : x(0) = \frac{C_1}{\sqrt{C_1^2 + C_2^2}}, \quad y(0) = \frac{C_2}{\sqrt{C_1^2 + C_2^2}}, \quad z(0) = 0,$$

which belongs to the Equator $E : x^2 + y^2 = 1$, $z = 0$ of the sphere S^2: $x^2 + y^2 + z^2 = 1$.

At $t \to +\infty \Rightarrow x(t) \to 0, x(t) \to 0, z(t) = -1$ (South pole).

At $t \to -\infty \Rightarrow x(t) \to 0, x(t) \to 0, z(t) = +1$ (North pole).

4.6.8. Implementation of dynamical systems on the m-dimensional hypersurfaces M^m in the Euclidean space $R^n(x_1, \ldots, x_n)$

Suppose that in the Euclidean space $R^n(x_1, \ldots, x_n)$ the m-dimensional hypersurface is given:

$$M^m: F_1(x_1, \ldots, x_n) = 0, \ldots, F_{r=n-m}(x_1, \ldots, x_n) = 0 \qquad (4.81)$$

where $1 \le m \le n - 1$. The number $1 \le r = n - m \le n - 1$ is called the co-dimension of the hypersurface M^m in the Euclidean space $R^n(x_1, \ldots, x_n)$.

We give any dynamical system in the space $R^n(x_1, \ldots, x_n)$:

$$\frac{dx_1}{dt} = P_1(x_1, \ldots, x_n), \ldots, \frac{dx_n}{dt} = P_n(x_1, \ldots, x_n), \qquad (4.82)$$

where $P_1 = P_1(x_1, \ldots, x_n)$, $P_n = P_n(x_1, \ldots, x_n)$ are continuous or smooth functions of the variables $x_1, \ldots, x_n$.

Next we consider the vector field $\vec{V} = (P_1, \ldots, P_n)$ in the space $R^n(x_1, \ldots, x_n)$. We project this vector field $\vec{V}$ to the hypersurface (4.81) $M^m: F_1(x_1, \ldots, x_n) = 0, \ldots, F_{r=n-m}(x_1, \ldots, x_n) = 0$ so that the resulting vector field $\vec{W} = (P_1^*, \ldots, P_n^*)$ becomes the vector field tangent to this hypersurface M^m. Then this vector field $\vec{W}$ will have the form:

$$\vec{W} = \vec{V} + \sum_{i=1}^{n-m} k_i \vec{N}_i, \qquad (4.83)$$

where $(\vec{W} \cdot \vec{N}_j) = 0, \vec{N}_j = (\frac{\partial F_j}{\partial x_1}, \ldots, \frac{\partial F_j}{\partial x_n})$ is the vector field orthogonal to each hypersurface $F_j(x_1, \ldots, x_n) = 0$, where $j = 1, 2, 3, \ldots, r = n - m$, k_i are real numbers (scalars), and $(\vec{W} \cdot \vec{N}_j)$ are scalar products of vectors. To find scalars k_i we scalarly multiply both sides of (4.83) by the vector $\vec{N}_j$ resulting in:

$$(\vec{W} \cdot \vec{N}_j) = (\vec{V} \cdot \vec{N}_j) + \sum_{i=1}^{n-m} k_i(\vec{N}_i \cdot \vec{N}_j) \tag{4.84}$$

where $(\vec{W} \cdot \vec{N}_j)$, $(\vec{N}_i \cdot \vec{N}_i) = |\vec{N}_i|^2$, $|\vec{N}_i|$ is a modulus of the vector $\vec{N}_i$. From (4.84) we obtain a system of linear equations for the scalars k_i ($i = 1, 2, \ldots, n - m$):

$$(\vec{V} \cdot \vec{N}_j) + \sum_{i=1}^{n-m} k_i(\vec{N}_i \cdot \vec{N}_j) = 0, \quad i = 1, 2, 3, \ldots, r = n - m. \tag{4.85}$$

In matrix form, the equation (4.85) can be rewritten as follows:

$$\begin{pmatrix} a_{11} & a_{12} & \cdots & a_{1r} \\ a_{21} & a_{22} & \cdots & a_{2r} \\ \vdots & \vdots & \cdots & \vdots \\ a_{r1} & a_{r2} & \cdots & a_{rr} \end{pmatrix} \begin{pmatrix} k_1 \\ k_2 \\ \vdots \\ k_r \end{pmatrix} = \begin{pmatrix} b_1 \\ b_2 \\ \vdots \\ b_r \end{pmatrix}, \tag{4.86}$$

where $r = n - m$.

Here we have:

$$\begin{cases} a_{11} = |\vec{N}_1|^2, \qquad a_{12} = (\vec{N}_1 \cdot \vec{N}_2), \ldots, a_{1r} = (\vec{N}_1 \cdot \vec{N}_r) \\ a_{21} = (\vec{N}_2 \cdot \vec{N}_1), \quad a_{22} = |\vec{N}_2|^2, \ldots, a_{2r} = (\vec{N}_2 \cdot \vec{N}_r) \\ \qquad\qquad\qquad\qquad \vdots \\ a_{r1} = (\vec{N}_r \cdot \vec{N}_1), \quad a_{r2} = (\vec{N}_r \cdot \vec{N}_2), \ldots, a_{rr} = |\vec{N}_r|^2 \\ b_1 = -(\vec{V} \cdot \vec{N}_1), \quad b_2 = -(\vec{V} \cdot \vec{N}_2), \ldots, b_r = -(\vec{V} \cdot \vec{N}_r) \end{cases} \tag{4.87}$$

Therefore, the vector field tangent to the hypersurface M^m: $F_1(x_1, \ldots, x_n) = 0, \ldots, F_{r=n-m}(x_1, \ldots, x_n) = 0$ has the form (4.83) $\vec{W} = \vec{V} + \Sigma_{i=1}^{n-m} k_i \vec{N}_i$. We write the vector field $\vec{W}$ in the coordinate form: $\vec{W} = (P_1^*, \ldots, P_n^*)$. As

$$\vec{V} = (P_1, \ldots, P_n), \quad \vec{N}_1 = \left(\frac{\partial F_1}{\partial x_1}, \ldots, \frac{\partial F_1}{\partial x_n}\right), \ldots, \vec{N}_r = \left(\frac{\partial F_r}{\partial x_1}, \ldots, \frac{\partial F_r}{\partial x_n}\right),$$

$$\tag{4.88}$$

then, by virtue of (4.83), the dynamical system on the hypersurface (4.81):

$$M^m: F_1(x_1, \ldots, x_n) = 0, \ldots, F_{r=n-m}(x_1, \ldots, x_n) = 0$$

will have the following form:

$$\begin{cases} \dfrac{dx_1}{dt} = P_1^* = P_1 + k_1 \dfrac{\partial F_1}{dx_1} + \cdots + k_r \dfrac{\partial F_r}{dx_1} \\ \qquad\qquad \vdots \\ \dfrac{dx_n}{dt} = P_n^* = P_n + k_1 \dfrac{\partial F_1}{dx_n} + \cdots + k_r \dfrac{\partial F_r}{dx_n} \end{cases}, \qquad (4.89)$$

where the scalars $k_1, \ldots, k_r$ are defined by the matrix equation (4.86).

References

[1] "Mathematical beauty", *Wikipedia, The Free Encyclopedia*, https://en.wikipedia.org/wiki/Mathematical_beauty (accessed November 28, 2015).

[2] Wenninger, R. M. *Polyhedron Models*. Moscow: Mir (1974) (Russian; translated from English).

[3] Stakhov, A. P., assisted by S. Olsen. *The Mathematics of Harmony. From Euclid to Contemporary Mathematics and Computer Science*. Singapore: World Scientific (2009).

[4] Coxeter, H. S. M. *Introduction to Geometry*. New York: John Wiley and Sons (1961).

[5] Vorobyov, N. N. *Fibonacci Numbers*. Moscow: Nauka (1978) (first edition, 1961) (Russian).

[6] Hoggat, Jr. V. E. *Fibonacci and Lucas Numbers*. Boston, MA: Houghton Mifflin (1969).

[7] Vajda, S. *Fibonacci & Lucas Numbers, and the Golden Section. Theory and Applications*. Ellis Horwood Limited (1989).

[8] Shevelev, I. *Shaping: Number, Shape, Art, Life*. Kostroma: Diar (1995) (Russian).

[9] Stakhov, A. P. *Introduction into Algorithmic Measurement Theory*. Moscow: Soviet Radio (1977) (Russian).

[10] Stakhov, A. P. *Codes of the Golden Proportion*. Moscow: Radio and Communications (1984) (Russian).

[11] Stakhov, A. P. "The generalized golden proportions and a new approach to geometric definition of a number", *Ukrainian Mathematical Journal* (2004) Vol. 56: 1143–1150 (Russian).

[12] Stakhov, A. P. "The golden section and modern harmony mathematics", *Applications of Fibonacci Numbers*. Kluwer Academic Publishers (1998) Vol. 7: 393–399.

[13] Stakhov, A. P. "The mathematics of harmony: clarifying the origins and development of mathematics", *Congressus Numerantium* (2008) Vol. CXCIII: 5–48.

[14] Stakhov, A. P. and Tkachenko, I. S. "Hyperbolic Fibonacci trigonometry", *Reports of the Ukrainian Academy of Sciences* (1993) Vol. 208, No. 7: 9–14 (Russian).

[15] Stakhov, A. P. and Rozin, B. N. "On a new class of hyperbolic function", *Chaos, Solitons & Fractals* (2004) Vol. 23: 379–389.

[16] Stakhov, A. P. and Rozin, B. N. "The golden section, Fibonacci series and new hyperbolic models of nature", *Visual Mathematics* (2006) Vol. 8, No. 3. http://www.mi.sanu.ac.rs/vismath/stakhov/index.html (accessed November 4, 2015).

[17] Stakhov, A. P. and Rozin, B. N. "The 'golden' hyperbolic models of universe", *Chaos, Solitons & Fractals* (2007) Vol. 34, Issue 2: 159–171.

[18] Stakhov, A. P. "Gazale's formulas, a new class of hyperbolic Fibonacci and Lucas Functions and the improved method of the 'golden' cryptography", Academy of Trinitarism, Moscow: Electronic number 77-6567, publication 14098 (2006) (Russian). http://www.trinitas.ru/rus/doc/0232/004a/02321 063.htm (accessed November 4, 2015).

[19] Stakhov, A. P. "A generalization of the Cassini formula", *Visual Mathematics* (2012) Vol. 14, No. 2. http://www.mi.sanu.ac.rs/vismath/stakhovsept2012/ cassini.pdf (accessed November 4, 2015).

[20] Stakhov, A. P. "On the general theory of hyperbolic functions based on the hyperbolic Fibonacci and Lucas functions and on Hilbert's Fourth Problem", *Visual Mathematics* (2013), Vol. 15, No. 1. http://www.mi.sanu.ac.rs/ vismath/2013stakhov/hyp.pdf (accessed November 4, 2015).

[21] Stakhov, A. P. "Non-Euclidean geometries. From the 'game of postulates' to the 'game of functions'", Academy of Trinitarism, Moscow: Electronic number 77-6567, publication 18048 (2013) (Russian). http://www.trinitas.ru/ rus/doc/0016/001d/00162125.htm (accessed November 4, 2015).

[22] Stakhov, A. P. "Hilbert's fourth problem: Searching for harmonic hyperbolic worlds of nature", *Journal of Applied Mathematics and Physics* (2013) Vol. 1, No. 3: 60–66.

[23] Aranson, S. Kh. "Dynamical systems on two-dimensional manifolds", in *Proceedings of V-th International Conference on the Nonlinear Oscillations. Qualitative Methods in the Theory of Nonlinear Oscillations.* Kiev: Institute of Mathematics, Ukrainian Academy of Sciences (1970) Vol. 2: 46–52 (Russian).

[24] Aranson, S. Kh. and Grines, V. Z. "On some invariants of dynamical systems on two-dimensional manifolds (necessary and sufficient conditions for the topological equivalence of transitive dynamical systems)", *Math. USSR-Sb* (1973) Vol. 19, No. 3: 366–393.

[25] Aranson, S. Kh. "On some arithmetic properties of dynamical systems on two-dimensional manifolds", *Soviet. Math. Dokl.* (1975) Vol. 16, No. 3: 605–609.

[26] Aranson, S. Kh. and Zuzhoma, E. V. "Classification of transitive foliations on the sphere with four singularities of spine type", *Selecta Mathematica Sovietica* (1990) Vol. 9, No. 2: 117–121.

[27] Aranson, S. Kh. and Grines, V. Z. "Classification of dynamical systems on two-dimensional manifolds", in *The 9th International Conference on Non-linear Oscillations. Qualitative Methods of Theory of Nonlinear Oscillations* (edited by Yu. A. Mitropolsky). Kiev: Naukova Dumka (1984) Vol. 2: 23–25 (Russian).

[28] Aranson, S. Kh. *Foliations with Singularities and Homeomorphisms with Invariant Foliations on Two-dimensional Manifolds.* Manuscript No. 4915-85, deposited at VINITI, Moscow (1985) (Russian).

[29] Aranson, S. Kh. and Grines, V. Z. "Topological classification of flows on closed two-dimensional manifolds", *Russian Math. Surveys* (1986) Vol. 41, No. 1: 183–208.

[30] Aranson, S. Kh. "Topological equivalence of fiberings with singularities and homeomorphisms with invariant fiberings on two-dimensional manifolds", *Russian Math. Surveys* (1986) Vol. 41, No. 3: 197–198.

[31] Aranson, S. Kh. *Topological Classification of Foliations with Singularities and Homeomorphisms with Invariant Foliations on Closed Surfaces. Part 1. Fiberings.* Moscow: deposited at VINITI, Moscow (1988), No. 6887-B 88 (Russian).

[32] Arnold, V. I., Il'yashenko, Yu. S., Anosov, D. V., Bronshtein, I. U., Aranson, S. Kh. and Grines, V. Z. *Dynamical Systems I, Encyclopaedia of Mathematical Sciences.* Berlin: Springer Verlag (1988) Vol. 1.

[33] Aranson, S. Kh. *Topological Classification of Foliations with Singularities and Homeomorphisms with Invariant Foliations on Closed Surfaces. Part 2. Homeomorphisms,* Moscow: deposited at VINITI (1989), No. 1043-B 89 (Russian).

[34] Aranson, S. Kh. "Global problems of the qualitative theory of dynamical systems on closed surfaces", *Dissertation for the degree of Doctor of Physical-Mathematical Sciences. Specializations: Differential Equations, Geometry and Topology* (1990) (Russian).

[35] Aranson, S. Kh. "Trajectories on non-orientable two-dimensional manifolds", *Math. USSR-Sb* (1969) Vol. 9: 297–313 (Russian).

[36] Aranson, S. Kh. and Grines, V. Z. "Topological classification of cascades on closed two-dimensional manifolds", *Russian Math. Surveys* (1990) Vol. 45, No. 1: 1–35.

[37] Aranson, S. Kh. "Topology of vector fields, foliations with singularities, and homeomorphisms with invariant foliations on closed surfaces", in *Proceedings of the Steklov Institute of Mathematics* (1993) Issue 193: 13–18.

[38] Aranson, S. Kh. and Zhuzhoma, E. V. "Quasi-minimal sets of foliations and one-dimensional basic sets of axiom A of diffeomorphisms of surfaces", *Dokl. Math* (1993) Vol. 47, No. 3: 448–450 (Russian).

[39] Aranson, S. Kh. and Zhuzhoma, E. V. "On trajectories of covering flows in the case of branched coverings for a sphere and projective plane", *Mathematics Notes* (1993) Vol. 43, No. 5–6, 463–468.

[40] Aranson, S. Kh., Grines, V. Z. and Zhuzhoma, E. V. "On the geometry and topology of foliations on surfaces", *Russian Math. Surveys* (1995) Vol. 50, No. 4: 731 (Joint sessions of the Petrovskii Seminar on Differential Equations and Mathematical Problems of Physics and of the Moscow Mathematical Society (seventeenth session, January 24–27, 1995).

[41] Anosov, D. V., Aranson, S. Kh., Grines, V. Z., Plykin, R. V., Safonov, A. V., Sataev, E. A., Shlyachkov, S. V., Solodov, V. V., Starkov, A. N. and Stepin, A. M. "Dynamical systems with hyperbolic behavior", in *Encyclopedia of Mathematical Sciences. Dynamical Systems IX*. Berlin: Springer (1995) Vol. 66.

[42] Aranson, S. Kh., Belitsky, G. R. and Zhuzhoma, E. V. *Introduction to the Qualitative Theory of Dynamical Systems on Surfaces*. USA: American Mathematical Society (1996) Vol. 153.

[43] Anosov, D. V., Aranson, S. Kh., Arnold, V. I., Bronshtein, I. U., Grines, V. Z. and Il'yashenko, Yu. S. *Ordinary Differential Equations and Smooth Dynamical Systems*. Berlin: Springer (1997).

[44] Aranson, S. Kh. and Zhuzhoma, E. V. "Qualitative theory of flows on surfaces (a review)", *Journal of Mathematical Sciences* (1998) Vol. 90, No. 3: 2051–2110.

[45] Aranson, S. Kh., Bronshtein, I. U., Zhuzhoma, E. V. and Nikolaev, I. V. "Qualitative theory of foliations on closed surfaces", *Journal of Mathematical Sciences* (1998) Vol. 90, No. 3: 2111–2149.

[46] Aranson, S. Kh., Zhuzhoma, E. V. and Telnykh, I. A. "Transitive and supertransitive flows on closed nonorientable surfaces", *Mathematics Notes* (1998) Vol. 63, No. 4: 549–552.

[47] Aranson, S. Kh. and Telnykh, I. A. *Topological dynamics super transitive flows on closed nonorientable surfaces*. Moscow: deposited at VINITI (1998), No. 1751-B 98 (Russian).

[48] Aranson, S. Kh. "Qualitative properties of foliations on closed surfaces", *Journal of Dynamical and Control Systems* (2000) Vol. 6, No. 1: 127–157.

[49] Aranson S. Kh., Medvedev, V. and Zhuzhoma, E. "Collapse and continuity of geodesic frameworks of surface foliations", in *Methods of Qualitative Theory of Differential Equations and Related Topics*. Providence, RI: American Mathematical Society (2000) Vol. 200, Series 2.

[50] Aranson, S. Kh., Gorelikova, I. A. and Zhuzhoma, E. V. "The influence of absolute on the local and smooth properties of foliations and homeomorphisms with invariant foliations on closed surfaces", *Dokl. Math* (2001) Vol. 64, No. 1: 25–28.

[51] Aranson, S. Kh. and Zhuzhoma, E. V. "Nonlocal properties of analytic flows on closed orientable surfaces", in *Proceedings of the Steklov Institute of Mathematics* (2004) Vol. 244: 2–17.

[52] Stakhov, A. P. and Aranson, S. Kh. "Hyperbolic Fibonacci and Lucas functions, 'golden' Fibonacci goniometry, Bodnar's geometry, and Hilbert's fourth problem. Part I. Hyperbolic Fibonacci and Lucas functions and 'golden' Fibonacci goniometry", *Applied Mathematics* (2011) Vol. 2, No. 1: 74–84.

[53] Stakhov, A. P. and Aranson, S. Kh. "Hyperbolic Fibonacci and Lucas functions, 'golden' Fibonacci goniometry, Bodnar's geometry, and Hilbert's fourth problem. Part II", *Applied Mathematics* (2011) Vol. 2, No. 2: 181–188.

[54] Stakhov, A. P. and Aranson, S. Kh. "Hyperbolic Fibonacci and Lucas functions, 'golden' Fibonacci goniometry, Bodnar's geometry, and Hilbert's fourth problem. Part III", *Applied Mathematics* (2011) Vol. 2, No. 3: 283–293.

[55] Stakhov, A. P and Aranson, S. Kh. "The mathematics of harmony and Hilbert's fourth problem: the way to the harmonic hyperbolic and spherical worlds of nature", Academy of Trinitarism, Moscow: Electronic number 77-6567, publication 18814 (2014) (Russian). http://www.trinitas.ru/rus/doc/0232/007a/02321016.htm (accessed November 4, 2015).

[56] Stakhov, A. P. and Aranson, S. Kh. *The Mathematics of Harmony and Hilbert's Fourth Problem. The Way to the Harmonic Hyperbolic and Spherical Worlds of Nature.* Germany: Lambert Academic Publishing (2014).

[57] Poincaré, H. *On Curves Defined by Differential Equations.* Moscow-Lenigrad: Gostechizdat (1947) (Russian; translated from French).

[58] Denjoy, A. "Sur les courbes par les equation differentielles a la surface du tore", *Journal Math Pures Appl* (1932) Vol. 11: 333–375.

[59] Kneser, H. "Regulare kurvenscharen auf ringflachen", *Math. Ann.* (1923) Vol. 91: 135–154.

[60] Nemytskii, V. V. and Stepanov, V. V. *Qualitative Theory of Differential Equations.* Moscow-Leningrad: Gostekhizdat (1949) (Russian).

[61] Anosov, D. V. "Geodesic flows on closed Riemannian manifolds with negative curvature", *Proceedings of the Steklov Institute of Mathematics* (1969) Vol. 90 (Russian).

[62] de Spinadel, V. W. "The metallic means and design", in *NEXUS II: Architecture and Mathematics* (1998).

[63] de Spinadel, V. W. "The metallic means family and multifractal spectra", *Nonlinear Analysis* (1999) Vol. 36: 721–745.

[64] de Spinadel, V. W. *From the Golden Mean to Chaos.* Nueva Libreria (first edition 1998); Nobuko (second edition 2004).

[65] Khinchin, A. Ya. *Continued Fractions.* Moscow-Lenigrad: Gostekhizdat (1935) (Russian).

[66] Khovansky, A. N. *The Application of Continued Fractions and Their Generalizations to the Approximate Analysis.* Moscow: Gostekhizdat (1956) (Russian).

[67] Hurwitz, A. *Mathematische Werke.* Germany: Basel (1932) Bd. 1: 321–383.

[68] Mayer, A. G. "Trajectories on orientable surfaces", *Mathematical Collection.* Moscow: Academy of Sciences of USSR (1943) Vol. 12: 71–84 (Russian).

[69] Markley, N. G. "The Poincare–Bendixson Theorem for the Klein bottle", *Amer. Math. Soc.* (1969) Vol. 135: 159–165.

[70] Markley, N. G. "On the number recurrent orbit closures", *Amer. Math. Soc.* (1970) Vol. 25, No. 2: 413–416.

[71] Anosov, D. V. "Roughness of geodesic flows on compact Riemannian manifolds of negative curvature", in *Reports of Academy of Sciences of USSR.* Moscow: Academy of Sciences of USSR (1962) Vol. 145, No. 4: 707–709

(Russian).

[72] Anosov, D. V. "Ergodic properties of geodesic flows on closed manifolds of negative curvature", in *Reports of Academy of Sciences of USSR*. Moscow: Academy of Sciences of USSR (1963) Vol. 151, No. 6: 1250–1252 (Russian).

[73] Anosov, D. V. and Sinai, Ya. G. "Some smooth ergodic systems", *Russian Math. Surveys* (1967) Vol. 22, Issue 5: 103–167.

[74] Sinai, Ya. G. "Markov's partitions and U-diffeomorphisms", *Functional Analysis and Its Applications* (1968) Vol. 2, Issue 1: 61–82 (Russian).

[75] Sinai, Ya. G. "Construction of Markov's partitions", *Functional Analysis and Its Applications*. (1968) Vol. 2, Issue 3: 245–253 (Russian).

[76] Nitezki, Z. *Differentiable Dynamics. An Introduction to the Orbit Structure of Diffeomorphisms*. Boston, MA: MIT Press Cambridge (1971).

[77] Kornfeld, I. P., Sinai, Ya. G. and Fomin, S. V. *Ergodic Theory*. Moscow: Nauka, Fizmatgiz (1980) (Russian).

[78] Arnold, V. I. and Avez, A. *Ergodic Problems of Classical Mechanics*. New York: Benjamin Publishing House (1968).

[79] Pilyugin, S. Y. "Spaces of dynamical systems", *NITs Regular Chaotic Dynamics*, Moskva-Izhevsk: Institute of Computer Sciences (2008) (Russian).

[80] Goryachenko, V. D. *Elements of the Theory of Oscillations*. Krasnoyarsk University Press (1995) (Russian).

[81] Aranson, S. Kh. and Grines, V. Z. "On topological classification of the supertransitive 2-webs on closed surfaces of Euler negative characteristic", University of Burgundy U.F.R. Preprint (2000), No. 224 (BP-47 870-21078 CEDEX Dijon-France).

[82] Aranson, S. Kh., Grines, V. Z. and Kaimanovich, V. A. "Classification of supertransitive 2-webs on surfaces", Bonn (Germany): Max-Plank-Inst. fur Mathematik, Preprint (2002) No. 365/91-02.

[83] Aranson, S. Kh., Grines, V. Z. and Kaimanovich, V. A. "Classification of supertransitive 2-webs on surfaces", *Journal of Dynamical and Control Systems* (2003) Vol. 9, Issue 4: 455–468.

[84] Aranson, S. Kh., Grines, V. and Zhuzhoma, E. "Using the Lobachevsky plane to study surface flows, foliations and 2-webs", in *Non-Euclidean Geometry in Modern Physics and Mathematics. Proceedings of the International Conference BGL-4 (Bolyai-Gauss-Lobachevsky)*. Kiev-Nizhny Novgorod (2004): 8–24 (Russian).

[85] Arnold, V. I. "Small denominators. I. On the maps of a circle onto itself", *Izvestija of Sciences of USSR Ser. Math* (1961) Vol. 25, No. 1: 21–86 (Russian).

[86] Aranson, S. Kh. and Zhuzhoma E. V. "On the topological classification of the Reeb foliation of co-dimension of one on a three-dimensional torus", in *Interuniversity Thematic Collection of Scientific Papers "Methods of Qualitative Theory of Differential Equations"*. Gorky State University (1978), pp. 41–64 (Russian).

[87] Aranson, S. Kh., Mamaev, V. and Zhuzhoma, E. "Asymptotic properties of co-dimension of one foliations and Anosov–Weil Problem", in *Proceedings of Geometric Study of Foliations. Intern. Symposium and Workshop on Geometric Study of Foliations* (Tokyo, November 1993). Singapore: World Scientific (1994), pp. 145–151.

[88] Aranson, S. Kh. "On the classification of Pfaff's differential equations", in *Interuniversity Thematic Collection of Scientific Papers "Methods of Qualitative Theory of Differential Equations"*, pp. 101–115, Gorky State University (1982) (Russian).

Chapter 5

The Basic Stages
of the Mathematical Solution
to the Fine-Structure Constant
Problem as a Physical
Millennium Problem

5.1. Physical Millennium Problems

The twenty-three Mathematical Problems presented in 1900 by the prominent mathematician David Hilbert at the International Congress of Mathematicians in Paris [1–3] are well known.

In May 2000, the Clay Mathematics Institute of Cambridge announced (in Paris, exactly as Hilbert did) the seven mathematical "Millennium Prize Problems", each with a bounty of $1 million [4].

Modern physicists decided not to lag behind the mathematicians in formulating major problems that are in need of solutions. As part of the Millennium Madness, they have proposed 10 Physics Problems for the Next Millennium [5]. These physical Millennium Problems were presented at the Strings 2000 Conference (held on July 10–15, 2000 at the University of Michigan). All participants at the Conference were invited to formulate the 10 most important unsolved problems in fundamental physics. Each participant was allowed to submit one candidate for consideration as a Millennium Problem. To qualify, the problem had to stoke our curiosity, be of high importance, well-defined and clearly stated.

The 10 best problems were selected at the end of the conference by a selection panel comprising:

- Michael Duff (University of Michigan)
- David Gross (Institute for Theoretical Physics, Santa Barbara)
- Edward Witten (Caltech & Institute for Advanced Studies)

The first physical Millennium Problem was formulated by the prominent physicist and Nobel Laureate in Physics 2004, David Gross (University of California, Santa Barbara), as follows:

"*Are all the (measurable) dimensionless parameters that characterize the physical universe calculable in principle or are some merely determined by historical or quantum mechanical accident and incalculable?*"

Let us analyze Gross's formulation of the above Millennium Problem:

(1) The first question is: what are "(measurable) dimensionless parameters that characterize the physical universe"?

(2) The second question concerns the essence of this Millennium Problem: are these "dimensionless parameters... calculable in principle" or are they incalculable (or non-calculable) and are some "merely determined by historical or quantum mechanical accident"?

In answering the first question, we immediately arrive at the main dimensionless constant, which is widely known in physics as the fine-structure constant α. As is highlighted in the Wikipedia article [6]:

"*In physics, the fine-structure constant, also known as Sommerfeld's constant, commonly denoted by α (the Greek letter alpha), is a fundamental physical constant characterizing the strength of the electromagnetic interaction between elementary charged particles. It is related to the elementary charge (the electromagnetic coupling constant) e, which characterizes the strength of the coupling of an elementary charged particle with the electromagnetic field, by the formula $4\pi\varepsilon_0 \hbar c \alpha = e^2$. Being a dimensionless quantity, it has the same numerical value in all systems of units. Arnold Sommerfeld introduced the fine-structure constant in 1916.*"

Note that the physical significances of all symbols appearing in the formula $4\pi\varepsilon_0 \hbar c \alpha = e^2$ are the following: ε_0 is an *electric constant*; $\hbar$ is *Dirac's constant*; c is the speed of light in a vacuum; e is an *elementary charge*.

Thus the problem, formulated by Gross, can be narrowed down to the fine-structure constant as the main dimensionless constant of physical world:

"*Is the fine-structure constant, characterizing the physical universe, calculable or non-calculable?*"

It should be noted that Gross's formulation of the Millennium Problem remains consistent even under our definition. We simply focus our attention on the primary dimensionless constant of the physical world, the fine-structure constant. The following question arises from such a modification of Gross's Millennium Problem: Is the fine-structure constant problem a purely physical (non-calculable) or physical-mathematical (calculable) problem?

In this chapter, we consider the problem of the fine-structure constant as a physical-mathematical problem. This means that, first of all, we will consider it as a physical problem. However, we are solving this problem by using mathematical methods. To model this problem, we use a special mathematical theory, the so-called Fibonacci special theory of relativity [7]. The Mathematics of Harmony [8] and the "golden" matrices [9], introduced by Stakhov in 2007, underlie the Fibonacci special theory of relativity.

We will study the variation problem of the fine-structure constant in dependence on the age of the Universe, beginning with the Big Bang. We have deduced the formula for this constant as a function of time T, starting from the moment of the Big Bang. This makes it possible to calculate the values of this constant for all stages of evolution of the Universe beginning with the Big Bang, including the Dark Ages, the Light Ages (the positive arrow of time), and finally the Black Hole (the negative arrow of time). Comparisons of theoretical calculations and experimental astronomical data show a very high degree of coincidence.

These results allow us to argue that we have obtained an original solution to Gross's Millennium Problem for the case of the fine-structure constant as the most important dimensionless constant of the physical world.

5.2. Classical Special Theory of Relativity

5.2.1. *Lorentz transformations and classical special theory of relativity*

The model of four-dimensional space-time, based on the transformation by Hendrik Antoon Lorentz, was used by Albert Einstein in 1905 [10] for the creation of the special theory of relativity (STR). The mathematical apparatus for transformations of coordinates and time between different frames of reference (for the purpose of conservation of the electromagnetic field equations) had been formulated previously by the French mathematician Henri Poincaré. He offered to call them Lorentz transformations, although Lorentz had only presented approximate formulas before [11].

The main difference between Poincaré and Einstein's approaches, disguised by the resemblance between their mathematical models, is that both scientists had different interpretations of the deep physical (more than just mathematical) nature of said models. All new effects, interpreted by Poincaré as dynamic properties of ether, were interpreted in Einstein's theory of relativity as objective properties of space and time. Einstein moved them from dynamics to kinematics. For more details about this, see articles [7, 11–18], and the articles "The special theory of relativity" and "Poincaré, Henri" in Wikipedia.

This theory necessitated a revision of all concepts of classical physics. The classical special theory of relativity (STR) is given by a system of two matrix transformations [7, 10–18]:

$$
\begin{pmatrix} ct \\ x \\ y \\ z \end{pmatrix} = \begin{pmatrix} \dfrac{1}{\sqrt{1-(\bar{v})^2}} & \dfrac{\bar{v}}{\sqrt{1-(\bar{v})^2}} & 0 & 0 \\ \dfrac{\bar{v}}{\sqrt{1-(\bar{v})^2}} & \dfrac{1}{\sqrt{1-(\bar{v})^2}} & 0 & 0 \\ 0 & 0 & 1 & 0 \\ 0 & 0 & 0 & 1 \end{pmatrix} \begin{pmatrix} ct' \\ x' \\ y' \\ z' \end{pmatrix},
$$

$$
\begin{pmatrix} ct \\ x \\ y \\ z \end{pmatrix} = \begin{pmatrix} \mathrm{ch}(\theta) & \mathrm{sh}(\theta) & 0 & 0 \\ \mathrm{sh}(\theta) & \mathrm{ch}(\theta) & 0 & 0 \\ 0 & 0 & 1 & 0 \\ 0 & 0 & 0 & 1 \end{pmatrix} \begin{pmatrix} ct' \\ x' \\ y' \\ z' \end{pmatrix} \tag{5.1}
$$

where:

(1) $c = \mathrm{const}$ [m sec^{-1}] is the speed of light in a vacuum.

(2) t [sec] is a time, ct [m] is a time coordinate, x [m] is a length, y [m] is a width, z [m] is a height.

(3) $\bar{v} = \frac{v}{c}$ is the normalized Lorentzian speed of the light source (dimensionless), $|\bar{v}| < 1 \Leftrightarrow |v| < c$.

(4) v [m sec^{-1}] is the Lorentzian speed of the light source (the speed of uniform motion of the light source along the x axis).

(5) $\theta(-\infty < \theta < +\infty)$ is a hyperbolic rotation angle (dimensionless).

(6) $\mathrm{sh}(\theta) = \frac{e^{\theta}-e^{-\theta}}{2}$ is a hyperbolic sine, $\mathrm{ch}(\theta) = \frac{e^{\theta}+e^{-\theta}}{2}$ is a hyperbolic cosine, $\mathrm{th}(\theta) = \frac{e^{\theta}-e^{-\theta}}{e^{\theta}+e^{-\theta}}$ is a hyperbolic tangent, $e = \lim_{n\to+\infty}(1+\frac{1}{n})^n \approx 2.71828$ is the number of Napier.

The first transformation in (5.1) is called the Lorentz transformation for the speed of the light source. The second transformation is called the

Lorentz transformation to hyperbolic rotation angle θ. Because the conjugated matrices in (5.1) coincide, the following relationships result:

$$\frac{\bar{v}}{\sqrt{1-(\bar{v})^2}} = \mathrm{sh}(\theta), \qquad \frac{1}{\sqrt{1-(\bar{v})^2}} = \mathrm{ch}(\theta), \qquad \bar{v} = \frac{v}{c} = \mathrm{th}(\theta), \qquad v = c \cdot \mathrm{th}(\theta).$$

$$(5.2)$$

5.2.2. *Einstein's postulates*

Einstein proposed two postulates as the starting points for the special relativity theory.

(1) The principle of relativity expresses the invariance of the laws of nature and their equations at the transitions from one inertial system to another. That is, all inertial reference systems (IRS) are indistinguishable in their properties, and therefore none can be selected as preferred.
(2) The principle of the independence of light speed from the light source claims that the speed of light c in a vacuum is the same in all directions and is not dependent on the speed v of the movement of the light source. This implies that the speed of light in a vacuum must be limited and constant in all inertial reference systems.

Figure 5.1 presents the graph of the function $\bar{v} = \frac{v}{c} = \mathrm{th}(\theta)$.

Adding to the spatial coordinates (x, y, z) the time coordinate ct [m], we obtain the *four-dimensional space-time* $D^4 = (ct, x, y, z)$. We supply the space $D^4 = (ct, x, y, z)$ with the alternating Minkowski metric, where dl is

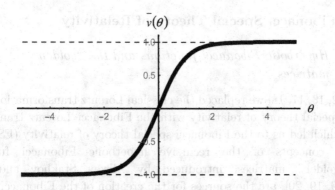

Figure 5.1. The graph of the normalized speed of the light source $\bar{v}(\theta) = \frac{v(\theta)}{c} = \mathrm{th}(\theta)$ (black curve), $\lim_{\theta \to \pm \infty} \bar{v}(\theta) = \pm 1$ are the limit values of the normalized speed of the light source (dashed lines).

the element of arc length, that is

$$(dl)^2 = [d(ct)]^2 - (dx)^2 - (dy)^2 - (dz)^2. \tag{5.3}$$

The Minkowski metric (5.3) has the remarkable property of invariance relative to the Lorentz transformation (5.1), that is

$$(dl)^2 = [d(ct)]^2 - (dx)^2 - (dy)^2 - (dz)^2$$
$$= [d(ct')]^2 - (dx')^2 - (dy')^2 - (dz')^2.$$

5.2.3. Relativistic effects of the classical special theory of relativity

A number of well-known consequences of the special theory of relativity (STR), the STR relativistic effects for moving light sources, compared with a stationary observer [11] follows from the invariance of the Minkowski metric relative to Lorentz transformations.

(1) Time dilation of the moving light source (time flows slowly for the moving light source in comparison with the stationary light source).

(2) Shortening the length of the moving light source (the length of a moving light source appears shorter to the fixed light source).

(3) Increasing the mass of the moving light source (the mass of the moving light source appears greater to the fixed light source).

(4) Increasing the energy of the moving light source (the energy of the moving light source appears greater to the stationary light source).

5.3. Fibonacci Special Theory of Relativity

5.3.1. Hyperbolic Fibonacci functions and the "golden" matrices

In [7, 12, 13, 17, 18] we replaced the classical Lorentz transformations (5.1) of the special theory of relativity with the Fibonacci–Lorentz transformations, which led us to the Fibonacci special theory of relativity (FSTR).

The concepts of the recursive hyperbolic Fibonacci functions and "golden" matrices, introduced by Alexey Stakhov and Boris Rozin [9, 19, 20], are the sources for the creation of the Fibonacci special theory of relativity. The hyperbolic Fibonacci functions are defined earlier as follows:

$$\text{Fibonacci hyperbolic sine: } sF(x) = \frac{\Phi^x - \Phi^{-x}}{\sqrt{5}} = \frac{2}{\sqrt{5}}\text{sh}(x \ln \Phi)$$

$$\text{Fibonacci hyperbolic cosine: } cF(x) = \frac{\Phi^x + \Phi^{-x}}{\sqrt{5}} = \frac{2}{\sqrt{5}}\text{ch}(x \ln \Phi)$$

$$\text{The golden ratio: } \Phi = \frac{1 + \sqrt{5}}{2} \approx 1.61803$$

$$\text{The basic relation: } [cF(x-1)]^2 - sF(x-2)sF(x) = 1. \qquad (5.4)$$

The "*golden*" *matrix* A:

$$
A = \begin{pmatrix} cF(x-1) & sF(x-2) \\ sF(x) & cF(x-1) \end{pmatrix}
$$

$$
= \begin{pmatrix} \dfrac{2}{\sqrt{5}}\text{ch}[(x-1) \cdot \ln \Phi] & \dfrac{2}{\sqrt{5}}\text{sh}[(x-2) \cdot \ln \Phi] \\ \dfrac{2}{\sqrt{5}}\text{sh}[x \cdot \ln \Phi] & \dfrac{2}{\sqrt{5}}\text{ch}[(x-1) \cdot \ln \Phi] \end{pmatrix}; \quad \det(A) = 1.
$$

$$(5.5)$$

5.3.2. *Fibonacci–Lorentz transformations and Fibonacci special theory of relativity*

The Fibonacci special theory of relativity, set forth in [7, 12, 13, 17, 18], is based on the following Fibonacci–Lorentz transformations:

$$
\begin{pmatrix} c_0 t \\ x \\ y \\ z \end{pmatrix} = \begin{pmatrix} \dfrac{1}{\sqrt{1-(\bar{v})^2}} & \dfrac{1}{\bar{c}(\psi)}\dfrac{\bar{v}}{\sqrt{1-(\bar{v})^2}} & 0 & 0 \\ \bar{c}(\psi)\dfrac{\bar{v}}{\sqrt{1-(\bar{v})^2}} & \dfrac{1}{\sqrt{1-(\bar{v})^2}} & 0 & 0 \\ 0 & 0 & 1 & 0 \\ 0 & 0 & 0 & 1 \end{pmatrix} \cdot \begin{pmatrix} c_0 t' \\ x' \\ y' \\ z' \end{pmatrix}
$$

$$(5.6)$$

$$
\begin{pmatrix} c_0 t \\ x \\ y \\ z \end{pmatrix} = \begin{pmatrix} cF(\psi-1) & sF(\psi-2) & 0 & 0 \\ sF(\psi) & cF(\psi-1) & 0 & 0 \\ 0 & 0 & 1 & 0 \\ 0 & 0 & 0 & 1 \end{pmatrix} \cdot \begin{pmatrix} c_0 t' \\ x' \\ y' \\ z' \end{pmatrix}.
$$

5.3.3.   The main postulate of the Fibonacci special theory of relativity

For the Fibonacci special theory of relativity, the light speed in a vacuum is not constant, but is determined by the function $c = \bar{c}(\psi) \cdot c_0$ [m sec^{-1}]. The postulate on the variability of light's speed in a vacuum c is consistent with data gathered by the astronomer John Webb (www.vokrugsveta.ru/telegraph/cosmos/1298). He found that the light, which is coming to us from the observed Universe, follows the principle of non-decreasing entropy; this means, that for this case the Second Law of Thermodynamics has been preserved and therefore the speed of light c should decrease with the increasing age of the Universe. James Franson's article [21] confirms this conclusion.

5.3.4.   The meaning of the symbols for the Fibonacci special theory of relativity

(1) Agreement on the symbols and $\frac{|\psi|}{\psi}$ and $\frac{|T|}{T}$. Throughout, the symbols $\frac{|\psi|}{\psi}$ and $\frac{|T|}{T}$ will be used in the usual sense only for $\{-\infty < \psi < 0\} \cup \{2 < \psi < +\infty\}$. For $\{0 < \psi < 2\}$ we assume that $\frac{|\psi|}{\psi}$ and $\frac{|T|}{T}$ is minus (-1).

(2) The dimensionless parameter ψ means the angle of the Fibonacci rotation, also known as the parameter of self-organization [17]. We will use both definitions for this parameter.

(3) $c_0 = \frac{c^*}{\Phi}$ [m sec^{-1}] = const means the normalized Lorentzian speed of light in a vacuum. Here c^* [m sec^{-1}] = const is the speed of light in a vacuum for the case of the classical special theory of relativity. For the modern period the accepted value for c^* is:

$$c^* \approx 2.99792458 \cdot 10^8 \text{ [m sec}^{-1}].$$

The value of c_0 does not depend on the speed of the movement of the light source or the observer, and is the same for all inertial reference systems. For the modern period we have:

$$c_0 = \frac{c^*}{\Phi} \approx 1.85281990 \cdot 10^8 \text{ [m sec}^{-1}].$$

(4) The parameter $c(\psi) = c_0 \cdot \bar{c}(\psi)$ [m sec^{-1}] is called the Fibonacci speed of light in a vacuum.

(5) The dimensionless parameter $\bar{c}(\psi) = \sqrt{\dfrac{\mathrm{sF}(\psi)}{\mathrm{sF}(\psi-2)}} = \dfrac{\sqrt{[\mathrm{cF}(\psi-1)]^2-1}}{|\mathrm{sF}(\psi-2)|}$ and is called the normalized Fibonacci speed of light in a vacuum.

(6) The parameter $v(\psi) = c(\psi) \cdot \bar{v}(\psi) = c_0 \cdot \dfrac{\mathrm{sF}(\psi)}{\mathrm{cF}(\psi-1)}$ [m sec^{-1}] is called the Fibonacci speed of the light source in a vacuum.

(7) The dimensionless parameter $\bar{v}(\psi) = \dfrac{1}{\bar{c}(\psi)} \cdot \dfrac{\mathrm{sF}(\psi)}{\mathrm{cF}(\psi-1)} = \dfrac{|\psi|}{\psi} \cdot \dfrac{\sqrt{[\mathrm{cF}(\psi-1)]^2 - 1}}{\mathrm{cF}(\psi-1)}$ is called the normalized Fibonacci speed of the light source in a vacuum.

5.3.5. *Properties of the normalized Fibonacci light speed*

For the normalized Fibonacci speed of light in a vacuum $\bar{c}(\psi) = \sqrt{\dfrac{\mathrm{sF}(\psi)}{\mathrm{sF}(\psi-2)}}$, the following properties are realized:

(1) For the values $\{-\infty < \psi < 0\}$ we have $\mathrm{sF}(\psi) < 0, \mathrm{sF}(\psi - 2) < 0$, but for the values $\{2 < \psi < +\infty\}$ we have $\mathrm{sF}(\psi) > 0, \mathrm{sF}(\psi - 2) > 0$. Therefore, $\{-\infty < \psi < 0\} \cup \{2 < \psi < +\infty\}$ results in $\dfrac{\mathrm{sF}(\psi)}{\mathrm{sF}(\psi-2)} > 0$ which means that $\bar{c}(\psi) = \sqrt{\dfrac{\mathrm{sF}(\psi)}{\mathrm{sF}(\psi-2)}} > 0$.

For such values of ψ, the following limit values are realized:

$$\bar{c}(\psi = -\infty) = \frac{1}{\Phi}, \qquad \bar{c}(\psi = 0 - 0) = 0,$$
$$\bar{c}(\psi = 2 + 0) = +\infty, \qquad \bar{c}(\psi = +\infty) = \Phi. \tag{5.7}$$

Hence, we obtain the following limit values for the Fibonacci speed of light in a vacuum:

$$\lim_{\psi \to +\infty} c = \lim_{\psi \to +\infty} \bar{c}(\psi) \cdot c_0 = \Phi \cdot \frac{c^*}{\Phi} = c^*,$$
$$\lim_{\psi \to -\infty} c = \lim_{\psi \to -\infty} \bar{c}(\psi) \cdot c_0 = \frac{1}{\Phi} \cdot \frac{c^*}{\Phi} = \frac{c^*}{\Phi^2}. \tag{5.8}$$

(2) For the values $\{0 < \psi < 2\}$ we have $\mathrm{sFs}(\psi) > 0, \mathrm{sFs}(\psi - 2) < 0$. Therefore, for the case $\{0 < \psi < 2\}$ we have $\dfrac{\mathrm{sFs}(\psi)}{\mathrm{sFs}(\psi-2)} < 0$. But then $\bar{c}(\psi) = i \cdot |\bar{c}(\psi)|$, where the absolute value $|\bar{c}(\psi)| = \sqrt{-\dfrac{\mathrm{sF}(\psi)}{\mathrm{sF}(\psi-2)}} > 0$.

For such values of ψ in the case of the absolute values $|\bar{c}(\psi)|$ the following limit values are realized:

$$|\bar{c}(\psi = 0 + 0)| = 0, \qquad |\bar{c}(\psi = 2 - 0)| = +\infty. \tag{5.9}$$

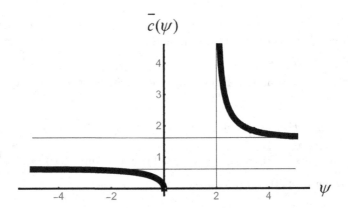

Figure 5.2.   The graph $\bar{c}(\psi) > 0$ for the case $\{-\infty < \psi < 0\} \cup \{2 < \psi < +\infty\}$ (the black curve).

(3) The values $\psi = 0$ and $\psi = 2$ correspond to significant bifurcation points (see Fig. 5.2).

5.3.6.   *Identical types of symbols for the Fibonacci special theory of relativity*

(1) The normalized Fibonacci speed of light in a vacuum $\bar{c}(\psi) = \sqrt{\dfrac{\mathrm{sFs}(\psi)}{\mathrm{sFs}(\psi-2)}}$ is derived directly from the equality of the conjugating matrices in (5.6). It follows that:

$$\begin{cases} \dfrac{1}{\sqrt{1-(\bar{v})^2}} = \mathrm{cF}(\psi-1), \quad \dfrac{1}{\bar{c}(\psi)} \dfrac{\bar{v}}{\sqrt{1-(\bar{v})^2}} = \mathrm{sF}(\psi-2) \\[4mm] \bar{c}(\psi)\dfrac{\bar{v}}{\sqrt{1-(\bar{v})^2}} = \mathrm{sF}(\psi) \end{cases} . \qquad (5.10)$$

Therefore:

$$\frac{\bar{v}}{\sqrt{1-(\bar{v})^2}} = \bar{c}(\psi) \cdot \mathrm{sF}(\psi-2) = \frac{1}{\bar{c}(\psi)} \cdot \mathrm{sF}(\psi)$$

$$\Rightarrow [\bar{c}(\psi)]^2 = \frac{\mathrm{sF}(\psi)}{\mathrm{sF}(\psi-2)} \Rightarrow \bar{c}(\psi) = \sqrt{\frac{\mathrm{sF}(\psi)}{\mathrm{sF}(\psi-2)}}.$$

(2) From $\bar{c}(\psi) = \sqrt{\dfrac{\mathrm{sF}(\psi)}{\mathrm{sF}(\psi-2)}}$, and $\mathrm{sF}(\psi)\mathrm{sF}(\psi-2) = [\mathrm{cF}(\psi-1)]^2 - 1$ (see (5.4)) we get an identical form for $\bar{c}(\psi)$. It follows then that:

$$\bar{c}(\psi) = \sqrt{\frac{sF(\psi)}{sF(\psi - 2)}} = \sqrt{\frac{sF(\psi)sF(\psi - 2)}{[sF(\psi - 2)]^2}};$$

$$\sqrt{\frac{sF(\psi)sF(\psi - 2)}{[sF(\psi - 2)]^2}} = \frac{\sqrt{[cF(\psi - 1)]^2 - 1}}{|sF(\psi - 2)|}.$$

From here we get the following identical form for $\bar{c}(\psi)$:

$$\bar{c}(\psi) = \frac{\sqrt{[cF(\psi - 1)]^2 - 1}}{|sF(\psi - 2)|}. \tag{5.11}$$

(3) From the relations:

$$\frac{1}{\sqrt{1 - (\bar{v})^2}} = cF(\psi - 1),$$

$$\frac{\bar{v}}{\sqrt{1 - (\bar{v})^2}} = \bar{c}(\psi) \cdot sF(\psi - 2) = \frac{1}{\bar{c}(\psi)} \cdot sF(\psi),$$

$$\bar{c}(\psi) = \sqrt{\frac{sF(\psi)}{sF(\psi - 2)}}$$

we arrive at an identical form for $\bar{v}(\psi)$ as follows:

(a) $\quad \dfrac{\bar{v}}{\sqrt{1 - (\bar{v})^2}} = \dfrac{1}{\bar{c}(\psi)} \cdot sF(\psi) = \sqrt{\dfrac{sF(\psi - 2)}{sF(\psi)}} \cdot sF(\psi)$

$$= \frac{|\psi|}{\psi} \cdot \sqrt{\frac{sF(\psi - 2)sF(\psi)^2}{sF(\psi)}}$$

$$= \frac{|\psi|}{\psi} \sqrt{sF(\psi - 2)sF(\psi)} = \frac{|\psi|}{\psi} \cdot \sqrt{[cF(\psi - 1)]^2 - 1}.$$

(b) $\quad \dfrac{\bar{v}}{\sqrt{1 - (\bar{v})^2}} = \bar{c}(\psi) \cdot sF(\psi - 2) = \sqrt{\dfrac{sF(\psi)}{sF(\psi - 2)}} \cdot sF(\psi - 2)$

$$= \frac{|\psi|}{\psi} \sqrt{\frac{sF(\psi) [sFs(\psi - 2)]^2}{sFs(\psi - 2)}}$$

$$= \frac{|\psi|}{\psi} \sqrt{sF(\psi)sF(\psi - 2)} = \frac{|\psi|}{\psi} \cdot \sqrt{[cFs(\psi - 1)]^2 - 1}.$$

(c) $\quad \dfrac{\frac{\bar{v}}{\sqrt{1-(\bar{v})^2}}}{\frac{1}{\sqrt{1-(\bar{v})^2}}} = \bar{v} = \dfrac{|\psi|}{\psi} \cdot \dfrac{\sqrt{[cF(\psi - 1)]^2 - 1}}{cF(\psi - 1)}.$

Hence, the following formula applies to the normalized Fibonacci speed of the source of light in a vacuum:

$$\bar{v}(\psi) = \frac{|\psi|}{\psi} \cdot \frac{\sqrt{[cF(\psi - 1)]^2 - 1}}{cF(\psi - 1)}. \tag{5.12}$$

Using (5.10), we rewrite (5.12) as follows:

$$\frac{1}{\sqrt{1 - (\bar{v})^2}} = cF(\psi - 1), \quad \bar{c}(\psi)\frac{\bar{v}}{\sqrt{1 - (\bar{v})^2}} = sF(\psi),$$

$$sF(\psi) = \bar{c}(\psi)\frac{\bar{v}}{\sqrt{1 - (\bar{v})^2}} = \bar{c}(\psi)\bar{v} \cdot cF(\psi - 1).$$

Hence we arrive at the following formula for the Fibonacci normalized speed of the light source in a vacuum:

$$\bar{v}(\psi) = \frac{1}{\bar{c}(\psi)} \cdot \frac{sF(\psi)}{cF(\psi - 1)}. \tag{5.13}$$

Then, the formula $v = \bar{c}(\psi)c_0 \cdot \bar{v}$ is transformed as follows. Because

$$v(\psi) = \bar{c}(\psi)c_0 \cdot \bar{v}(\psi) \quad \text{and} \quad \bar{v}(\psi) = \frac{1}{\bar{c}(\psi)} \cdot \frac{sF(\psi)}{cF(\psi - 1)},$$

we then have:

$$v(\psi) = \bar{c}(\psi)c_0 \cdot \bar{v}(\psi) = \bar{c}(\psi)c_0 \cdot \frac{1}{\bar{c}(\psi)} \cdot \frac{sF(\psi)}{cF(\psi - 1)} = c_0 \cdot \frac{sF(\psi)}{cF(\psi - 1)}.$$

Thus, we get the following formula for the Fibonacci speed of the light source in a vacuum $v(\psi)$:

$$v(\psi) = c_0 \cdot \frac{sF(\psi)}{cF(\psi - 1)}. \tag{5.14}$$

Figure 5.3 shows the graphs $\bar{c}(\psi)$ and $\bar{v}(\psi)$.

5.3.7. *The physical interpretation of the mathematical model of the Fibonacci special theory of relativity: bifurcation points, material Universe, Dark and Light Ages, Black Hole and anti-matter*

In this section, we use the works [7, 12, 13, 17, 18]. Let us interpret Fig. 5.2 from the physical point of view:

(1) The point $\psi = 0$ is the first point of bifurcation and corresponds to the point of the singularity, the Big Bang.

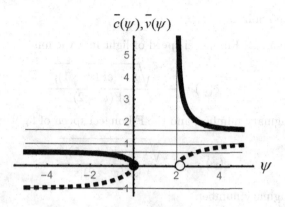

Figure 5.3. The graph $\bar{c}(\psi)$ (the black curves) and the graph $\bar{v}(\psi)$ (the dashed curves), where the parameter ψ is within the limits $\{-\infty < \psi < 0\} \cup \{2 < \psi < +\infty\}$. The symbols ● and ○ correspond to the bifurcation points $\psi = 0$ and $\psi = 2$, respectively.

(2) The domain $\{0 < \psi < +\infty\}$ (the right half of the graph in Fig.5.2) corresponds to the Material Universe; this domain is divided by the second bifurcation point $\psi = 2$ into two sub-domains:

   (a) The sub-domain $\{0 < \psi < 2\}$ or the Dark Ages, when there was nothing to illuminate the Universe. In the sub-domain $\{0 < \psi < 2\}$ the normalized Fibonacci speed of light in a vacuum $\bar{c}(\psi)$ is an imaginary magnitude, although the Material Universe began to evolve and elementary particles started to form.

   (b) The second point of bifurcation $\psi = 2$ corresponds to the beginning of the transition from the Dark Ages to the Light Ages.

   (c) The sub-domain $\{2 < \psi < +\infty\}$ means the Light Ages, when the first proto-stars ignited and the Light Universe began and has continued to evolve up to the present day.

(3) The domain $\{-\infty < \psi < 0\}$ (the left half of the graph in Fig. 5.2) corresponds to the Black Hole, which consists of anti-matter.

5.3.8. *Fibonacci–Lorentz transformations for the Dark Ages* $\{0 < \psi < 2\}$

For this case we get that $\frac{|\psi|}{\psi} = -1$ and $[cF(\psi - 1)]^2 - 1 < 0$. Because for all values $\{0 < \psi < 2\}$ the following non-equality is realized: $[cF(\psi - 1)]^2 - 1 < 0$, then for the Dark Ages the following properties are realized:

(1) The normalized Fibonacci speed of light in a vacuum

$$\bar{c}(\psi) = -i \cdot \frac{\sqrt{1 - [cF(\psi - 1)]^2}}{|sF(\psi - 2)|}$$

is an imaginary number, and the Fibonacci speed of light in a vacuum:

$$c(\psi) = -i \cdot \frac{\sqrt{1 - [cF(\psi - 1)]^2}}{|sF(\psi - 2)|} \cdot c_0$$

is an imaginary number.

(2) The normalized Fibonacci speed of the light source in a vacuum

$$\bar{v}(\psi) = -i \cdot \frac{\sqrt{[1 - cF(\psi - 1)]^2}}{cF(\psi - 1)}$$

is an imaginary number, and the Fibonacci speed of the light source in a vacuum:

$$v(\psi) = c_0 \cdot \frac{sF(\psi)}{cF(\psi - 1)} \; [\text{m sec}^{-1}]$$

is a real number.

Previously we presented in Figs. 5.1 and 5.2 the graphs of the real dimensionless functions $\bar{c}(\psi), \bar{v}(\psi)$ for the Black Hole $\{-\infty < \psi < 0\}$ and for the Light Ages $\{2 < \psi < +\infty\}$.

For the Dark Ages $\{0 < \psi < 2\}$, the functions $\bar{c}(\psi), \bar{v}(\psi)$ are imaginary and have the following forms $\bar{c}(\psi) = |\bar{c}(\psi)| \cdot i, \bar{v}(\psi) = -|\bar{v}(\psi)| \cdot i$, respectively. That is why Figs. 5.4 and 5.5 represent the graphs of the modules $|\bar{c}(\psi)|$ and $|\bar{v}(\psi)|$ for the Dark Ages, which are positive dimensionless real functions.

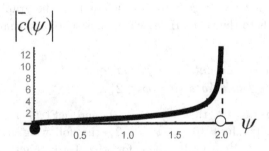

Figure 5.4.   The graph of the module $|\bar{c}(\psi)|$ on the interval $\{0 < \psi < 2\}$.

Table 5.1. Numerical characteristics $\bar{c}(\psi)$ and $\bar{v}(\psi)$ depending on the parameter of self-organization ψ for the range $\{-\infty < \psi < +\infty\}$.

The dimensionless magnitudes	ψ	$\bar{c}(\psi)$	$\bar{v}(\psi)$
The Black Hole (the negative arrow of time) $\{-\infty < \psi < 0\}$	$-\infty$	0.618	-1
	-5	0.615	-0.992
	-1	0.5	-0.666
	0	0	0
The Dark Ages $\{0 < \psi < 2\}$	0	0	0
	0.5	$0.555 \cdot i$	$-0.424 \cdot i$
	1	$1 \cdot i$	$-0.5 \cdot i$
	1.5	$1.798 \cdot i$	$-0.424 \cdot i$
	$2 - 0$	$\infty \cdot i$	0
The Light Ages (the positive arrow of time) $\{2 < \psi < +\infty\}$	$2 + 0$	$+\infty$	0
	3	2	0.666
	$+\infty$	1.618	1

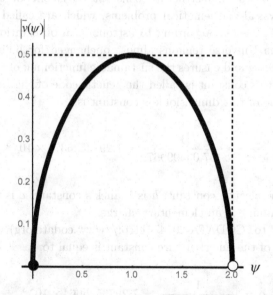

Figure 5.5. The graph of the module $|\bar{v}(\psi)|$ on the interval $\{0 < \psi < 2\}$.

We will try to establish the interconnection between the dimensionless parameter of self-organization and the age of the Universe at any given time T [billion years] beginning with the Big Bang by using the mathematical model of the Fibonacci special theory of relativity. This allows for a full interpretation of this model of the evolution of the Universe not only qualitatively but also quantitatively as for both the positive arrow of time $(T > 0)$ and the negative arrow of time $(T < 0)$. To solve it, we need to study the problem of the dependence of the dimensionless constant (the fine-structure constant) on the age of the Universe, starting with the Big Bang. For example, for the present time, according to modern studies, the age of the Universe is $T_{present} = 13.81$ [billion years] (the data of WMAP for 2012 [22]) or the more accurate value $T_{present} = 13.81 \pm 0.06$ [billion years] (the data of PLANCK for 2013 [23]), and the value of the fine-structure constant is equal to $\alpha = 0.0072973525376$.

5.4.   The Fine-Structure Constant α and Its Relationship with the Evolution of the Universe

5.4.1.   *The fine-structure constant*

The problem of the fine-structure constant α, is one of the 10 most important physical-mathematical problems, which are called Millennium Problems [5, 6, 24–29]. According to astronomical observations, the constant α for the Universe remains almost unchanged for billions of years after the Big Bang and ensures the sustainable functioning of the Universe, and therefore this constant is called the genetic code of the Universe. The numerical value of this dimensionless constant is:

$$\alpha = \frac{e^2}{2\varepsilon_0 c \cdot h} = \frac{1}{137.035999679} = 7.2973525376 \times 10^{-3} \qquad (5.15)$$

where ε_0 is the electric constant; h is Planck's constant; c is the speed of light in a vacuum; e is an elementary charge.

According to CODATA-2014 (http://www.codata.org), the recommended value of the fine-structure constant is equal to:

$$\alpha = \frac{1}{137.035999139} = 7.2973525664 \times 10^{-3}.$$

The problem of the fine-structure constant can be included in Gross's formulation of the First Physics Millennium Problem [5, 6, 24–29].

5.4.2. The significances of the symbols for the fine-structure constant α

(1) Electric constant (or dielectric permeability of the vacuum)

$$\varepsilon_0 = 8.854187817629 \times 10^{-12} \left[\frac{A^2 \cdot (\sec)^4}{m^3 \cdot kg} \right] \qquad (5.16)$$

where A (ampere) is a unit amperage (SI).

(2) Planck's constant

$$h = 6.626068959046 \left[\frac{kg \cdot m^2}{\sec} \right]. \qquad (5.17)$$

(3) Dirac's constant (or the reduced Planck's constant)

$$\bar{h} = \frac{h}{2\pi} = 1.054571628 \times 10^{-34} \left[\frac{kg \cdot m^2}{\sec} \right].$$

(4) Elementary charge e:

$$e = 1.6022176487 \times 10^{-19} \, [A \cdot \sec]. \qquad (5.18)$$

5.4.3. Excursus on the discovery of the fine-structure constant

The fine-structure constant α is a fundamental physical constant, which characterizes the power of the electromagnetic interaction. It does not characterize separate physical bodies, but the physical properties of our world as a whole.

For the first time in 1916, the constant α was described by the German physicist Arnold Sommerfeld as a measure of the relativistic corrections in the description of atomic spectral lines in the model of the Danish physicist Niels Bohr. In other words, the fine-structure constant α describes the splitting of atomic levels on a few close sublevels (multiplets) due to the effects of the special theory of relativity. American physicist Richard Feynman made a similar observation about the constant α. A more detailed description of α can be found in James Carter's book [27].

5.4.4.   The hypotheses and experiments about the variability of the fine-structure constant, depending on the age of the Universe

There is a long history to the question of whether the fine-structure constant is really a fundamental physical constant, that is, always unchanged or has it possibly changed during the tenure of the Universe. In 1995, the prominent Russian physicist Lev Landau predicted that this constant can vary depending on time: "However, such changes cannot be very large, otherwise they would have already 'emerged' in relatively simple experiments."

In the late 1990s, new data from astronomical observations appeared. Astronomer John Webb and his colleagues [30] found a tiny change in the wavelength of the light from distant quasars. Simulation of quasar light showed that 10 to 12 billion years ago, the value of the constant α was greater than its current value.

More detailed observations of quasars, made in April 2004 by using the 8.2-meter telescope in Paranal Observatory in Chile, showed that over 10 billion years ago, the possible value of the constant α could not be more than 6×10^{-7} of the present value of α.

5.4.5.   Hilbert's Sixth Problem and recommendations for the solution of physical problems by means of harmony between experience and thought

It should be noted that already in 1900, David Hilbert in his paper, "Mathematical problems" [1–3], presented before the Second International Congress of Mathematicians in Paris (August 8, 1900), had formulated the Sixth Problem "Mathematical treatment of the axioms of physics". In the Sixth Problem he drew attention not only to the study of physical phenomena and constants by using experiments, but also to the creation of rigorous mathematical theories, which were directed towards the solution of physical problems by employing the "realm of pure thought".

Hilbert stated the Sixth Problem for mathematicians as follows:

> "The investigations on the foundations of geometry suggest the problem:
> To treat in the same manner, by means of axioms, those physical sciences in which mathematics plays an important part.
>   Further, the mathematician has the duty to test exactly in each instance whether the new axioms are compatible with the previous ones. The physicist, as his theories develop, often finds himself forced by the results of his experiments to make new hypotheses, while he depends, with

respect to the compatibility of the new hypotheses with the old axioms, solely upon these experiments or upon a certain physical intuition, a practice which in the rigorously logical building up of a theory is not admissible. The desired proof of the compatibility of all assumptions seems to me also of importance, because the effort to obtain such proof always forces us most effectually to an exact formulation of the axioms."

In the Introductory part of his paper, Hilbert states:

"In the meantime, while the creative power of pure reason is at work, the outer world again comes into play, forces upon us new questions from actual experience, opens up new branches of mathematics, and while we seek to conquer these new fields of knowledge for the realm of pure thought, we often find the answers to old unsolved problems and thus at the same time advance most successfully the old theories. And it seems to me that the numerous and surprising analogies and that apparently pre-arranged harmony which the mathematician so often perceives in the questions, methods and ideas of the various branches of his science, have their origin in this ever-recurring interplay between thought and experience."

Thus, *a priori*, even Hilbert himself points out that the Sixth Problem is too vague and virtually impracticable and, therefore, a mathematical statement of the axioms of physics is inherently unsound. That is why Hilbert's Sixth Problem, "Mathematical treatment of the axioms of physics" has not attracted much attention from mathematicians.

However, the underlying importance of Hilbert's Sixth Problem is the fact that Hilbert suggests solving physical problems by the creation of rigorous mathematical theories. That is, Hilbert's Sixth Problem is physical, even though its solution is considered to be mathematical.

5.4.6. Formula by Nikolai Kosinov interconnecting three major dimensionless constants: the fine-structure constant α, the number π and the golden ratio Φ

The findings of Ukrainian physicist Nikolai Kosinov [24, 28, 29] became an important breakthrough in the "understanding of the geometrical status of the fine-structure constant, and also that all dimensionless parameters, which characterize the micro-world and Universe, are calculable in principle."

In 2000, Kosinov [24] found a simple and beautiful relationship, which links the dimensionless constants: the fine-structure constant α, the number π and the golden ratio $\Phi = \frac{1+\sqrt{5}}{2}$. This formula is as follows:

$$\alpha = 10^{-\frac{43}{20}} \times \pi^{\frac{1}{260}} \times \Phi^{\frac{7}{130}}. \tag{5.19}$$

The newly calculated value of the fine-structure constant was obtained in [24] on the basis of formula (5.19):

$$\alpha = 7.297351997351997377362 \times 10^{-3}. \tag{5.20}$$

Thus, the absolute error $\Delta\alpha$ between the true value of α (see (5.15)) and the calculated value of α (see (5.20)) equals the following:

$$\Delta\alpha = |7.2973525376 \times 10^{-3} - 7.2973519973 \times 10^{-3}|$$

$$= 5.403 \times 10^{-10} = 0.0000000005403. \tag{5.21}$$

In this case, the relative error $\frac{\Delta\alpha}{\alpha}$ is equal to the value:

$$\frac{\Delta\alpha}{\alpha} = \frac{5.403 \times 10^{-10}}{7.2973525376 \times 10^{-3}} = 7.40406 \times 10^{-8} = 0.0000000740406,$$
$$\tag{5.22}$$

that is, Kosinov's formula (5.19) coincides numerically with the value (5.15) with pinpoint accuracy.

5.4.7. Postulate of the dependence of the fine-structure constant upon the age of the Universe for the mathematical model of the Fibonacci special theory of relativity

The presence in the formula (5.19) of the golden ratio, which is known to be the indicator of mathematical harmony, beginning with the ancient Greeks, led us to the use of the Fibonacci special theory of relativity, based on the golden ratio. The theoretical study of the variations of the fine-structure constant depends on the age of the Universe, starting with the Big Bang up to the present time $T_{present}$, which according to the data of WMAP [22] $T_{present} = 13.75 \pm 0.13$ [billion years] or according to the specified data of PLANCK [23] $T_{present} = 13.81 \pm 0.06$ [billion years].

To this end, we introduce for consideration the following postulate:

The fine-structure constant α depends on the time T [billion years], counted from the moment of the Big Bang ($T = 0$) for the

Dark and Light Ages $(T > 0)$, and for the Black Hole $(T < 0)$ by the formula:

$$\alpha = 10^{-\frac{43}{20}} \times \pi^{\frac{1}{260}} \times |\bar{c}(\psi)|^{\frac{7}{130}}, \quad \psi = \lambda_0 \cdot T. \tag{5.23}$$

5.4.8. *The significances of symbols in the formula for the fine-structure constant, depending on the age of the Universe*

(1) T [billion years] is the time counted from the moment of the Big Bang.

(2) $\lambda_0 \left[\dfrac{1}{\text{billion years}}\right] = \text{const} > 0$ is a weight coefficient.

(3) $\bar{c}(\psi) = \sqrt{\dfrac{\text{sF}(\psi)}{\text{sF}(\psi-2)}} = \dfrac{\sqrt{[\text{cF}(\psi-1)]^2 - 1}}{|\text{sF}(\psi-2)|}$ (dimensionless) is the normalized Fibonacci speed of light in a vacuum, $|\bar{c}(\psi)|$ is the module of the normalized Fibonacci speed of light in a vacuum, ψ is a parameter of self-organization.

Note that $\bar{c}(\psi) = |\bar{c}(\psi)|$ for the range $\{-\infty < \psi < 0\} \cup \{2 < \psi < +\infty\}$, and $\bar{c}(\psi) = i \cdot |\bar{c}(\psi)|$ for the range $\{0 < \psi < 2\}$, where $i = \sqrt{-1}$ is the imaginary unit. The values $\psi = 0$ and $\psi = 2$ are the bifurcation values for $\bar{c}(\psi)$. For ψ and $|\bar{c}(\psi)|$ we have the following limits:
$\{2 < \psi < +\infty\}$ (the Light Ages) for the range $\{\Phi < |\bar{c}(\psi)| < +\infty\}$,
$\{0 < \psi < 2\}$ (the Dark Ages) for the range $\{0 < |\bar{c}(\psi)| < +\infty\}$,
$\{-\infty < \psi < 0\}$ (the Black Hole) for the range $\{0 < |\bar{c}(\psi)| < \Phi^{-1}\}$.

(4) $\text{sF}(x) = \dfrac{\Phi^x - \Phi^{-x}}{\sqrt{5}} = \dfrac{2}{\sqrt{5}}\text{sh}(x \cdot \ln \Phi)$ (dimensionless) is the hyperbolic Fibonacci sine, $\text{cF}(x) = \dfrac{\Phi^x + \Phi^{-x}}{\sqrt{5}} = \dfrac{2}{\sqrt{5}}\text{ch}(x \cdot \ln \Phi)$ (dimensionless) is the hyperbolic Fibonacci cosine.

(5) $\Phi = \dfrac{1+\sqrt{5}}{2} \approx 1.61803$ (dimensionless) is the golden ratio.

5.4.9. *The procedure for finding the numerical values of the weighting factor in the postulate of the dependence of the fine-structure constant on the age of the Universe*

(1) Experimentally: As a result of astronomical observations, or other experiences we find for the Light Ages a supporting experiment for the case $(T_0 > 0, \alpha_0 > 0)$, where T_0 [billion years] > 0 is the fixed value of the time T, measured from the moment of the Big Bang and $\alpha_0 > 0$ is the value of the fine-structure constant at the moment of time T_0.

(2) From the formula (5.23) for the case $\alpha_0 > 0$, we calculate the corresponding value of the normalized Fibonacci speed of light in a vacuum:

$$\bar{c}(\psi_0) = |\bar{c}(\psi_0)| = \sqrt[14]{10^{559} \cdot \pi^{-1}} \cdot \sqrt[7]{\alpha_0^{130}}$$

$$= 7.817371000127518 \cdot 10^{39} \cdot \sqrt[7]{\alpha_0^{130}}. \tag{5.24}$$

Here the following inequality is fulfilled:

$$\{\Phi < |\bar{c}(\psi_0)| < +\infty\},$$

because, by hypothesis, the experiment should be conducted for the Light Ages of the Universe $\{2 < \psi < +\infty\}$, when our Universe began to be illuminated by stars.

(3) From the formulas (5.23) and (5.24) for the case $\bar{c}(\psi_0)$, we calculate the corresponding value of the parameter of self-organization ψ_0 by the formula:

$$\psi_0 = 0.5 \cdot \left[\log_\Phi \left(\frac{1 - |\bar{c}(\psi_0)|^2 \cdot \Phi^2}{1 - |\bar{c}(\psi_0)|^2 \cdot \Phi^{-2}} \right) \right]. \tag{5.25}$$

The value of $\psi_0 > 2$ should be verified experimentally, as we are in the Light Ages and according to the mathematical model of the Fibonacci special theory of relativity, ψ_0 corresponds to the range $\{2 < \psi < +\infty\}$.

(4) For the cases $T_0 > 0$ and $\psi_0 > 2$, we calculate the unknown value of the weighting factor λ_0 by the formula:

$$\lambda_0 = \frac{\psi_0}{T_0} \left[\frac{1}{\text{billion years}} \right]. \tag{5.26}$$

5.4.10. *Temporary restrictions of the different periods of the existence of the Universe following from the supporting experiment*

From the formula (5.26) for any value of T in the range $-\infty < T < +\infty$, we can calculate the corresponding value of the self-organization parameter ψ from the formula:

$$\psi = \lambda_0 \cdot T \text{ (dimensionless)}. \tag{5.27}$$

Conversely, for any value of the parameter of self-organization in the range $-\infty < \psi < +\infty$, we can calculate the corresponding value of the time in the range $-\infty < T < +\infty$ from the formula:

$$T = \frac{1}{\lambda_0} \cdot \psi \; [\text{billion years}]. \tag{5.28}$$

Hence, for the Fibonacci special theory of relativity, we get the following **time intervals** T [billion years] for the different periods of the existence of the Universe:

(1) For the Light Ages $\{2 < \psi < +\infty\}$ we have $\{\frac{2}{\lambda_0} < T < +\infty\}$.
(2) For the Dark Ages $\{0 < \psi < 2\}$ we have $\{0 < T < \frac{2}{\lambda_0}\}$.
(3) For the Black Hole $\{-\infty < \psi < 0\}$ we have $\{-\infty < T < 0\}$.
(4) For the bifurcation point of $\psi = 0$ (the Big Bang as the transition from the Dark Ages to the Black Hole) we have $T = 0$.
(5) For the bifurcation point $\psi = 2$ (the transition from the Dark Ages to the Light Ages) we have $T = \frac{2}{\lambda_0}$.

5.4.11. A quantitative description of the Universe's time evolution based on the supporting experiment

We recall the basic formulas, which describe the evolution of the Universe based on the Fibonacci special theory of relativity, depending on the parameter of self-organization ψ, $\{-\infty < \psi < +\infty\}$. These formulas are:

$$\begin{cases} \bar{c}(\psi) = \sqrt{\dfrac{\text{sF}(\psi)}{\text{sF}(\psi-2)}}, \quad c(\psi) = c_0 \cdot \bar{c}(\psi) \; [\text{m sec}^{-1}], \\[2mm] \bar{v}(\psi) = \dfrac{|\psi|}{\psi} \cdot \dfrac{\sqrt{[\text{cF}(\psi-1)]^2 - 1}}{\text{cF}(\psi-1)} \\[2mm] v(\psi) = c_0 \cdot \dfrac{\text{sF}(\psi)}{\text{cF}(\psi-1)} \; [\text{m sec}^{-1}], \quad c_0 = \frac{c^*}{\Phi} \approx 1.8528199 \; [\text{m sec}^{-1}] \end{cases} \tag{5.29}$$

The following procedure is a quantitative description of the Universe's time evolution on the basis of the supporting experiment:

(1) For the Light Ages we carry out the supporting experiment ($T_0 > 0, \alpha_0 > 0$).
(2) We calculate the positive supporting numerical coefficient by the formula $\lambda_0 = \frac{\psi_0}{T_0}[\frac{1}{\text{billion years}}]$.
(3) We postulate that for any T in the range $\{-\infty < T < +\infty\}$ the following relation is valid: $\psi = \lambda_0 \cdot T$.
(4) For any T in the range $\{-\infty < T < +\infty\}$, the values $\psi = \lambda_0 \cdot T$ are substituted into the relation (5.29). The following formulas resulting

from the Fibonacci special theory of relativity, give not only the complete qualitative, but also quantitative information on the evolution of the Universe:

$$
\begin{cases}
\bar{c}(T) = \sqrt{\dfrac{\mathrm{sF}(\lambda_0 T)}{\mathrm{sF}(\lambda_0 T - 2)}}, \quad c(T) = c_0 \cdot \bar{c}(T) \ [\mathrm{m}\ \mathrm{sec}^{-1}], \\[3mm]
\bar{v}(T) = \dfrac{|T|}{T} \cdot \dfrac{\sqrt{[\mathrm{cF}(\lambda_0 T - 1)]^2 - 1}}{\mathrm{cF}(\lambda_0 T - 1)} \\[3mm]
v(T) = c_0 \cdot \dfrac{\mathrm{sF}(\lambda_0 T)}{\mathrm{cF}(\lambda_0 T - 1)} [\mathrm{m}\ \mathrm{sec}^{-1}], \quad c_0 = \frac{c^*}{\Phi} \approx 1.8528199\,[\mathrm{m}\ \mathrm{sec}^{-1}]
\end{cases}
$$

$$(5.30)$$

5.4.12.   Limit values of the fine-structure constant for the Light Ages, the Dark Ages, and the Black Hole

It follows from the formulas (5.29) and (5.30) that even in the absence of the supporting experiments for the Light Ages $\{2 < \psi < +\infty\}$ and the Black Hole $\{-\infty < \psi < 0\}$ we have the following limit relations, respectively:

$$
\lim_{T \to +\infty} \bar{c}(\psi(T)) = \lim_{\psi \to +\infty} \bar{c}(\psi) = \Phi, \quad \lim_{T \to -\infty} \bar{c}(\psi(T)) = \lim_{\psi \to -\infty} \bar{c}(\psi) = \Phi^{-1}.
$$

$$(5.31)$$

These arguments lead us to the following conclusions:

(1) For the Light Ages $\{2 < \psi < +\infty\}$ (the positive arrow of time), the function $\bar{c}(\psi) = |\bar{c}(\psi)| = \sqrt{\dfrac{\mathrm{sF}(\psi)}{\mathrm{sF}(\psi - 2)}}$, when changing ψ through its limit range from 2 to $+\infty$, decreases monotonically from $\bar{c}(\psi = 2) = +\infty$ to $\bar{c}(\psi = +\infty) = \Phi$. Consequently, the number of $\bar{c}(+\infty) = \Phi$ which is the lower bound of the $\bar{c}(\psi)$ in the range $\{2 < \psi < +\infty\}$.

But then, according to (5.23), the fine-structure constant $\alpha(\psi) = 10^{-\frac{43}{20}} \times \pi^{\frac{1}{260}} \times |\bar{c}(\psi)|^{\frac{7}{130}}$, when changing ψ through its limit range from 2 to $+\infty$, decreases monotonically from the value

$$
\alpha(\psi = 2) = 10^{-\frac{43}{20}} \times \pi^{\frac{1}{260}} \times |\bar{c}(\psi = 2)|^{\frac{7}{130}}
$$
$$
= 10^{-\frac{43}{20}} \times \pi^{\frac{1}{260}} \times |(+\infty)|^{\frac{7}{130}} = +\infty,
$$

to the value

$$
\alpha(\psi = +\infty) = 10^{-\frac{43}{20}} \times \pi^{\frac{1}{260}} \times |\bar{c}(\psi = +\infty)|^{\frac{7}{130}}
$$
$$
= 10^{-\frac{43}{20}} \times \pi^{\frac{1}{260}} \times \Phi^{\frac{7}{130}} = 0.007297351997377362.
$$

Consequently, the lower bound of the fine-structure constant $\alpha(\psi)$ for the Light Ages is equal to

$$\alpha(\psi = +\infty) = 10^{-\frac{43}{20}} \times \pi^{\frac{1}{260}} \times \Phi^{\frac{7}{130}} = 0.007297351997377362, \quad (5.32)$$

which corresponds to Kosinov's formula (5.19). Due to the limit relations (5.31), the lower bound of the fine-structure constant $\alpha(T)$ for the Light Ages, even in the absence of a supporting experiment, is equal to

$$\alpha(T = +\infty) = 10^{-\frac{43}{20}} \times \pi^{\frac{1}{260}} \times \Phi^{\frac{7}{130}} = 0.007297351997377362.$$
$$(5.33)$$

(2) For the Black Hole $\{-\infty < \psi < 0\}$ when changing ψ through its limit range from 0 to $-\infty$ (the negative arrow of time), the function $\bar{c}(\psi) = |\bar{c}(\psi)| = \sqrt{\dfrac{sF(\psi)}{sF_{(\psi-2)}}}$ increases monotonically from $\bar{c}(\psi = 0) = 0$ to $\bar{c}(\psi = -\infty) = \Phi^{-1}$. Consequently, the number of $\bar{c}(\psi = -\infty) = \Phi^{-1}$ is the upper bound of the function $\bar{c}(\psi)$ when changing limit ψ from 0 to $-\infty$. But then, according to $\bar{c}(\psi) = |\bar{c}(\psi)| = \sqrt{\dfrac{sF(\psi)}{sF_{(\psi-2)}}}$, the fine-structure constant (5.23) when changing ψ through its limit range from 0 to $-\infty$ increases monotonically from the value

$$\alpha(\psi = 0) = 10^{-\frac{43}{20}} \times \pi^{\frac{1}{260}} \times |\bar{c}(\psi = 0)|^{\frac{7}{130}} = 10^{-\frac{43}{20}} \times \pi^{\frac{1}{260}} \times |0|^{\frac{7}{130}} = 0$$

to the value

$$\alpha(\psi = -\infty) = 10^{-\frac{43}{20}} \times \pi^{\frac{1}{260}} \times |\bar{c}(\psi = -\infty)|^{\frac{7}{130}}$$
$$= 10^{-\frac{43}{20}} \times \pi^{\frac{1}{260}} \times |\Phi^{-1}|^{\frac{7}{130}} = 0.0069288144971348135.$$

Due to the limit relations (5.31), the upper bound of the fine-structure constant $\alpha(T)$ for the Black Hole, even without the supporting experiment, at changing of T from 0 to $-\infty$, is the number:

$$\alpha(T = -\infty) = 10^{-\frac{43}{20}} \times \pi^{\frac{1}{260}} \times |\Phi^{-1}|^{\frac{7}{130}} = 0.0069288144971348135.$$
$$(5.34)$$

(3) For the Dark Ages $\{0 < \psi < 2\}$, the normalized Fibonacci speed of light in a vacuum $\bar{c}(\psi) = i \cdot |\bar{c}(\psi)|$, that is, $\bar{c}(\psi)$ is an imaginary quantity. In order to find the values of the fine-structure constant we use the formula (5.23):

$$\alpha(\psi) = 10^{-\frac{43}{20}} \times \pi^{\frac{1}{260}} \times |\bar{c}(\psi)|^{\frac{7}{130}},$$

where instead of $\bar{c}(\psi)$ we take the module $|\bar{c}(\psi)|$, then at the interval $\{0 < \psi < 2\}$ the fine-structure constant $\alpha = \alpha(\psi)$ is a real number.

The limit values for $\alpha(\psi)$ are the following:

$$\alpha(\psi = 0) = 10^{-\frac{43}{20}} \times \pi^{\frac{1}{260}} \times |\bar{c}(\psi) = 0|^{\frac{7}{130}}$$

$$= 10^{-\frac{43}{20}} \times \pi^{\frac{1}{260}} \times 0^{\frac{7}{130}} = 0,$$

$$\alpha(\psi = 2) = 10^{-\frac{43}{20}} \times \pi^{\frac{1}{260}} \times |\bar{c}(\psi) = 2|^{\frac{7}{130}}$$

$$= 10^{-\frac{43}{20}} \times \pi^{\frac{1}{260}} \times (+\infty)^{\frac{7}{130}} = +\infty.$$

The fine-structure constant $\alpha(\psi)$, when increasing ψ through the range from 0 to 2, *increases monotonically* from the value $\alpha(\psi = 0) = 0$ to $\alpha(\psi = 2) = +\infty$. With regard to the limits of the fine-structure constant α for the Dark Ages, starting at the moment of the Big Bang, for which $T = 0$, then the first limit value for $T = 0$ does not depend on the supporting experiment $(T_0 > 0, \alpha_0 > 0)$ and is equal to $\alpha(T = 0) = 0$.

As for the other limit value, when $\alpha(\psi = 2) = +\infty$, the value of the time $T^* = T(\psi = 2)$ depends on the reliability of the supporting experiment. As the supporting experiment should be carried out for the Light Ages and determines the values $(T_0 > 0, \alpha_0 > 0)$. From here the positive supporting numerical coefficient is given by the formula $\lambda_0 = \frac{\psi_0}{T_0} > 0$.

Next, we use the relation $\psi = \lambda_0 \cdot T$ for any values of ψ and T. From here, we get the formula $T = \frac{\psi}{\lambda_0}$, see the formula (5.28). But then the interval $\{0 < \psi < 2\}$ for the Dark Ages in terms of the time T [billion years] is given by $\left\{0 < T < \frac{2}{\lambda_0}\right\}$, that is, it depends upon the supporting experiment. For the endpoints $\psi = 0$ and $\psi = 2$, which are the bifurcation points, we get $T(\psi = 0) = 0$ [billion years] and $T(\psi = 2) = \frac{2}{\lambda_0}$ [billion years], respectively.

5.4.13.   *Graphs, tables, and calculations*

Graphs.   The use of the Fibonacci special theory of relativity as a quantitative description for the evolution of the Universe, based upon the given supporting experiment (T_0, α_0) and applied to the Light Ages $\{2 < \psi < +\infty\}$, is demonstrated by the fact that when we change the parameter of self-organization ψ from 2 to $+\infty$ the derivative $\frac{d\alpha(\psi)}{d\psi} < 0$. Note also that for the Black Hole $\{2 < \psi < +\infty\}$ the derivative $\frac{d\alpha(\psi)}{d\psi} < 0$. For the Dark Ages the derivative $\frac{d\alpha(\psi)}{d\psi} > 0$.

The derivation of the derivative $\frac{d\alpha}{d\psi}$. Let us prove that the derivative $\frac{d\alpha}{d\psi} < 0$ for the function

$$\alpha(\psi) = 10^{-\frac{43}{20}} \times \pi^{\frac{1}{260}} \times |\bar{c}(\psi)|^{\frac{7}{130}},$$

where

$$|\bar{c}(\psi)| = \sqrt{\frac{sF(\psi)}{sF(\psi - 2)}},$$

if $\{-\infty < \psi < 0\} \cup \{2 < \psi < +\infty\}$, and that the derivative $\frac{d\alpha}{d\psi} > 0$ for the function $|\bar{c}(\psi)| = \sqrt{-\frac{sF(\psi)}{sF(2-\psi)}}$ if $\{0 < \psi < 2\}$.

The derivative $\frac{d\alpha}{d\psi}$ for the interval $\{2 < \psi < +\infty\}$ is calculated as follows. Let us introduce the following designations:

$$k = 10^{-\frac{43}{20}} \cdot \pi^{\frac{1}{260}}; \quad \Phi = \frac{1 + \sqrt{5}}{2}; \quad \alpha = k \cdot |\bar{c}(\psi)|^{\frac{7}{130}};$$

$$|\bar{c}(\psi)| = \sqrt{\frac{\Phi^{\psi} - \Phi^{-\psi}}{\Phi^{\psi-2} - \Phi^{2-\psi}}}.$$

Then we have:

$$\frac{d\alpha}{d\psi} = \frac{7k \cdot \ln \Phi \left[-\frac{(\Phi^{\psi} - \Phi^{-\psi}) \cdot (\Phi^{\psi-2} + \Phi^{2-\psi})}{(\Phi^{\psi-2} - \Phi^{2-\psi})^2} + \frac{\Phi^{\psi} + \Phi^{-\psi}}{\Phi^{\psi-2} - \Phi^{2-\psi}} \right]}{260 \left(\frac{(\Phi^{\psi} - \Phi^{-\psi})}{(\Phi^{\psi-2} - \Phi^{2-\psi})} \right)^{\frac{253}{260}}}.$$

The sign of the derivative $\frac{d\alpha}{d\psi}$ coincides with the sign of the following expression:

$$(\Phi^{-\psi} - \Phi^{\psi}) \cdot (\Phi^{\psi-2} + \Phi^{2-\psi}) + (\Phi^{-\psi} + \Phi^{\psi}) \cdot (\Phi^{\psi-2} - \Phi^{2-\psi})$$

$$= -2 \cdot (\Phi^2 - \Phi^{-2}) < 0.$$

That is, in this case, the sign $\frac{d\alpha}{d\psi} = -1$. Similarly the derivative $\frac{d\alpha}{d\psi} < 0$ is calculated for the interval $\{-\infty < \psi < 0\}$ and the derivative $\frac{d\alpha}{d\psi} > 0$ is calculated for the interval $\{0 < \psi < 2\}$.

The points $\psi = 0, \psi = 2$ are the points of bifurcation. In the bifurcation points $\psi = 0, \psi = 2$, the derivative $\frac{d\alpha}{d\psi}$ is not defined and it is necessary to define it on the left and on the right of the bifurcation points.

To illustrate the above statements, we confine ourselves to the graph of the function (5.23) $\alpha(\psi) = 10^{-\frac{43}{20}} \times \pi^{\frac{1}{260}} \times |\bar{c}(\psi)|^{\frac{7}{130}}$ on an infinite interval $\{-\infty < \psi < \infty\}$ (see Fig. 5.6).

We also need to find the explicit dependence of the parameter of self-organization of ψ for the given values of the module $|\bar{c}(\psi)|$ of the normalized Fibonacci speed of light in a vacuum $\bar{c}(\psi)$. The normalized Fibonacci speed of light in a vacuum has the form:

$$\bar{c}(\psi) = |\bar{c}(\psi)| \text{ at } \{-\infty < \psi < 0\} \cup \{2 < \psi < +\infty\}$$

and $\bar{c}(\psi) = i \cdot |\bar{c}(\psi)|$ at $\{0 < \psi < 2\}$.

$$\alpha(\psi)$$

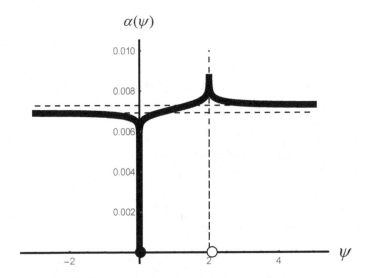

Figure 5.6.   The graph of the fine-structure constant (black curve) at $\{-\infty < \psi < \infty\}$.

As $\bar{c}(\psi) = \sqrt{\frac{\Phi^\psi - \Phi^{-\psi}}{\Phi^{\psi-2} - \Phi^{-(\psi-2)}}}$, then $[\bar{c}(\psi)]^2 = \sigma \cdot \frac{\Phi^\psi - \Phi^{-\psi}}{\Phi^{\psi-2} - \Phi^{-(\psi-2)}}$, where $\sigma = 1$ at $\{-\infty < \psi < 0\} \cup \{2 < \psi < +\infty\}$ and $\sigma = -1$ at $\{0 < \psi < 2\}$. Hence we obtain the formula of dependence of ψ from $\bar{c}(\psi)$, which has the following form:

$$\psi = 0.5 \left[\log_\Phi \left(\frac{\sigma - |\bar{c}(\psi)|^2 \cdot \Phi^2}{\sigma - |\bar{c}(\psi)|^2 \cdot \Phi^{-2}} \right) \right], \tag{5.35}$$

where $\log_\Phi(x)$ is the logarithm with the base Φ from the variable x. Recall that $\log_\Phi(x) = \frac{\ln(x)}{\ln(\Phi)}$. Figure 5.7 shows the graph of the formula (5.35) for any value of ψ in the range $\{-\infty < \psi < +\infty\}$.

The wide black curve in the upper part of Fig. 5.7 corresponds to the Light Ages, for which $\{\Phi < |\bar{c}(\psi)| < +\infty\} \Leftrightarrow \{2 < \psi < +\infty\}, \sigma = 1$.

The thin black curve in the upper part of Fig. 5.7 corresponds to the Dark Ages, for which $\{0 < |\bar{c}(\psi)| < +\infty\} \Leftrightarrow \{0 < \psi < 2\}, \sigma = -1$.

The wide black curve in the lower part of Fig. 5.7 corresponds to the Black Hole, for which $\{0 < |\bar{c}(\psi)| < \Phi^{-1}\} \Leftrightarrow \{-\infty < \psi < 0\}, \sigma = 1$.

Tables.   *The Black Hole* $\{-\infty < \psi < 0\}$. The formulas for numerical calculations:

$$\alpha(\psi) = 10^{-\frac{43}{20}} \times \pi^{\frac{1}{260}} \times |\bar{c}(\psi)|^{\frac{7}{130}}, \quad \bar{c}(\psi) = |\bar{c}(\psi)|$$

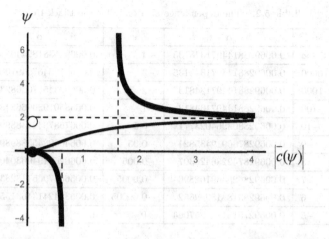

Figure 5.7.   The graph of the dependence of the parameter of self-organization ψ from the module $|\bar{c}(\psi)|$ for the normalized Fibonacci speed of light in a vacuum $\bar{c}(\psi)$.

$$|\bar{c}(\psi)| = \sqrt{\frac{\Phi^{\psi} - \Phi^{-\psi}}{\Phi^{\psi-2} - \Phi^{-(\psi-2)}}}$$

$$= \mathrm{Exp}\left[\frac{130}{7}\ln(\alpha) + \frac{559}{14}\ln(10) - \frac{1}{14}\ln(\pi)\right]$$

$$\psi = 0.5\left[\log_{\Phi}\left(\frac{1 - |\bar{c}(\psi)|^2 \cdot \Phi^2}{1 - |\bar{c}(\psi)|^2 \cdot \Phi^{-2}}\right)\right],$$

where the function e^x is denoted by $\mathrm{Exp}(x)$.

The Dark Ages $\{0 < \psi < 2\}$. The formulas for numerical calculations:

$$\alpha = 10^{-\frac{43}{20}} \times \pi^{\frac{1}{260}} \times |\bar{c}(\psi)|^{\frac{7}{130}}, \quad \bar{c}(\psi) = i \cdot |\bar{c}(\psi)|$$

$$|\bar{c}(\psi)| = \sqrt{\frac{\Phi^{\psi} - \Phi^{-\psi}}{\Phi^{\psi-2} - \Phi^{-(\psi-2)}}} = \mathrm{Exp}\left[\frac{130}{7}\cdot\ln(\alpha) + \frac{559}{14}\ln(10) - \frac{1}{14}\ln(\pi)\right],$$

$$\psi = 0.5\left[\log_{\Phi}\left(\frac{1 + |\bar{c}(\psi)|^2 \cdot \Phi^2}{1 + |\bar{c}(\psi)|^2 \cdot \Phi^{-2}}\right)\right].$$

The Light Ages $\{2 < \psi < +\infty\}$. The formulas for numerical calculations:

$$\alpha = 10^{-\frac{43}{20}} \times \pi^{\frac{1}{260}} \times |\bar{c}(\psi)|^{\frac{7}{130}}, \quad \bar{c}(\psi) = |\bar{c}(\psi)|,$$

Table 5.2.   The dependence of α on ψ for the Black Hole.

ψ	α	ψ	α
$-\infty$	0.0069288144971348135	-4	0.00692538187440185
-10000	0.0069288144971348135	-3	0.006919646794970036
-1000	0.006928814497134813	-2	0.006903455705682327
-100	0.006928814497134813	-1	0.006850192996684811
-10	0.006928803964023771	-0.5	0.006768784213889069
-9	0.006928786919435824	-0.05	0.006406984558898941
-8	0.006928742286426307	-0.005	0.006026519400304021
-7	0.006928625369162896	-0.0005	0.005664706511828369
-6	0.006928318818909652	-0.00005	0.0053242413463155105
-5	0.0069275131121957064	0	0

Table 5.3.   The dependence of ψ on $|\bar{c}(\psi)|$ for the Black Hole.

$|\bar{c}(\psi)|$	ψ	$|\bar{c}(\psi)|$	ψ
$0.618033988749895 = \Phi^{-1}$	$-\infty$	0.6123724356957945	-4
0.618033988749894827	-10000	0.6030226891555273	-3
0.6180339887498948	-1000	0.5773502691896257	-2
0.6180339887498948	-100	0.5	-1
0.6180165405913052	-10	0.40044657145607854	-0.5
0.617988307121335	-9	0.14437305931408442	-0.05
0.6179143806533247	-8	0.04631539686890963	-0.005
0.6177207681213422	-7	0.014667481696040147	-0.0005
0.6172133998483676	-6	0.0046389387118200036	-0.00005
0.6158817620514397	-5	0	0

$$|\bar{c}(\psi)| = \sqrt{\frac{\Phi^{\psi} - \Phi^{-\psi}}{\Phi^{\psi-2} - \Phi^{-(\psi-2)}}} = \operatorname{Exp}\left[\frac{130}{7} \cdot \ln(\alpha) + \frac{559}{14} \ln(10) - \frac{1}{14} \ln(\pi)\right],$$

$$\psi = 0.5 \left[\log_{\Phi}\left(\frac{1 - |\bar{c}(\psi)|^2 \cdot \Phi^2}{1 - |\bar{c}(\psi)|^2 \cdot \Phi^{-2}}\right)\right].$$

Calculations. According to current data, the period of the Dark Ages $\{0 < \psi < 2\}$ ended 550 [million years] after the moment of the Big Bang, when the first stars, quasars, galaxies, clusters and super clusters of galaxies

Table 5.4.   The dependence of α on ψ for the Dark Ages.

ψ	α	ψ	α
0	0	1.15	0.007173203111587392
0.00000005	0.004420677984437741	1.295	0.007236786535409398
0.0000005	0.004703400500624787	1.3995	0.007286414039484436
0.000005	0.0050042047632669956	1.49995	0.007339081525719611
0.00005	0.005324250600866376	1.599995	0.007399158790659007
0.0005	0.005664804976045141	1.6999995	0.007471875746442799
0.005	0.006027567016514314	1.79999995	0.007568544855416565
0.05	0.006418132522321802	1.899999995	0.007725409677896242
0.5	0.006889391594589547	1.9999999995	0.012947368238045895
1	0.007110696049622987	2	$+\infty$

Table 5.5.   The dependence of ψ on $|\bar{c}(\psi)|$ for the Dark Ages.

$|\bar{c}(\psi)|$	ψ	$|\bar{c}(\psi)|$	ψ
0	0	1.176495436910126	1.15
0.00014669849260547704	0.00000005	1.3860143882604832	1.295
0.000463901433387913	0.0000005	1.5735816664444588	1.3995
0.001466987270482024	0.000005	1.798780757504753	1.49995
0.004639088462289398	0.00005	2.0928249016653453	1.599995
0.014672217221849427	0.0005	2.5096386981179037	1.6999995
0.04646514788007947	0.005	3.186330402711177	1.79999995
0.14911029919188895	0.05	4.663866193655711	1.899999995
0.5558929702514211	0.5	68167.01784975648	1.9999999995
1	1	$+\infty$	2

began to form. It was then that hydrogen was re-ionized by the light of stars and quasars. For more detailed information we present a Chronology of the Big Bang (Table 5.8).

In this publication, for the Fibonacci special theory of relativity we employ the following strategy:

(1) We use the value of time T, measured from the Big Bang to the present time, the same as in Table 5.8, i.e. $T = 13.7$ [billion years]. Until now there is no clear consensus on the value of time T, because in addition

Table 5.6.   The dependence of α on ψ for the Light Ages.

ψ	α	ψ	α
2	$+\infty$	5	0.007307020113639036
2.0000005	0.01075009435860901	6	0.007300968990174425
2.000005	0.010103904514440983	7	0.0072987228593058675
2.00005	0.009496563927388494	8	0.0072978740776367965
2.0005	0.008925793102351615	9	0.007297551190335574
2.005	0.008389917123235882	10	0.007297428049702034
2.05	0.007891699729461058	100	0.007297351997377362
2.5	0.0074698788899746965	1000	0.007297351997377362
3	0.007381105661489203	10000	0.007297351997377362
4	0.007324157706770783	$+\infty$	0.007297351997377362

Table 5.7.   The dependence of ψ on $|\bar{c}(\psi)|$ for the Light Ages.

$|\bar{c}(\psi)|$	ψ	$|\bar{c}(\psi)|$	ψ
$+\infty$	2	1.6583123951777001	5
2155.6310252322273	2.0000005	1.632993161855452	6
681.6713738379511	2.000005	1.6236882817719775	7
215.56654703247312	2.00005	1.620185174601965	8
68.17802951613534	2.0005	1.618854426800759	9
21.59109211198998	2.005	1.618347187425374	10
6.926500032284402	2.05	1.618033988749895	100
2.4972120409568324	2.5	1.6180339887498953	1000
2	3	1.61803398874989	10000
1.7320508075688774	4	1.618033988749895 = Φ	$+\infty$

to $T = 13.7$ [billion years], other options have been suggested including $T = 13.73,\ 13.75,\ 13.798,\ 13.81$ [billion years] [21–29].

This once again confirms the opinion, expressed by Hilbert in 1900 that, "*there is no completely precise data in the experimental sciences.*" This data always has some uncertainty. In order to achieve authentic harmony with reality, we need to create a rigorous mathematical model. According to Hilbert [1–3]:

"*It remains to discuss briefly what general requirements may be justly laid down for the solution of a mathematical problem. I should*

Table 5.8. Chronology of the Big Bang (compiled from Wikipedia [31]).

Time T from the moment of the Big Bang	Age	Event	To the present moment
0	Singularity	Big Bang	13.7 billion years
$0 - 10^{-43}$ sec	Beginning Dark Ages. Planck epoch	Birth of particles	13.7 billion years
$10^{-43} - 10^{-35}$ sec	The era of Grand Unification	Separation of gravity from the united electroweak and strong interactions. Possible birth of monopoles. The destruction of the Grand Unification.	13.7 billion years
$10^{-35} - 10^{-32}$ sec	The inflationary epoch	The radius of the Universe increases exponentially on many orders. The structure of the primary quantum fluctuations by means of swelling gives the beginning of the large-scale structure of the universe. Secondary heating and Baryogenesis.	13.7 billion years
$10^{-32} - 10^{-12}$ sec	Electroweak epoch	The universe is filled with a quark-gluon plasma, leptons, photons, W- and Z-bosons, the Higgs boson. Breaking supersymmetry.	13.7 billion years
$10^{-12} - 10^{-6}$ sec	Quark era	Electroweak symmetry is broken. Now all four fundamental interactions exist separately. Quarks have not yet merged into hadrons. The universe is filled with a quark-gluon plasma, leptons and photons.	13.7 billion years
$10^{-6} - 100$ sec	Hadron era	Hadronization. The annihilation of baryon-antibaryon pairs. Thanks to CP-violation there remains a small excess of baryons over antibaryons (about $1:10^9$).	13.7 billion years
100 seconds–3 minutes	Lepton era	The annihilation of lepton-antileptons pairs. The collapse of the neutrons. Substance becomes transparent to neutrinos.	13.7 billion years

(Continued)

Table 5.8.   (Continued)

Time T from the moment of the Big Bang	Age	Event	To the present moment
3 minutes – 380,000 years	Proton era	Nucleosynthesis of helium, deuterium, traces of lithium-7 (20 minutes). Substance starts to dominate over radiation (70,000 s) which leads to changes in the expansion of the Universe. At the end of the era (380,000 years) there is recombination of hydrogen and the Universe becomes transparent to photons of thermal radiation.	13.7 billion years
380,000 years – 550 million years	The Dark Ages. The end of the Dark Ages	The Universe is filled with hydrogen and helium, the relic radiation of atomic hydrogen at 21 cm. The stars, quasars and other bright sources are absent.	13.15 billion years
550 million years – 1 billion years	Beginning Light Ages. Reionization	Forming the first stars (stars population III), quasars, galaxies, clusters and superclusters of galaxies. Reionization of hydrogen by the light of stars and quasars.	12.7 billion years
1 billion years – 8.9 billion years	The era of the substance	The formation of interstellar clouds, which gave rise to the Solar System.	4.8 billion years
8.9 billion years – 9.1 billion years		The formation of the Earth and other planets of the Solar System, general solidification.	4.6 billion years

say first of all, this: that it shall be possible to establish the correctness of the solution by means of a finite number of steps based upon a finite number of hypotheses which are implied in the statement of the problem and which must always be exactly formulated. This requirement of logical deduction by means of a finite number of processes is simply the requirement of rigor in reasoning. Indeed the requirement of rigor, which has become proverbial in mathematics, corresponds to a universal philosophical necessity of our understanding; and, on the other hand, only by satisfying this requirement do the thought content and the suggestiveness of the problem attain their full effect."

(2) As the initial value of the fine-structure constant α, we use the value given by Kosinov's formula [24, 28, 29]:

$$\alpha = 10^{-\frac{43}{20}} \times \pi^{\frac{1}{260}} \times |\Phi|^{\frac{7}{130}} = 0.007297351997377362.$$

Next, we will consider two examples of experiments and their quantitative comparison with numerical data, obtained by using the mathematical model of the Fibonacci special theory of relativity. This problem primarily involves numerical comparisons of the theoretical and experimental results, related to the value of time T, which is measured from the moment of the Big Bang, and also with modifications of the fine-structure constant, depending on time T. We note that in the framework of the classical special theory of relativity, which postulates the constancy of the speed of light in a vacuum, such a comparison is, in principle, impossible.

Example 1. (Experiment for the bifurcation point $\psi = 2$.) For the observable Universe at $\psi_0 = 2$ (this bifurcation point corresponds to the transition from the Dark Ages to the Light Ages), the fine-structure constant $\alpha = +\infty$. Therefore, the supporting experiment in the case (T_0, α_0) is impossible. However, it was determined experimentally (see Table 5.8), that the Dark Ages ended at $T_0 = 550$ [million years] $= 0.55$ [billion years]. Then we get the following values for the supporting numerical coefficient λ_0 and the *inverse supporting numerical coefficient* λ_0^{-1}:

$$\lambda_0 = \frac{\psi_0}{T_0} = \frac{2}{0.55} = 3.63636363 \left[\frac{1}{\text{billion years}}\right],$$

$$\lambda_0^{-1} = 0.275 \text{ [billion years]}. \tag{5.36}$$

Therefore, for any T [billion years] we get:

$$\psi = \lambda_0 \cdot T = \frac{2}{0.55} \cdot T \text{ [dimensionless]}, \tag{5.37}$$

and for any ψ we have:

$$T = \lambda_0^{-1} \cdot \psi = 0.275 \text{ [billion years]}. \tag{5.38}$$

In particular, for $T = 13.7$ [billion years] we get:

$$\psi = \lambda_0 \cdot T = \frac{2}{0.55} \cdot T = 3.63636 \cdot 13.7 = 49.8182, \tag{5.39}$$

$$\bar{c}(\psi) = |\bar{c}(\psi)| = \sqrt{\frac{\Phi^\psi - \Phi^{-\psi}}{\Phi^{\psi-2} - \Phi^{-(\psi-2)}}}$$

$$= 1.618033988749895 \text{ [dimensionless]}, \tag{5.40}$$

$$\alpha = 10^{-\frac{43}{20}} \times \pi^{\frac{1}{260}} \times |\bar{c}(\psi)|^{\frac{7}{130}}$$

$$= 0.00729735199737362 \text{ [dimensionless]}. \tag{5.41}$$

Example 2. As noted above the quasar observations, made in April 2004 by using a UVES spectrograph on one of the four 8.2-meter telescopes at Paranal Observatory in Chile, showed that 10 billion years ago the possible value of α could not have been more than 6×10^{-7} of α. Since we assumed that the current age of the Universe, measured from the moment of the Big Bang is $T = 13.7$ [billion years], then 10 billion years ago this age was equal to $T_0 = 13.7 - 10 = 3.7$ [billion years]. For modern times $T = 13.7$ [billion years] and the fine-structure constant $\alpha = 0.007297351997377362$. Then, according to the Paranal Observatory in Chile, when the age of the Universe was $T_0 = 3.7$ [billion years] the fine-structure constant α_0 satisfied the non-equalities:

$$\alpha_{\min} = 0.007297351997377362 < \alpha_0 \leq \alpha_{\max}$$

$$= 0.007297351997377362 + \Delta\alpha = 0.00729735637885605, \tag{5.42}$$

where $\Delta\alpha = 6 \cdot 10^{-7} \cdot \alpha = 4.37841198426416 \cdot 10^{-9}$.

According to the data, obtained in Example 1,

$$\lambda_0 = \frac{\psi_0}{T_0} = 3.636363636363636363 \left[\frac{1}{\text{billion years}} \right].$$

This implies that the parameter of self-organization ψ_0, corresponding to the past age of the Universe $T_0 = 3.7$ [billion years] is equal to the value

$$\psi_0 = \lambda_0 T_0 = 3.636363636363636363 \cdot 3.7$$

$$= 13.454545454545455 \text{ [dimensionless]}. \tag{5.43}$$

Because the resulting value $\psi_0 > 2$, the given supporting experiment (T_0, α_0) is verified for the Light Ages $\{2 < \psi < +\infty\}$. As mentioned above, the following relation is valid for the Fibonacci special theory of relativity during the Light Ages:

$$\bar{c}(\psi) = |\bar{c}(\psi)| = \sqrt{\frac{\Phi^{\psi} - \Phi^{-\psi}}{\Phi^{\psi-2} - \Phi^{-(\psi-2)}}},$$

where $\bar{c}(\psi)$ is the normalized Fibonacci speed of light in a vacuum. By substituting $\psi_0 = 13.454545454545455$ into the formula for $|\bar{c}(\psi)|$, we get $|\bar{c}(\psi_0)| = 1.618045254395514$. By substituting this value $|\bar{c}(\psi_0)|$ into the formula $\alpha = 10^{-\frac{43}{20}} \times \pi^{\frac{1}{260}} \times |\bar{c}(\psi)|^{\frac{7}{130}}$, we get $\alpha_0 = 0.007297354733194072$. The above arguments result in the following non-equalities:

$$\alpha_{\min} = 0.007297351997377362 < \alpha_0 = 0.007297354733194072$$

$$< \alpha_{\max} = 0.0072973563757885605. \tag{5.44}$$

From the non-equalities (5.44) we obtain the following deviations:

(1) $\Delta\alpha_{\min} = |\alpha_{\min} - \alpha_0| = 2.735816710328076 \cdot 10^{-9}$ is the absolute deviation of $\alpha_{\min}$ from α_0, $\frac{\Delta\alpha_{\min}}{\alpha_{\min}} = 3.7490540559251047 \cdot 10^{-7}$ is the relative deviation of $\alpha_{\min}$ from α_0.

(2) $\Delta\alpha_{\max} = |\alpha_{\max} - \alpha_0| = 4.378411198706356 \cdot 10^{-9}$ is the absolute deviation of $\alpha_{\max}$ from α_0, $\frac{\Delta\alpha_{\max}}{\alpha_{\max}} = 6 \cdot 10^{-7}$ is the relative deviation of $\alpha_{\max}$ from α_0.

5.4.14. *Reconciliation of the theoretical results of the Fibonacci special theory of relativity with experimental data*

(1) Thus, we see that indeed the inequalities (5.44) are realized:

$$\alpha_{\min} < \alpha_0 \leq \alpha_{\max}.$$

Moreover, the difference between the experimental data for $\alpha_{\min}, \alpha_{\max}$ (*"experience"*), obtained as a result of astronomical observations, and the theoretical data α_0 (*"thinking"*), obtained with the Fibonacci special theory of relativity, matched to the ninth decimal place for the absolute deviation and to the seventh decimal place for the relative deviation.

(2) Consequently the mathematical model of the Fibonacci special theory of relativity is fully consistent with the following experimental data:

(a) The ending of the Dark Ages happened 550 million years after the Big Bang (see Table 5.8).

(b) 10 billion years ago, the possible value of the fine-structure constant α could not be more than 0.6 millionth part (6×10^{-7}) of α (observational data of Paranal Observatory in Chile).

5.5. Quantitative Results of the Fibonacci Special Theory of Relativity from the Onset of the Big Bang $T = 0$ to any Time T [Billion Years]

5.5.1.   *The Light Ages ($T > 0.55$ [billion years])*

(1) The fine-structure constant α [dimensionless] with increasing time T decreases from $\alpha = +\infty$ to

$$\alpha = \frac{1}{137.03600982375468} = 0.007297351997377362.$$

(2) The speed of light in a vacuum c [m sec^{-1}] with increasing time T decreases from $c = +\infty$ to $c = 2.99792458 \cdot 10^8$ (the speed of light in a vacuum for the classical special theory of relativity).

(3) The speed of the light source in a vacuum v [m sec^{-1}] with increasing time T increases from $v = 0$ to $v = 2.99792458 \cdot 10^8$.

5.5.2.   *The Dark Ages ($0 < T < 0.55$ [billion years])*

(1) The fine-structure constant α [dimensionless] with increasing time T increases from $\alpha = 0$ to $\alpha = +\infty$.

(2) The speed of light in a vacuum c [m sec^{-1}] is an imaginary quantity.

(3) The speed of the light source in a vacuum v [m sec^{-1}] is an imaginary quantity.

5.5.3.   *The Black Hole ($T < 0$)*

(1) The fine-structure constant α [dimensionless] with decreasing time T increases from

$$\alpha = 0 \text{ to } \alpha = \frac{1}{144.32483369464106} = 0.0069288144971348135.$$

(2) The speed of light in a vacuum c [m sec^{-1}] with decreasing time T increases from $c = 0$ to $c = 1.1451052938512468 \cdot 10^8$.

(3) The speed of the light source in a vacuum v [m sec^{-1}] with decreasing time T decreases from $v = 0$ to $v = -1.1451052938512468 \cdot 10^8$.

5.6. Advantages of the Fibonacci Special Theory of Relativity in Comparison with the Classical Special Theory of Relativity

The Fibonacci special theory of relativity is a theory capable of solving the variability of the fine-structure constant and provides a strong match between the theoretical data and the latest experimental data. Based on the Mathematics of Harmony [8] and the "golden" matrices [9], it allows one to derive, depending on the value of T (from the beginning of the Big Bang), the qualitative and quantitative (numerical) information regarding the following calculable variables: the fine-structure constant $\alpha = \alpha(T)$, the speed of light in a vacuum $c = c(T)$, the speed of the light source $v = v(T)$ for not only the Light Ages, but also for the Dark Ages and the Black Hole (the negative arrow of time). This theory reveals a kind of genetic code for the Universe. Perhaps it is no wonder that the physical-mathematical problem of the fine-structure constant has been referred to as a Millennium Problem.

Knowledge of the dependence of the fine-structure constant $\alpha = \alpha(T)$ on the age of Universe T is one of the most mysterious physical-mathematical problems of modern science. Even minor deviations, as recent experiments show, can dramatically alter our views of the evolution of the Universe.

Einstein's postulate about the constancy of the speed of light in a vacuum is the primary postulate of the classical special theory of relativity. In this connection, the classical special theory of relativity does not have the same capabilities as the Fibonacci special theory of relativity. Employing the formulas of the classical special theory of relativity it is fundamentally impossible to calculate the values of the fine-structure constant, depending on the different ages of the Universe. However, as mentioned above this is the primary feature of the Fibonacci special theory of relativity.

Another feature exhibited by the Fibonacci special theory of relativity is universal harmony, between the mathematical formula and observable data. Hilbert's vaguely stated Sixth Problem appears to be reducible to the issue of harmony between mathematical theory and physical experiment. Hilbert states [1–3]:

> *"I should say first of all, this: that it shall be possible to establish the correctness of the solution by means of a finite number of steps based upon a finite number of hypotheses which are implied in the statement of the problem and which must always be exactly formulated. This requirement of logical deduction by means of a finite number of processes is simply the requirement of rigor in reasoning.*
>
> *In the meantime, while the creative power of pure reason is at work, the outer world again comes into play, forces upon us new questions from actual experience, opens up new branches of mathematics, and while we seek to conquer these new fields of knowledge for the realm of pure thought, we often find the answers to old unsolved problems and thus at the same time advance most successfully the old theories. And it seems to me that the numerous and surprising analogies, and that apparently pre-arranged harmony which the mathematician so often perceives in the questions, methods and ideas of the various branches of his science, have their origin in this ever-recurring interplay between thought and experience."*

Although Hilbert's Sixth Problem is called a *"Mathematical statement of the axioms of physics"*, Hilbert recognized that the creation of a system of axioms in physics which would fully satisfy the properties of consistency, independence and completeness, is virtually impossible.

Therefore, *a priori*, as follows logically from Hilbert's above-cited words, the only possibility that remains is to create a mathematical theory of a physical problem, and then to check its reliability by using a finite number of experiments. Our Fibonacci special theory of relativity satisfies the basic requirements of Hilbert's Sixth Problem as it harmonizes mathematical theory and experimental data, marrying a thought experiment to the observed phenomena by resolving the puzzling anomaly of the variability of the fine-structure constant.

We present this final chapter as a solution to the Millennium Problem posed by Gross, who states this problem as follows:

> *"Are all the (measurable) dimensionless parameters that characterize the physical universe calculable in principle or are some merely determined by historical or quantum mechanical accident and incalculable?"*

We have reformulated this problem and focused on the fine-structure constant (Sommerfeld's constant), the main fundamental dimensionless physical constant, which characterizes the strength of the electromagnetic interaction between elementary charged particles, as follows:

> "*Is the fine-structure constant, which characterizes the physical universe, calculable or non-calculable?*"

We developed a theory of the fine-structure constant that fits this restated problem. We confirmed the coincidence of our theoretical calculations with the astronomical observations to a very high degree of accuracy. Taking into consideration the above arguments, we claim that the scientific results, presented in this chapter, is the mathematical solution to the most important Millennium Problem in Physics, formulated by Gross in 2000.

5.7. The Ratio of the Proton Mass M to the Electron Mass m Depending on the Universe Evolution

Besides the fine-structure constant α, there are other important dimensionless constants in physics.

In [28, 29, 32], the so-called **Kosinov principle** has been formulated: **all dimensionless physical constants are derived from the constants of α and π.**

Note that the constant of π does not depend on the age of the Universe. The fine-structure constant α, as is shown above, depends on the age of the Universe T [billion years]. The dependence α from T has the following form:

$$\alpha(\psi) = 10^{-\frac{43}{20}} \times \pi^{\frac{1}{260}} \times |\bar{c}(\psi)|^{\frac{7}{130}}, \psi = \lambda_0 \cdot T,$$

$\lambda_0 = \frac{2}{0.55} = 3.63636363\ldots$ [billion years]$^{-1}$, $T = 0.275 \cdot \psi$ [billion years],

$$\bar{c}(\psi) = \sqrt{\frac{\Phi^\psi - \Phi^{-\psi}}{\Phi^{\psi-2} - \Phi^{-(\psi-2)}}}, \quad \psi = 0.5 \left[\log_\Phi \left(\frac{\sigma - |\bar{c}(\psi)|^2 \cdot \Phi^2}{\sigma - |\bar{c}(\psi)|^2 \cdot \Phi^{-2}} \right) \right],$$

where $\sigma = 1$ at $\{-\infty < \psi < 0\} \cup \{2 < \psi < +\infty\}$ and $\sigma = -1$ at $\{0 < \psi < 2\}$.

Based on the above result, which has been proved earlier in this chapter, and the **Kosinov principle**, we can formulate the following **Stakhov–Aranson principle: all dimensionless physical constants, based on the fine-structure constant, are changing depending on the age of the Universe, starting from the moment of the Big Bang.**

It is known that all atoms consist of protons, electrons and neutrons. The protons are positive particles, the electrons are negative particles, and the neutrons are neutral particles (that is, they do not carry an electrical charge). The ratio $\frac{M}{m}$ of the proton mass M to the electron mass m is one of the most important dimensionless physical constants.

The protons are involved in thermonuclear reactions, which are the main source of energy, generated by stars. Electron shells of atoms consist of electrons. Most of the chemical properties of an atom are determined by the structure of the outer electron shells. The movement of free electrons causes such phenomena as electric currents in conductors and vacuum.

It is clear that the important task is to establish the dependence of the ratio $\frac{M}{m}$ of the proton mass M to the electron mass m on the age of the Universe T.

But for this we must first find empirical formulas that link the ratio $\frac{M}{m}$ with the fine-structure constant α. We begin to find a solution to this task from the assumption that $\alpha = $ const and this constant is not dependent on T; besides, these formulas should be fairly well harmonized with the modern values of the physical constants M and m.

Many research studies suggested such empirical formulas. The most important results in this area have been obtained by Nikolai Kosinov in the works [24, 28, 33] and also by Natalia Miskinova and Boris Shvilkin in the works [34–37].

We use the empirical formula proposed by Miskinova and Shvilkin in the works [34–37]. This formula connects the fine-structure constant and the ratio $\frac{M}{m}$ and has the following form:

$$\alpha^{-1} = \frac{\pi \left[\sqrt{\frac{M}{m} + 1} - \frac{1}{\pi\sqrt{2}} \right]}{1 + \frac{1}{\pi\sqrt{2}} \cdot \frac{m}{M}}, \tag{5.45}$$

where M is the proton mass, m is the electron mass.

For the recently measured value of the mass ratio $\frac{M}{m} = 1836.15267247$, the calculated value $\alpha^{-1} = 137.03603899$ differs at eighth digits [37]: $\Delta\alpha^{-1} = 0.00003932$.

We get from (5.45):

$$\alpha = \frac{1 + \frac{1}{\frac{M}{m}}}{\pi \left(\sqrt{\frac{M}{m} + 1} - \frac{1}{\pi\sqrt{2}} \right)}. \tag{5.46}$$

We denote $x = \sqrt{\frac{M}{m}} > 0$. Then, the ratio $\frac{M}{m} = x^2$. After simple transformations we obtain the following equation:

$$Ax^3 + Bx^2 - 1 = 0, \tag{5.47}$$

where $A = \sqrt{2}\pi^2\alpha, B = \pi[(\sqrt{2}\pi - 1)\alpha - \sqrt{2}]$. Since $\alpha \geq 0$, then for this situation the equation (5.47) always has one real root $x_1 \geq 0$ and two complex-conjugate roots x_2, x_3. Then, we get $\frac{M}{m} = x_1^2$.

We find the algorithm for obtaining the value of the ratio $\frac{M}{m} = x_1^2$. Let us give the value of the fine-structure constant $\alpha > 0$. We introduce the following designations:

$$\lambda = -\frac{1}{\alpha\lambda} + 1 - \frac{\sqrt{2}}{2\pi}, \quad \mu = -\frac{\sqrt{2}}{2\alpha\pi^2}, \quad p = -\frac{\lambda^2}{3},$$

$$q = \frac{2}{27}\lambda^3 + \mu, \quad Q = \frac{p^3}{27} + \frac{q^2}{4}.$$

Then, we get:

$$\frac{M}{m} = x_1^2 = \left(\sqrt[3]{-\frac{q}{2} + \sqrt{Q}} + \sqrt[3]{-\frac{q}{2} - \sqrt{Q}} - \frac{\lambda}{3} \right)^2. \tag{5.48}$$

Changing the value of the ratio $\frac{M}{m}$ depending on the age of the Universe has direct relation to the disruption of fusion reactions on the Sun and the stars and to the other above-mentioned effects.

Table 5.9 shows the numerical characteristics of T [billion years], α and $\frac{M}{m}$ depending on the parameter of self-organization ψ for the range $\{-\infty < \psi < +\infty\}$.

Figure 5.8 shows the graph of the function $\frac{M}{m}(\alpha)$ for the range: $0 \leq \alpha \leq 10$. It explains the fact, presented in Table 5.9, that for the case $\alpha \to 0 + 0$, the value of the ratio $\frac{M}{m} \to +\infty$, and for the case $\alpha \to +\infty$, the value of $\frac{M}{m} \to 0 + 0$. Here, the function $\frac{M}{m} = \frac{M}{m}(\alpha)$ decreases monotonically with increasing α.

5.8.   General Conclusions

(1) Discussing the history of mathematics and the development of new mathematical ideas and theories, we should draw particular attention to the central role of Euclid's *Elements*. Academician Andrey Kolmogorov identifies several stages in the development of mathematics [38]. According to Kolmogorov, the modern period in mathematics

Table 5.9.   Numerical characteristics T [billion years], α and $\frac{M}{m}$ depending on the parameter of self-organization ψ for the range $\{-\infty < \psi < +\infty\}$.

ψ	T [billion years]	α	$\dfrac{M}{m}$
The Black Hole (the negative arrow of time) $\{-\infty < \psi < 0\}$			
$-\infty$	$-\infty$	0.0069288144971348135	2040.344104791557
-5	-1.375	0.0069275131121957064	2041.1237432867479
-1	-0.275	0.006850192996684811	2088.2500646387352
$0 - 0$	0	0	$+\infty$
The Dark Ages $\{0 < \psi < 2\}$			
$0 + 0$	0	0	$+\infty$
0.5	0.1375	0.006889391594589547	2064.1592455987684
1	0.275	0.007110696049622987	1935.5798368691815
1.5	0.4125	0.007339081525719611	1814.9606105036796
$2 - 0$	0.55	$+\infty$	0
The Light Ages (the positive arrow of time) $\{2 < \psi < +\infty\}$			
$2 + 0$	0.55	$+\infty$	0
3	0.825	0.007381105661489203	1793.984114748964
$+\infty$	$+\infty$	0.007297351997377362	1836.1518766721745

began in the 19th century. According to him, "expanding the content of mathematics became the most significant feature of 19th century mathematics." At the same time, according to Kolmogorov [38], the creation of Lobachevsky's "imaginary geometry" became:

> "a remarkable example of the theory that has arisen as a result of the internal development of mathematics. ... It is the example of geometry, which overcame a belief in the permanence of axioms, as a consecrated MILLENNIUM development of mathematics, and comprehended the possibility of creating significant new mathematical theories."

As we know, "hyperbolic geometry" in its origins dates back to Euclid's Fifth Postulate. For several centuries, from Ptolemy to

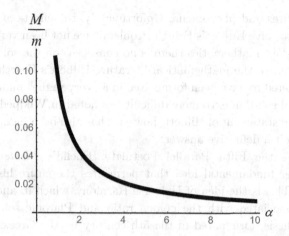

Figure 5.8. Graph of the function $\frac{M}{m}(\alpha)$ for the range $0 \leq \alpha \leq 10$.

Proclus, mathematicians tried to prove this postulate. An initial brilliant solution to this problem was given by Russian mathematician Nikolai Lobachevsky during the first half of the 19th century. This marked the beginning of the contemporary stage in the development of mathematics.

(2) At the interface of the 19th and 20th centuries, the eminent mathematician David Hilbert formulated 23 mathematical problems, which greatly stimulated the development of mathematics throughout the 20th century. One of these (Hilbert's Fourth Problem) refers directly to non-Euclidean geometry. Hilbert presented for mathematicians the following fundamental problem [1]:

> "*The more general question now arises: whether from other suggestive standpoints geometries may not be devised which, with equal right, stand next to Euclidean geometry.*"

Hilbert's quote contains the formulation of a very important scientific problem, which is of fundamental importance not only for mathematics, but also for all theoretical natural sciences: are there non-Euclidean geometries, which are close to Euclidean geometry and are interesting from "other suggestive standpoints"? If we consider it in the context of the theoretical natural sciences, then Hilbert's Fourth Problem aims to search for new non-Euclidean worlds of Nature, which are close to Euclidean geometry and reflect some properties of Nature's structures and phenomena. Unfortunately, the efforts of mathematicians to solve Hilbert's Fourth Problem have not been very successful. In modern mathematics there is no consensus on the solution to this problem. In the mathematical literature Hilbert's Fourth Problem is considered to have been formulated in a "very vague" manner, making its final solution extremely difficult. As noted in Wikipedia [39], "the original statement of Hilbert, however, has also been judged too vague to admit a definitive answer."

(3) Besides the Fifth Parallel Postulate, Euclid's *Elements* contains another fundamental idea that permeates the entire history of science. This is the idea of Universe Harmony, which in ancient Greece was associated with the golden ratio and Platonic solids. Proclus' hypothesis, formulated in the 5th century by the Greek philosopher and mathematician Proclus Diadochus (412–485), contains an unexpected view of Euclid's *Elements*. According to Proclus, Euclid's main goal was to build a complete theory of the regular polyhedra (Platonic solids). This theory was outlined by Euclid in Book XIII, that is, in the concluding book of the *Elements*, which in itself indirectly confirms Proclus' hypothesis. To solve this problem, Euclid included all the necessary mathematical data in the *Elements*. He then used this data to solve the main problem — the creation of a complete theory of the Platonic solids. The most curious (yet tell-tale) thing is that he had introduced the golden ratio early on in Book II for its later use in the creation of the geometric theory of the dodecahedron.

(4) Starting from Euclid, the golden ratio and Platonic solids run like a "red thread" throughout the history of mathematics and the natural sciences. In modern science, Platonic solids have become a source for significant scientific discoveries, particularly of fullerenes (Nobel Prize in Chemistry, 1996) and quasi-crystals (Nobel Prize in Chemistry, 2011). The publication of Stakhov's 2009 book, *The Mathematics of Harmony. From Euclid to Contemporary Mathematics and Computer Science* [8] is a reflection of this very important trend in the development of modern science (including mathematics): the revival of the "harmonic ideas" of Pythagoras, Plato and Euclid.

(5) Argentinean mathematician Vera W. de Spinadel's metallic proportions [40–42], which are a generalization of the classic golden ratio, are a new class of mathematical constants of fundamental theoretical and practical importance. Besides Spinadel, several researchers from various countries and continents (e.g. French mathematician

Midhat Gazale[43], American mathematician Jay Kappraff [44], Russian engineer Alexander Tatarenko [45], Armenian philosopher and physicist Hrant Arakelyan [46], Russian researcher Victor Shenyagin [47], Ukrainian physicist Nikolai Kosinov [48], Spanish mathematicians Falcon Sergio and Plaza Angel [49]) also independently developed de Spinadel's works [40–42]. All of this confirms the fact that the appearance of new (harmonic) mathematical constants has been maturing in modern mathematics.

(6) In this book we have applied the Mathematics of Harmony [8] and Spinadel's metallic proportions to the qualitative theory of dynamical systems. Of greatest interest here is the connection of Anosov's automorphism with Spinadel's metallic proportions [40–42]. We outlined the prospects of the applications of this approach in such important areas as the qualitative theory of dynamical systems as applied to small denominators, Reeb foliations, Pfaff's equations, etc.

(7) The new classes of hyperbolic functions, based on the golden ratio and Fibonacci numbers (Fibonacci hyperbolic functions) (see Stakhov and Tkachenko [50] and Stakhov and Rozin [19, 51]), have become one of the most important results of the Mathematics of Harmony [8], having direct relevance to hyperbolic geometry. The λ-Fibonacci hyperbolic functions, based on Spinadel's metallic proportions became an important step in the derivation of a general theory of "harmonic" hyperbolic functions [20, 52]. It is important to emphasize here that in contrast to classic hyperbolic functions, the λ-Fibonacci hyperbolic functions, based on the golden and metallic proportions, have recursive properties, because they are fundamentally connected with Fibonacci numbers and their generalizations — λ-Fibonacci numbers. This fundamental mathematical fact allows us to place the λ-Fibonacci hyperbolic functions into a new and unique class of hyperbolic functions called recursive hyperbolic functions.

(8) The research of Ukrainian architect Oleg Bodnar was a significant step in the development of hyperbolic geometry. He showed in [53] that a special kind of hyperbolic geometry, based on "golden" hyperbolic functions, has a wide distribution in Nature's flora and fauna, and underlies the botanic phenomenon of phyllotaxis (pine cones, cacti, pineapples, sunflower heads, etc.). His discovery demonstrates that hyperbolic geometry is much more common in Nature than previously thought. Perhaps Nature itself is the epitome of Bodnar's geometry. It is important to emphasize that Bodnar's geometry is based on

the "golden" recursive hyperbolic functions, the basis of which is the golden ratio.

(9) From this point of view, the original solution to Hilbert's Fourth Problem, based upon Stakhov's Mathematics of Harmony [8] — in particular on the metallic proportions [40–42] and λ-Fibonacci hyperbolic functions [20, 52], is of special significance for mathematics and all the sciences. This is one of the main results of this book along with our previous works in this area [12–18]. These works demonstrate that there is an infinite number of new hyperbolic geometries, which "with equal right, stand next to Euclidean geometry" (David Hilbert). It is important to emphasize that the recursive Fibonacci hyperbolic functions underlying our original solution to Hilbert's Fourth Problem led to the creation of a new class of hyperbolic geometries, called recursive hyperbolic geometries, which have the common title **The "Golden" Non-Euclidean Geometry**.

(10) This solution to Hilbert's Fourth Problem places the search for new (harmonic) worlds of Nature in the center of the natural sciences (physics, chemistry, biology, genetics and so on). In this regard, we should draw special attention to the fact that the "silver" recursive hyperbolic geometry, based on the silver proportion $\Phi_2 = 1 + \sqrt{2} \approx 2.41$, is the closest to Lobachevsky's geometry, based on the classical hyperbolic functions with the base $e \approx 2.71$. Its distance to Lobachevsky's geometry is equal to $\bar{\rho}_{12} \approx 0.1677$, which is the smallest amongst all the distances for Lobachevsky's metric forms. We may predict that the "silver" recursive hyperbolic geometry, based on the "silver" hyperbolic functions with the base $\Phi_2 = 1 + \sqrt{2} \approx 2.41$, will soon be found in Nature and they along with Bodnar's geometry, based upon the "golden" recursive hyperbolic functions with the base $\Phi = \frac{1+\sqrt{5}}{2} \approx 1.618$, will become the most important recursive hyperbolic geometries of the physical and biological worlds of Nature.

(11) We have attempted to extend the approach, used in [12–18], into the realm of spherical geometry. To solve this problem, a new class of elementary functions has been introduced called λ-Fibonacci spherical functions having recursive properties. This class of functions allows for the solution to Hilbert's Fourth Problem with respect to recursive spherical geometry.

(12) The subject of this book is of considerable interest from the standpoint of the history of mathematics and the prospects for its further development in close association with the theoretical natural sciences. This

study unites the ancient golden ratio, described in Euclid's *Elements*, with both spherical geometry and Lobachevsky's hyperbolic geometry. This unexpected union led to our original solution to Hilbert's Fourth Problem and to the creation of a new class of non-Euclidean geometry called recursive non-Euclidean geometry.

(13) Viewing recursive non-Euclidean geometry and Bodnar's geometry through Dirac's Principle of Mathematical Beauty, we believe that if Hilbert were alive today, he would have given his preference to solving the Fourth Problem in terms of the Mathematics of Harmony [8]. This "harmonic" solution unites Leibniz's "pre-established harmony" with Dirac's Principle of Mathematical Beauty as the foundation of a physical theory. The new solution to Hilbert's Fourth Problem and the creation of the new recursive non-Euclidean geometries can give vast opportunities for natural sciences in discovering applications of new non-Euclidean geometries in Nature.

(14) After an original solution to Hilbert's Fourth Problem, we then focused our attention towards the Millennium Problems. Amongst these was a formidable challenge to determine whether any of physics' dimensionless constants are calculable. The main dimensionless constant that underlies the whole physical world is the fine-structure constant. Through employing the Mathematics of Harmony, recursive Fibonacci hyperbolic functions and "golden" matrices, we derived the Fibonacci Special Theory of Relativity [7] connecting the fine-structure constant with time. The 2004 quasar observations from the Paranal Observatory in Chile is confirmation that the fine-structure constant was different 10 billion years ago. Employing our theory we were able to match the value observed at Paranal to the 9th decimal place. The close approximation of our mathematical theory to the astronomical observational data confirms that our theory is valid. This original solution can be considered to be one of the most important theoretical results ever obtained through the intersection of mathematical theory and experimental physics. This provides an answer to Gross's Millennium Problem.

(15) The fine-structure constant determines the majority of dimensionless physical constants. The ratio $\frac{M}{m}$ of the proton mass M to the electron mass m is one of them. By using the dependence of the fine-structure constant on time $T\alpha = \alpha(T)$, we have proved that the ratio $\frac{M}{m}(\alpha)$ also varies depending on the age of the Universe T.

References

[1] "Mathematical problems", lecture by Professor David Hilbert. http://aleph0.clarku.edu/~djoyce/hilbert/problems.html#prob4 (accessed November 4, 2015).

[2] Winton, M. "Mathematical problems", lecture delivered before the International Congress of Mathematicians at Paris in 1900 by Professor David Hilbert. http://www.mat.ucm.es/catedramdeguzman/old/01historias/haci-aelfuturo/Hilbertproblems1900/hilbertproblems.html (accessed November 4, 2015).

[3] Aleksandrov, P. S. (General Editor). *Hilbert's Problems*. Moscow: Publishing House "Science" (1969).

[4] "Millennium prize problems", *Wikipedia, The Free Encyclopedia*, https://en.wikipedia.org/wiki/Millennium_Prize_Problems (accessed November 4, 2015).

[5] "'Millennium Madness' — Physics problems for the next millennium". http://www.theory.caltech.edu/~preskill/millennium.html (accessed November 4, 2015).

[6] "Fine-structure constant", *Wikipedia, The Free Encyclopedia*, https://en.wikipedia.org/wiki/Fine-structure_constant (accessed November 4, 2015).

[7] Stakhov, A. P. and Aranson, S. Kh. "'Golden' Fibonacci Goniometry, Fibonacci–Lorentz Transformations, and Hilbert's Fourth Problem," *Congressus Numerantium* (2008), 193: 119–156.

[8] Stakhov, A. P., assisted by S. Olsen. *The Mathematics of Harmony. From Euclid to Contemporary Mathematics and Computer Science*. Singapore: World Scientific (2009).

[9] Stakhov, A. "The 'golden' matrices and a new kind of cryptography", *Chaos, Solitons & Fractals* (2007), Vol. 32, Issue 3: 1138–1146.

[10] Einstein, A. "On the electrodynamics of moving bodies", *Annals of Physics* (1905), 322(10): 891–921.

[11] Dubrovin, B. A., Novikov, S. P. and Fomenko, A. T. *Modern Geometry: Methods and Applications*. Moscow: Nauka (1965) (Russian).

[12] Stakhov, A. P. and Aranson, S. Kh. "The 'golden' Fibonacci goniometry, Fibonacci–Lorentz transformations and Hilbert's Fourth Problem", Academy of Trinitarism, Moscow: Electronic number 77-6567, publication 147816 (2008) (Russian).

[13] Stakhov, A. P. and Aranson, S. Kh. "The 'golden' Fibonacci goniometry, Hilbert's fourth problem, Fibonacci–Lorentz transformations and the 'golden' interpretation of the special theory of relativity", Academy of Trinitarism, Moscow: Electronic number 77-6567, publication 15225 (2009) (Russian).

[14] Stakhov, A. P. and Aranson, S. Kh. "Hyperbolic Fibonacci and Lucas functions, 'golden' Fibonacci goniometry, Bodnar's geometry, and Hilbert's fourth problem. Part I. Hyperbolic Fibonacci and Lucas functions and 'golden' Fibonacci goniometry", *Applied Mathematics* (2011) Vol. 2, No. 1: 74–84.

[15] Stakhov, A. P. and Aranson, S. Kh. "Hyperbolic Fibonacci and Lucas functions, 'golden' Fibonacci goniometry, Bodnar's geometry, and Hilbert's fourth problem. Part II", *Applied Mathematics* (2011) Vol. 2, No. 2: 181–188.

[16] Stakhov, A. P. and Aranson, S. Kh. "Hyperbolic Fibonacci and Lucas functions, 'golden' Fibonacci goniometry, Bodnar's geometry, and Hilbert's fourth problem. Part III", *Applied Mathematics* (2011) Vol. 2, No. 3: 283–293.

[17] Stakhov, A. P. and Aranson, S. Kh. "Fibonacci–Lorentz transformations and the 'golden' interpretation of the special theory of relativity on the four-dimensional torus (the closed model)", International Club of the Golden Section: Canada (2010) (Russian).

[18] Stakhov, A. P., Aranson, S. Kh. and Khanton, I. V. "The golden Fibonacci goniometry, the resonance structure of the genetic code of DNA, Fibonacci–Lorentz transformations and other applications. Part III. Fibonacci–Lorentz transformations and their relationship with the 'golden' universal genetic code", Academy of Trinitarism, Moscow: Electronic number 77-6567, publication 14782 (2008) (Russian).

[19] Stakhov, A. P. and Rozin, B. N. "On a new class of hyperbolic functions", *Chaos, Solitons & Fractals* (2005) Vol. 23, No. 2.

[20] Stakhov, A. P. "Gazale's formulas, a new class of hyperbolic Fibonacci and Lucas Functions and the improved method of the 'golden' cryptography", Academy of Trinitarism, Moscow: Electronic number 77-6567, publication 14098 (2006) (Russian). http://www.trinitas.ru/rus/doc/0232/004a/02321063.htm (accessed November 4, 2015).

[21] Franson, J. D. "Apparent correction to the speed of light in a gravitational potential", *New Journal of Physics* (2014) Vol. 16, No. 6.

[22] Jarosik, N., *et al.* Seven-Year Wilkinson Microwave Anisotropy Probe (WMAP) Observations: Sky Maps, Systematic Errors, and Basic Results (2011). http://lambda.gsfc.nasa.gov/product/map/dr4/map_bibliography.cfm (accessed November 4, 2015).

[23] Planck Collaboration. *Planck 2013 results. XVI. Cosmological parameters.* http://arxiv.org/abs/1303.5076 (accessed November 4, 2015).

[24] Kosinov, N. V. *Connection of three important constants* (2000) (Russian). http://www.roman.by/r-25512.html (accessed November 4, 2015).

[25] Johnson, G. "10 Physics Questions to Ponder for a Millennium or Two", *New York Times*, August 15, 2000.

[26] Gross, D. "Millennium Madness: Physics Problems for the Next Millennium", Strings 2000 Conference at University of Michigan, July 10–15, 2000.

[27] Carter, J. *The Other Theory of Physics.* Washington, 1994.

[28] Kosinov, N. V. "The physical equivalent of the number 'Pi' and the geometric equivalent of the fine-structure constant 'alpha'" (Russian). http://kosinov.314159.ru/kosinov10.htm (accessed November 4, 2015).

[29] Kosinov, N. V. "Five fundamental constants of vacuum, lying in the base of all physical laws, constants and formulas", *Physical Vacuum and Nature* (2000) 4 (Russian).

[30] Webb J. "Our latest measurement of variny alpha". http://www.phys.unsw. edu.au/~jkw/alpha/Welcome.html (accessed November 4, 2015).

[31] "The chronology of the universe", *Wikipedia, The Free Encyclopedia.* https://en.wikipedia.org/wiki/Chronology_of_the_universe (accessed November 4, 2015).

[32] Kosinov, N. V. "The constant physical and cosmological bases theory", *Physical Vacuum and Nature* (2002) No. 5: 69–104 (Russian).

[33] Kosinov N. V. "Emanation of matter vacuum and laws of structure", *Physical Vacuum and Nature* (1999) No 1: 82–104 (Russian).

[34] Miskinova, N. A. and Shvilkin B. N. "On the relationship between the fine-structure constant and the ratio of the masses of the proton and electron", *Science and Technology of Russia* (2002a) No. 4 (55): 19 (Russian).

[35] Miskinova, N. A. and Shvilkin, B. N. "On the question of the fine-structure constant", *Science and Technology of Russia* (2002b) No. 5/6 (56/57): 20 (Russian).

[36] Miskinova, N. A. and Shvilkin, B. N. "On the empirical formula of fine-structure constant", *Proceedings of the Russian Universities. Physics* (2004) No. 11: 97–98 (Russian).

[37] Miskinova, N. A. and Shvilkin B. N. "The relationship between the fundamental physical constants", *Science and Technology of Russia* (2012) Vol. 91, No 3: 25–32 (Russian).

[38] Kolmogorov, A. N. *Mathematics in its Historical Development.* Moscow: Nauka (1991) (Russian).

[39] "Hilbert's fourth problem", *Wikipedia. The Free Encyclopedia.* http://en. wikipedia.org/wiki/Hilbert's_fourth_problem (accessed November 4, 2015).

[40] de Spinadel, V. W. "The metallic means and design", in *NEXUS II: Architecture and Mathematics* (1998).

[41] de Spinadel, V. W. "The metallic means family and multifractal spectra", *Nonlinear Analysis* (1999) Vol. 36: 721–745.

[42] de Spinadel, V. W. *From the Golden Mean to Chaos.* Nueva Libreria (first edition 1998); Nobuko (second edition 2004).

[43] Gazale, M. J. *Gnomon. From Pharaohs to Fractals.* Princeton, NJ: Princeton University Press (1999).

[44] Kappraff, J. *Connections. The Geometric Bridge Between Art and Science. Second Edition.* Singapore, New Jersey, London, Hong Kong: World Scientific (2001).

[45] Tatarenko, A. "The golden T_m-harmonies and D_m-fractals is an essence of soliton-similar m-structure of the world", Academy of Trinitarism, Moscow: Electronic number 77-6567, publication 12691 (2005) (Russian). http:// www.trinitas.ru/rus/doc/0232/009a/02320010.htm (accessed November 4, 2015).

[46] Arakelyan, H. *The Numbers and Magnitudes in Modern Physics.* Yerevan: Armenian Academy of Sciences (1989) (Russian).

[47] Shenyagin, V. P. "Pythagoras, or how every man creates its own myth. The fourteen years after the first publication of the quadratic mantissa

s-proportions", Academy of Trinitarism, Moscow: Electronic number 77-6567, publication 17031, 27.11.2011 (Russian), http://www.trinitas.ru/rus/doc/0232/013a/02322050.htm (accessed November 4, 2015).

[48] Kosinov, N. V. "The golden ratio, golden constants, and golden theorems", Academy of Trinitarism, Moscow: Electronic number 77-6567, publication 14379 (2007) (Russian). http://www.trinitas.ru/rus/doc/0232/009a/02321049.htm (accessed November 4, 2015).

[49] Falcon, S. and Plaza, A. "On the Fibonacci k-numbers", *Chaos, Solitons & Fractals* (2007) Vol. 32, Issue 5: 1615–1624.

[50] Stakhov, A. P. and Tkachenko, I. S. "Hyperbolic Fibonacci trigonometry", *Reports of the Ukrainian Academy of Sciences* (1993) Vol. 208, No. 7: 9–14 (Russian).

[51] Stakhov, A. P. and Rozin, B. N. "The golden section, Fibonacci series and new hyperbolic models of nature", *Visual Mathematics* (2006) Vol. 8, No. 3.

[52] Stakhov, A. P. "On the general theory of hyperbolic functions based on the hyperbolic Fibonacci and Lucas functions and on Hilbert's Fourth Problem", *Visual Mathematics* (2013), Vol. 15, No. 1. http://www.mi.sanu.ac.rs/vismath/2013stakhov/hyp.pdf (accessed November 4, 2015).

[53] Bodnar, O. Y. *The Golden Section and Non-Euclidean Geometry in Nature and Art.* Lvov: Publishing House "Svit" (1994) (Russian).

From the "Golden" Geometry to the Multiverse

A.1. Conception of Multiverse

In a lecture delivered on June 10, 2007 at the Lebedev Physical Institute of the Russian Academy of Sciences (Moscow), Andrey Linde, a professor at Stanford University (USA), gave theoretical justification to the fact, that after the Big Bang there were formed not only our observed Universe, but also an infinite number of other Universes called the *Many Faces Universe* (or *Multiverse*) with their individual laws of evolution [1] (see also [2]).

The issue of the Multiverse is one of the most important conceptions of modern cosmology. It has indirect experimental confirmation. American astronomers have found in space dark flow, in addition to dark energy and dark matter. According to their hypothesis, dark flow is caused by the existence of other Universes that interact with our own. For example, the astronomer Alexander Kashlinsky from the Goddard Space Flight Center of NASA came to these conclusions in his paper "A measurement of large-scale peculiar velocities of clusters of galaxies: results and cosmological implications" [3].

The existence of the Multiverse is a subject of discussion among physicists. This idea is actively used, for example, in string theory. The assumption of the existence of the Multiverse is also used in the interpretation of quantum mechanics. Many scientists such as Stephen Hawking [4], Brian Greene [5], Alexander Vilenkin [6] and others are among the supporters of the idea of the Multiverse.

A.2.　The Conceptions and Theories used in this Study

A.2.1.　*Self-similarity principle and Fibonacci numbers in patterns of nature*

The SELF-SIMILARITY PRINCIPLE plays a fundamental role in the formation of the structures of the physical and biological world at all levels of its realizations, studied in theoretical natural sciences (including theoretical physics, chemistry, crystallography, biology, botany, genetics and so on).

In mathematics, this principle is expressed by the recurrence relations and has direct relation to patterns of nature [7]. The article [7] emphasizes the role of the most widely used recurrence numerical sequence, Fibonacci numbers, in the patterns of nature:

> "*In 1202, Leonardo Fibonacci (1170–1250) introduced the Fibonacci number sequence to the western world with his book Liber Abaci. Fibonacci gave an (unrealistic) biological example, on the growth in numbers of a theoretical rabbit population. In 1917, D'Arcy Wentworth Thompson (1860–1948) published his book On Growth and Form. His description of phyllotaxis and the Fibonacci sequence, the mathematical relationships in the spiral growth patterns of plants, is classic. He showed that simple equations could describe all the apparently complex spiral growth patterns of animal horns and mollusc shells.*"

We can mention the names of many prominent scholars and thinkers, such as Ernst Haeckel (1834–1919), Alan Turing (1912–1954), Benoît Mandelbrot (1924–2010), Aristid Lindenmayer (1925–1989) and others, who believed in the self-organization of nature and the self-similarity principle as the main principle of natural structures and processes.

A.2.2.　*The λ-Fibonacci numbers and metallic proportions*

In the late 20th and early 21st centuries, researchers of many countries have begun to explore the generalized Fibonacci numbers, given by the following recurrence relation:

$$F_\lambda(n + 2) = \lambda F_\lambda(n + 1) + F_\lambda(n);$$
$$F_\lambda(0) = 0, \quad F_\lambda(1) = 1 \quad (\lambda = 1, 2, 3, \ldots). \tag{A.1}$$

These number sequences are called λ-Fibonacci numbers.

The French mathematician of Egyptian origin Midhat Gazale [8] and Argentinean mathematician Vera W. de Spinadel [9–11] made the most contribution to the development of the λ-Fibonacci numbers and their applications. Vera W. de Spinadel introduced a new class of mathematical constants Φ_λ, called metallic means or metallic proportions:

$$\Phi_\lambda = \frac{\lambda + \sqrt{4 + \lambda^2}}{2} (\lambda = 1, 2, 3, \ldots). \tag{A.2}$$

For the case $\lambda = 1$ the metallic proportions Φ_λ ($\lambda = 1, 2, 3, \ldots$) are reduced to the classic golden ratio $\Phi = \frac{1 + \sqrt{5}}{2}$. The number of the metallic proportions is theoretically infinite, because each $\lambda = 1, 2, 3, \ldots$ creates its own metallic proportion. Thus, Spinadel's metallic proportions (A.2) extend *ad infinitum* the number of new mathematical constants, similar to the golden ratio, that relate to a new class of recurrence sequences (A.1).

A.2.3. Gazale's formulas and Gazale's hypothesis

Midhat Gazale in his book [8] has described two important scientific results. First of all, he introduced the so-called Gazale's formulas that allow us to express the λ-Fibonacci numbers (A.1) through the metallic proportions (A.2). Gazale's formulas are a generalization of the well-known Binet's formulas that allow us to express the classic Fibonacci numbers through the golden ratio. The formulation of Gazale's hypothesis is the second scientific result of Gazale. In his book [8] Gazale formulates this hypothesis as follows:

> **The λ-Fibonacci numbers ($\lambda = 1, 2, 3, \ldots$) "play a key role in the study of self-similarity".**

This means that the λ-Fibonacci numbers, according to Gazale's hypothesis, underlies the SELF-SIMILARITY PRINCIPLE. Gazale's hypothesis is quite an unexpected look at the SELF-SIMILARITY PRINCIPLE, which allows us to widely use mathematical apparatus for modeling of self-similar structures and systems of Nature at all levels of its organization, including the Multiverse.

A.2.4. The recursive λ-Fibonacci hyperbolic functions, Hilbert's Fourth Problem and the "golden" geometry

Introduction of the recursive λ-Fibonacci hyperbolic functions [12–14] is an important step in the extension of the universal SELF-SIMILARITY

PRINCIPLE in mathematics and theoretical natural sciences. The original solution to Hilbert's Fourth Problem (the Millennium Problem in Geometry), based on the recursive λ-Fibonacci hyperbolic functions, is one more confirmation of the effectiveness of the recursive λ-Fibonacci hyperbolic functions for the solution of complicated mathematical problems. This approach leads us to the recursive λ-Fibonacci hyperbolic geometries based on λ-Fibonacci numbers (A.1) and Spinadel's metallic proportions (A.2). These new recursive hyperbolic geometries have the common title of the "Golden" Geometry.

A.2.5.   Bodnar's geometry

The research of Ukrainian architect Oleg Bodnar is a brilliant example of the "golden" geometry. Bodnar showed in [15] that a special kind of hyperbolic geometry, based on the recursive ($\lambda = 1$)-Fibonacci hyperbolic functions, has wide distribution in Nature's flora and fauna, and underlies the botanic phenomenon of phyllotaxis (pine cones, cacti, pineapples, sunflower heads, etc.). Bodnar's discovery demonstrates that hyperbolic geometry is much more common in Nature than previously thought. Perhaps Nature itself is the epitome of Bodnar's geometry. It is important to emphasize that Bodnar's geometry is based on the recursive ($\lambda = 1$)-Fibonacci hyperbolic functions with the base of the golden ratio. Bodnar's geometry [15] is a brilliant confirmation of the effectiveness of the application of the universal SELF-SIMILARITY PRINCIPLE and Gazale's hypothesis in nature.

A.2.6.   Kosinov's formula for the fine-structure constant α

The expression of the fine-structure constant α through the basic physical constants (electric constant; Dirac's constant; the speed of light in a vacuum; the elementary charge) are well known in theoretical physics. In 2000, the Ukrainian physicist Nikolai Kosinov [16, 17] found a simple and beautiful relationship, which links the dimensionless constants: the fine-structure constant α, the number π and the golden ratio $\Phi = \frac{1+\sqrt{5}}{2}$. This formula is as follows:

$$\alpha = 10^{-\frac{43}{20}} \times \pi^{\frac{1}{260}} \times \Phi^{\frac{7}{130}}. \tag{A.3}$$

The methodological importance of Kosinov's formula (A.3) is in the fact that this formula links the fine-structure constant α (the main constant of physical world) with the golden ratio (the main Greek "harmonious" constant of the Universe). This means that Kosinov's formula (A.3) connects the Greek's "idea of harmony" (the golden ratio) with the physical world (the fine-structure constant).

A.3.   Mathematical Models of the Multiverse

A.3.1.   *General considerations and questions*

For the development of the conception of the Multiverse [1, 2], we attempt to unite the following original scientific conceptions and models:

— Principle of self-similarity, which has relation to all structures of Nature and theoretical natural sciences;

— Gazale's hypothesis on the key role of the λ-Fibonacci numbers in the study of self-similarity [8];

— Spinadel's metallic proportions as a new class of mathematical constants [9–11];

— Recursive λ-Fibonacci hyperbolic functions [12–14] as a new class of hyperbolic functions;

— Bodnar's geometry [15] as a new hyperbolic geometry of botanic phyllotaxis phenomenon;

— Kosinov's formula (A.3), which connects the fine-structure constant α with the golden ratio;

— Fibonacci special theory of relativity [18], a new special theory of relativity based on modern astronomical data on the speed of light in a vacuum;

— the original solution to Hilbert's Fourth Problem and, following from it, the conception of the "golden" geometry;

— the fine-structure constant problem as a Physical Millennium Problem (see Chapter 5).

By using the general SELF-SIMILARITY PRINCIPLE and its mathematical interpretation in the form of the λ-Fibonacci numbers and Spinadel's metallic proportions, we attempt to outline the ways to answer the following questions:

(1) What is the λ-Fibonacci special theory of relativity?

(2) What are the λ-Universes and how are they changing along with the development of the Multiverse since the Big Bang?

(3) What are the fine-structure λ-constants for the λ-Universes?

(4) How have the main physical λ-constants of the λ-Universes changed with the age of the Universe since the Big Bang?

A.3.2.   The λ-Fibonacci–Lorenz transformations and λ-Fibonacci special theory of relativity

The origin of the λ-Fibonacci special theory of relativity goes back to the article [18] and is based on the following λ-Fibonacci–Lorenz transformations ($\lambda = 1, 2, 3, \ldots$):

$$
\begin{pmatrix} c_\lambda^{(0)} t \\ x \\ y \\ z \end{pmatrix} = \begin{pmatrix} \dfrac{1}{\sqrt{1 - (\bar{v}_\lambda)^2}} & \dfrac{1}{\bar{c}_\lambda(\psi)}\dfrac{\bar{v}_\lambda}{\sqrt{1 - (\bar{v}_\lambda)^2}} & 0 & 0 \\ \bar{c}_\lambda(\psi)\dfrac{\bar{v}_\lambda}{\sqrt{1 - (\bar{v}_\lambda)^2}} & \dfrac{1}{\sqrt{1 - (\bar{v}_\lambda)^2}} & 0 & 0 \\ 0 & 0 & 1 & 0 \\ 0 & 0 & 0 & 1 \end{pmatrix} \cdot \begin{pmatrix} c_\lambda^{(0)} t' \\ x' \\ y' \\ z' \end{pmatrix}
$$

$$
\begin{pmatrix} c_\lambda^{(0)} t \\ x \\ y \\ z \end{pmatrix} = \begin{pmatrix} \mathrm{cF}_\lambda(\psi - 1) & \mathrm{sF}_\lambda(\psi - 2) & 0 & 0 \\ \mathrm{sF}_\lambda(\psi) & \mathrm{cF}_\lambda(\psi - 1) & 0 & 0 \\ 0 & 0 & 1 & 0 \\ 0 & 0 & 0 & 1 \end{pmatrix} \cdot \begin{pmatrix} c_\lambda^{(0)} t' \\ x' \\ y' \\ z' \end{pmatrix} \tag{A.4}
$$

A.3.3.   The meaning of the symbols for the λ-Fibonacci special theory of relativity

(1) The dimensionless parameter ψ is called the parameter of self-organization.

(2) t [sec] is a time, x [m] is a length, y [m] is a width, z [m] is a height.

(3) $c_\lambda^{(0)} = \frac{c_\lambda^*}{\Phi_\lambda}$ [m sec^{-1}] means the normalized λ-Lorentzian speed of light in a vacuum. If the value $\lambda = 1$, then $c_{\lambda=1}^*$ [m sec^{-1}] = const is the speed of light in a vacuum for the case of the classical special theory of relativity. For the modern period the accepted value for $c_{\lambda=1}^* = 2.99792458 \cdot 10^8$ [m sec^{-1}]. The value of $c_{\lambda=1}^{(0)}$ does not depend on the speed of the movement of the light source or the observer, and is the same for all inertial reference systems. For the modern period we have: $c_{\lambda=1}^{(0)} = 1.8528199 \cdot 10^8$ [m sec^{-1}], where $\Phi_{\lambda=1} = \Phi$ is the golden ratio.

(4) $\mathrm{sF}_\lambda(x) = \frac{\Phi_\lambda^x - \Phi_\lambda^{-x}}{\sqrt{4+\lambda^2}} = \frac{2}{\sqrt{4+\lambda^2}}\mathrm{sh}(x \ln \Phi_\lambda)$ is the λ-Fibonacci hyperbolic sine, $\mathrm{cF}_\lambda(x) = \frac{\Phi_\lambda^x - \Phi_\lambda^{-x}}{\sqrt{4+\lambda^2}} = \frac{2}{\sqrt{4+\lambda^2}}\mathrm{ch}(x \ln \Phi_\lambda)$ is the λ-Fibonacci hyperbolic cosine.

(5) The dimensionless parameter $\bar{c}_\lambda(\psi) = \sqrt{\dfrac{sF_\lambda(\psi)}{sF_\lambda(\psi-2)}}$ is the normalized λ-Fibonacci speed of light in a vacuum.

(6) $\bar{v}_\lambda(\psi) = \dfrac{1}{\bar{c}_\lambda(\psi)} \cdot \dfrac{sF_\lambda(\psi)}{cF_\lambda(\psi-1)}$ is the normalized λ-Fibonacci speed of the light source in a vacuum.

(7) $c_\lambda(\psi) = c_\lambda^{(0)} \cdot \bar{c}_\lambda(\psi)$ [m sec^{-1}] is the λ-Fibonacci speed of light in a vacuum.

(8) $v_\lambda(\psi) = c_\lambda(\psi) \cdot \bar{v}_\lambda(\psi) = c_\lambda^{(0)} \cdot \dfrac{sF_\lambda(\psi)}{cF_\lambda(\psi-1)}$ [m sec^{-1}] is the λ-Fibonacci speed of the light source in a vacuum.

A.3.4. The main postulate of the λ-Fibonacci special theory of relativity

We postulate the following hypothesis:

"For every $\lambda = 1, 2, 3, 4, \ldots$ in the Multiverse there exists the Universe, the evolution of which in time t and in space (x, y, z) is described theoretically by the physical-mathematical model, which is based on the λ-Fibonacci–Lorenz transformations (A.4) and is called the λ-Fibonacci special theory of relativity."

We name such a Universe the λ-Universe. In accordance with the terminology of the metallic proportions [9–11], when $\lambda = 1$, this λ-Universe (our observed Universe) is called the "golden" Universe; for the cases $\lambda = 2, \lambda = 3$ and $\lambda = 4$ these λ-Universes are called "silver", "bronze" and "copper" Universes, respectively.

For every λ-Universe ($\lambda = 1, 2, 3, 4, \ldots$), the speed of light in a vacuum c_λ, is not a constant value and is determined by the function $c_\lambda(\psi) = c_\lambda^{(0)} \cdot \bar{c}_\lambda(\psi)$ [m sec^{-1}].

For our observed ("golden") ($\lambda = 1$)-Universe, the postulate about the variability of the speed of light in a vacuum corresponds to the data compiled by John Webb (www.vokrugsveta.ru/telegraph/cosmos/1298). He found that the light, which is coming to us from the observed Universe, follows the principle of non-decreasing entropy; this means, that for this case the Second Law of Thermodynamics has been preserved and therefore the speed of light $c_{\lambda=1}$ should decrease with the increasing age of the Universe. James Franson's (Johns Hopkins University, USA) article [19] confirms this conclusion.

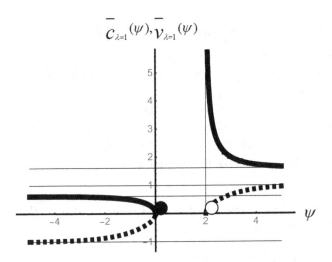

Figure A.1. The graph $\bar{c}_{\lambda=1}(\psi) > 0$ (the black curves) and the graph $\bar{v}_{\lambda=1}(\psi)$ (the dashed curves) for c ($\lambda = 1$) ("Observable" Universe, "Golden" Universe), where the parameter ψ is within the limits $\{-\infty < \psi < 0\} \cup \{2 < \psi < +\infty\}$. The symbols ● and ○ correspond to the bifurcation points $\psi = 0$ and $\psi = 2$, respectively.

A.3.5. Visual representation

A visual representation of the behavior of functions $\bar{c}_{\lambda=1}(\psi), \bar{v}_{\lambda=1}(\psi)$ and $\bar{c}_{\lambda=2}(\psi), \bar{v}_{\lambda=2}(\psi)$ can be seen in Figs. A.1 and A.2.

A.4.   Fundamental Physical Constants of the λ-Universes

A.4.1. Assumptions and definitions

We assume that the Multiverse, which appeared after the Big Bang, consists of many Universes, which begun to develop according to the universal SELF-SIMILARITY PRINCIPLE, based on the λ-Fibonacci numbers (Gazale's hypothesis) and Spinadel's metallic proportions. Each of these Universes is called the λ-Universe ($\lambda = 1, 2, 3, \ldots$).

There are specific physical laws in each λ-Universe. These physical laws are determined by their fine-structure λ-constant α_λ and other dimensionless λ-constants.

The physical world of the λ-Universe consists of elementary particles such as protons, electrons, neutrons and so on. The λ-masses of protons and electrons and their ratios (λ-ratios) are equal to special values. The

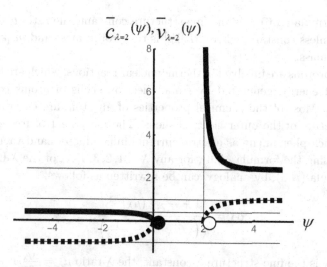

Figure A.2. The graph $\bar{c}_{\lambda=2}(\psi) > 0$ (the black curves) and the graph $\bar{v}_{\lambda=2}(\psi)$ (the dashed curves) for the case ($\lambda = 2$) ("Silver" Universe), where the parameter ψ is within the limits $\{-\infty < \psi < 0\} \cup \{2 < \psi < +\infty\}$. The symbols ● and ○ correspond to the bifurcation points $\psi = 0$ and $\psi = 2$, respectively.

basic dimensionless λ-constants of the λ-Universe change depending on the age of the Universe.

A.4.2.   The formula for the fine-structure λ-constants α_λ

Here we use Kosinov's formula (A.3). In order to get the expression for the fine-structure λ-constants we replace the classic golden ratio in the formula (A.3) with the metallic proportion Φ_λ ($\lambda = 1, 2, 3, \ldots$):

$$\alpha_\lambda = 10^{-\frac{43}{20}} \times \pi^{\frac{1}{260}} \times \Phi_\lambda^{\frac{7}{130}}, \qquad (A.5)$$

where $\Phi_\lambda = \frac{\lambda + \sqrt{4+\lambda^2}}{2}$ ($\lambda = 1, 2, 3, \ldots$) is Spinadel's metallic proportion [8–10].

The fine structure λ-constant α_λ for each λ-Universe characterizes the strength of the electromagnetic interaction. This constant for each λ-Universe characterizes its general physical properties.

A.4.3.   The formula for the λ-constant $\mu_\lambda = \frac{M_\lambda}{m_\lambda}$

Here we use, for $\lambda = 1$, the formula of Natalia Miskinova and Boris Shvilkin [20–23],

$$\alpha = \frac{1 + \frac{1}{\pi\sqrt{2}} \cdot (\mu)^{-1}}{\pi(\sqrt{\mu} + 1 - \frac{1}{\pi\sqrt{2}})}. \qquad (A.6)$$

In the formula (A.6) α is the fine-structure constant, the ratio $\mu = \frac{M}{m}$ is a dimensionless constant, where M [kg] is the proton mass and m [kg] is the electron mass.

The protons are involved in thermonuclear reactions, which are the main source of energy, generated by stars. Electron shells of atoms consist of electrons. Most of the chemical properties of an atom are determined by the structure of the outer electron shells. The movement of free electrons causes such phenomena as electric currents in conductors and vacuums.

By using the formula (A.6), for any $\lambda = 1, 2, 3, 4, \ldots$ of the λ-Universe, the formula (A.6), by analogy, can be rewritten as follows:

$$\alpha_\lambda = \frac{1 + \frac{1}{\pi\sqrt{2}} \cdot (\mu_\lambda)^{-1}}{\pi\left(\sqrt{\mu_\lambda} + 1 - \frac{1}{\pi\sqrt{2}}\right)}, \tag{A.7}$$

where α_λ is the fine-structure λ-constant, the λ-ratio $\mu_\lambda = \frac{M_\lambda}{m_\lambda}$ is a dimensionless constant, M_λ [kg] is the proton mass and m_λ [kg] is the electron mass for the given λ-Universe.

After simple transformations, we obtain the dependence $\mu_\lambda = \mu_\lambda(\alpha_\lambda)$ in the form:

$$\mu_\lambda = \frac{M_\lambda}{m_\lambda} = \frac{(A+B)^2}{9}, \tag{A.8}$$

where

$$A = -\frac{b}{a} + \frac{b^2}{a} \cdot \sqrt[3]{\frac{2}{\omega}},$$

$$B = \frac{1}{a} \cdot \sqrt[3]{\frac{\omega}{2}}, \quad \omega = 27a^2 - 2b^3 + 3 \cdot \sqrt{81a^4 - 12a^2b^3},$$

$$a = \pi^2\sqrt{2} \cdot \alpha_\lambda, \quad b = \pi \cdot [(\pi\sqrt{2} - 1)\alpha_\lambda - \sqrt{2}].$$

A visual representation of the behavior of constants α_λ and $\mu_\lambda = \frac{M_\lambda}{m_\lambda}$ ($\lambda = 1, 2, 3, 4$) is given in Table A.1.

A.4.4.   The variability of the fundamental physical constants α_λ and $\mu_\lambda = \frac{M_\lambda}{m_\lambda}$ depending on the age of the λ-Universe since the Big Bang

By analogy with the observed (golden) ($\lambda = 1$)-Universe (see Chapter 5), we postulate for other λ-Universes the following relationship α_λ and $\mu_\lambda = \frac{M_\lambda}{m_\lambda}$

Table A.1. Numerical values of the constants α_λ and $\mu_\lambda = \frac{M_\lambda}{m_\lambda}$ for the cases $\lambda = 1, 2, 3, 4$.

λ	α_λ	$\mu_\lambda = \dfrac{M_\lambda}{m_\lambda}$
1	0.007297351997377362	1836.1518766721751
2	0.007456295536924501	1757.339324899759
3	0.007583186851874606	1697.9660956279272
4	0.00768549168052453	1652.2359394410366

from ψ and therefore, *a priori*, from the cosmological time since the Big Bang:

$$\alpha_\lambda(\psi) = 10^{-\frac{43}{20}} \times \pi^{\frac{1}{260}} \times |\bar{c}_\lambda(\psi)|^{\frac{7}{130}}. \tag{A.9}$$

where $|\bar{c}_\lambda(\psi)| = \sqrt{\dfrac{\mathrm{sF}_\lambda(\psi)}{\mathrm{sF}_\lambda(\psi-2)}} > 0$, if $\{-\infty < \psi < 0\} \cup \{2 < \psi < +\infty\}$ (Black Hole + Light Ages), and $|\bar{c}_\lambda(\psi)| = \sqrt{-\dfrac{\mathrm{sF}_\lambda(\psi)}{\mathrm{sF}_\lambda(\psi-2)}} > 0$, if $\{0 < \psi < 2\}$ (Dark Ages).

If we substitute (A.9) in (A.8), we obtain the dependence $\mu_\lambda = \frac{M_\lambda}{m_\lambda}$ from ψ. A visual representation of the behavior of the functions $\alpha_{\lambda=1}(\psi), \mu_{\lambda=1}(\psi)$, and $\alpha_{\lambda=2}(\psi), \mu_{\lambda=2}(\psi)$, is given in Figs. A.3–A.6 and Tables A.2 and A.3.

We understand the hypothetical character of the above mathematical and physical models of the Multiverse, based on the Mathematics of Harmony. However, if we can obtain new astronomical confirmations of the conception of the Multiverse [1, 2], then the above questions will sooner or later appear in the forefront of modern theoretical physics.

In any case, we are confident that these hypothetical patterns of the Multiverse, based on the SELF-SIMILARITY PRINCIPLE and the λ-Fibonacci numbers, can attract the attention of mathematicians and physicists and can become a fruitful source for new scientific results in this area.

A.5. The Mathematics of Harmony as an Essential Part of Mathematical Physics

The above results touch on the role of the Mathematics of Harmony [20] in the development of Mathematical Physics. Taking into consideration the

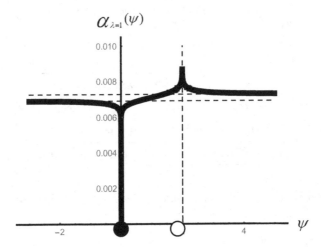

Figure A.3.   The graph $\alpha_{\lambda=1}(\psi) > 0$ for the case $\lambda = 1$ ("Observable" Universe, "Golden" Universe). The symbols ● and ○ correspond to the bifurcation points $\psi = 0$ and $\psi = 2$, respectively.

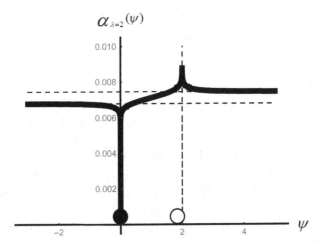

Figure A.4.   The graph $\alpha_{\lambda=2}(\psi) > 0$ for the case $\lambda = 2$ ("Silver" Universe). The symbols ● and ○ correspond to the bifurcation points $\psi = 0$ and $\psi = 2$, respectively.

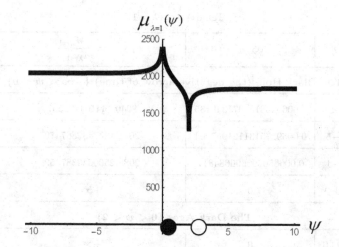

Figure A.5. The graph $\mu_{\lambda=1}(\psi) > 0$ for the case $\lambda = 1$ ("Observable" Universe, "Golden" Universe). The symbols ● and ○ correspond to the bifurcation points $\psi = 0$ and $\psi = 2$, respectively.

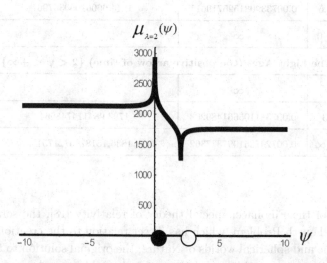

Figure A.6. The graph $\mu_{\lambda=2}(\psi) > 0$ for the case $\lambda = 2$ ("Silver" Universe). The symbols ● and ○ correspond to the bifurcation points $\psi = 0$ and $\psi = 2$, respectively.

Table A.2. $(\lambda = 1)$.

ψ	$\alpha_{\lambda=1}$	$\mu_{\lambda=1} = \dfrac{M_{\lambda=1}}{m_{\lambda=1}}$
The Black Hole (the negative arrow of time) $\{-\infty < \psi < 0\}$		
$-\infty$	0.0069288144971348135	2040.344104791557
-5	0.0069275131121957064	2041.1237432867479
-1	0.006850192996684811	2088.2500646387352
0–0	0	$+\infty$
The Dark Ages $\{0 < \psi < 2\}$		
0+0	0	$+\infty$
0.5	0.006889391594589547	2064.1592455987684
1	0.007110696049622987	1935.5798368691815
1.5	0.007339081525719611	1814.9606105036796
2–0	$+\infty$	0
The Light Ages (the positive arrow of time) $\{2 < \psi < +\infty\}$		
2+0	$+\infty$	0
3	0.007381105661489203	1793.984114748964
$+\infty$	0.007297351997377362	1836.1518766721751

creation of the Fibonacci special theory of relativity [18], the solution to Hilbert's Fourth Problem, which has direct relation to the creation of new hyperbolic and spherical worlds of Nature, the original solution to the fine-structure constant problem and the ratio $\frac{M}{m}$ problem, and especially the original ideas for the simulation of the Multiverse, we can rightfully claim that the Mathematics of Harmony [24] is a new branch of Mathematical Physics. This conclusion opens up interesting prospects for the creation of new mathematical theories of physical science based on the Mathematics of Harmony.

Table A.3. ($\lambda = 2$).

ψ	$\alpha_{\lambda=2}$	$\mu_{\lambda=2} = \dfrac{M_{\lambda=2}}{m_{\lambda=2}}$
The Black Hole (the negative arrow of time) $\{-\infty < \psi < 0\}$		
$-\infty$	0.006781115107325703	2131.731301384076
-5	0.006781088760702097	2131.7481406927013
-1	0.0067477576363425265	2153.210675513261
0–0	0	$+\infty$
The Dark Ages $\{0 < \psi < 2\}$		
0+0	0	$+\infty$
0.5	0.006858283163744207	2083.2439396253926
1	0.007110696049622987	1935.5798368691815
1.5	0.007372398762625624	1798.3005704875468
2–0	$+\infty$	0
The Light Ages (the positive arrow of time) $\{2 < \psi < +\infty\}$		
2+0	$+\infty$	0
3	0.007493155657785296	1739.779108505952
$+\infty$	0.0074562955369245	1757.3393248997609

References

[1] Linde, A. "The many faces of the universe", Third public lectures on physics, at the Lebedev Physical Institute of the Russian Academy of Sciences (Moscow) (2007). http://elementy.ru/lib/430484 (accessed January 10, 2015).

[2] Linde, A. "Inflation in supergravity and string theory: Brief history of the multiverse", presented at the State of the Universe Conference at the University of Cambridge, UK (2012). www.ctc.cam.ac.uk/stephen70/talks/swh70_linde.pdf (accessed January 10, 2015).

[3] Kashlinsky, A., Atrio-Barandela, F., Kocevski, D. and Ebeling, H. "A measurement of large-scale peculiar velocities of clusters of galaxies: results and cosmological implications", *Astrophys. J.* (2008) 686: 49–52.

[4] Hawking, S. *Cosmology from the top down, in Universe or Multiverse?* Cambridge University Press (2007): 91.

[5] Greene, B. *The Elegant Universe. Superstrings, Hidden Dimensions and the Quest for the Ultimate Theory,* Editorial URSS (2007) (translated from English).

[6] Vilenkin, A. *Many Worlds in One: The Search for Other Universes.* Macmillan (2007).

[7] "Pattern in nature", *Wikipedia, The Free Encyclopedia,* https://en.wikipedia.org/wiki/Patterns_in_nature (accessed January 4, 2016).

[8] Gazale, M. J. *Gnomon. From Pharaohs to Fractals.* Princeton, NJ: Princeton University Press (1999).

[9] de Spinadel, V. W. "The metallic means and design", in *NEXUS II: Architecture and Mathematics* (1998).

[10] de Spinadel, V. W. "The metallic means family and multifractal spectra", *Nonlinear Analysis* (1999) Vol. 36: 721–745.

[11] de Spinadel, V. W. *From the Golden Mean to Chaos.* Nueva Libreria (first edition 1998); Nobuko (second edition 2004).

[12] Stakhov, A. P. and Rozin, B. N. "On a new class of hyperbolic function", *Chaos, Solitons & Fractals* (2004) Vol. 23: 379–389.

[13] Stakhov, A. P. "Gazale's formulas, a new class of hyperbolic Fibonacci and Lucas Functions and the improved method of the 'golden' cryptography", Academy of Trinitarism, Moscow: Electronic number 77-6567, publication 14098 (2006) (Russian). http://www.trinitas.ru/rus/doc/0232/004a/02321063.htm (accessed November 4, 2015).

[14] Stakhov, A. P. "On the general theory of hyperbolic functions based on the hyperbolic Fibonacci and Lucas functions and on Hilbert's Fourth Problem", *Visual Mathematics* (2013), Vol. 15, No. 1. http://www.mi.sanu.ac.rs/vismath/2013stakhov/hyp.pdf (accessed November 4, 2015).

[15] Bodnar, O. Y. *The Golden Section and Non-Euclidean Geometry in Nature and Art.* Lvov: Publishing House "Svit" (1994) (Russian).

[16] Kosinov, N. V. "Connection of three important constants", (2000) (Russian). http://www.roman.by/r-25512.html (accessed November 4, 2015).

[17] Kosinov, N. V. "The physical equivalent of the number 'Pi' and the geometric equivalent of the fine-structure constant 'alpha'" (Russian). http://math-english.ru/constanta/kosinov/kosinov10.htm (accessed November 4, 2015).

[18] Stakhov, A. P. and Aranson, S. Kh. "The 'golden' Fibonacci goniometry, Fibonacci–Lorentz transformations, and Hilbert's fourth problem", *Congressus Numerantium* (2008) Vol. 193: 119–156 (International Journal, Canada, USA).

[19] Franson, J. D. "Apparent correction to the speed of light in a gravitational potential", *New Journal of Physics* (2014) Vol. 16, No. 6. http://iopscience.iop.org/1367-2630/16/6/065008 (accessed November 4, 2015).

[20] Miskinova, N. A, and Shvilkin B. N. "On the relationship between the fine-structure constant and the ratio of the masses of the proton and electron", *Science and Technology of Russia* (2002a) No. 4 (55): 19 (Russian).

[21] Miskinova, N. A. and Shvilkin, B. N. "On the question of the fine-structure constant", *Science and Technology of Russia* (2002b) No. 5/6 (56/57): 20 (Russian).

[22] Miskinova, N. A. and Shvilkin, B. N. "On the empirical formula of fine-structure constant", *Proceedings of the Russian Universities. Physics* (2004), No. 11: 97–98 (Russian).

[23] Miskinova, N. A. and Shvilkin B. N. "The relationship between the fundamental physical constants", *Science and Technology of Russia* (2012) Vol. 91, No. 3: 25–32 (Russian).

[24] Stakhov, A. P., assisted by S. Olsen. *The Mathematics of Harmony. From Euclid to Contemporary Mathematics and Computer Science.* Singapore: World Scientific (2009).

Index

Printed in the United States
By Bookmasters